Lecture Notes in Mathematics

Editors:
J.-M. Morel, Cachan
B. Teissier, Paris

For further volumes:
http://www.springer.com/series/304

FONDAZIONE
CIME
ROBERTO CONTI
CENTRO INTERNAZIONALE MATEMATICO ESTIVO
INTERNATIONAL MATHEMATICAL SUMMER CENTER

Fondazione C.I.M.E., Firenze

C.I.M.E. stands for *Centro Internazionale Matematico Estivo*, that is, International Mathematical Summer Centre. Conceived in the early fifties, it was born in 1954 in Florence, Italy, and welcomed by the world mathematical community: it continues successfully, year for year, to this day.

Many mathematicians from all over the world have been involved in a way or another in C.I.M.E.'s activities over the years. The main purpose and mode of functioning of the Centre may be summarised as follows: every year, during the summer, sessions on different themes from pure and applied mathematics are offered by application to mathematicians from all countries. A Session is generally based on three or four main courses given by specialists of international renown, plus a certain number of seminars, and is held in an attractive rural location in Italy.

The aim of a C.I.M.E. session is to bring to the attention of younger researchers the origins, development, and perspectives of some very active branch of mathematical research. The topics of the courses are generally of international resonance. The full immersion atmosphere of the courses and the daily exchange among participants are thus an initiation to international collaboration in mathematical research.

C.I.M.E. Director
Pietro ZECCA
Dipartimento di Energetica "S. Stecco"
Università di Firenze
Via S. Marta, 3
50139 Florence
Italy
e-mail: zecca@unifi.it

C.I.M.E. Secretary
Elvira MASCOLO
Dipartimento di Matematica "U. Dini"
Università di Firenze
viale G.B. Morgagni 67/A
50134 Florence
Italy
e-mail: mascolo@math.unifi.it

For more information see CIME's homepage: http://www.cime.unifi.it

CIME activity is carried out with the collaboration and financial support of:

- INdAM (Istituto Nazionale di Alta Matematica)

- MIUR (Ministero dell'Universita' e della Ricerca)

This course is partially supported by GDR-GDRE on CONTROLE DES EQUATIONS AUX DERIVEES PARTIELLES (CONEDP).

Fatiha Alabau-Boussouira • Roger Brockett
Olivier Glass • Jérôme Le Rousseau
Enrique Zuazua

Control of Partial Differential Equations

Cetraro, Italy 2010

Editors:
Piermarco Cannarsa
Jean-Michel Coron

 Springer

Fatiha Alabau-Boussouira
Université Paul Verlaine-Metz
LMAM
Ile du Saulcy
Metz
France

Roger Brockett
Harvard University
Engineering and Applied Sciences
Oxford St. 33
Cambridge Massachusetts
Maxwell Dworkin
USA

Olivier Glass
Université Paris-Dauphine
CEREMADE
Place du Maréchal de Lattre de
Tassigny
Paris
France

Jérôme Le Rousseau
Université d'Orléans
Laboratoire MAPMO
Orléans
France

Enrique Zuazua
Basque Center for Applied Mathematics
Bizkaia Technology Park
Derio
Spain

ISBN 978-3-642-27892-1   e-ISBN 978-3-642-27893-8
DOI 10.1007/978-3-642-27893-8
Springer Heidelberg Dordrecht London New York

Lecture Notes in Mathematics ISSN print edition: 0075-8434
                    ISSN electronic edition: 1617-9692

Library of Congress Control Number: 2012934256

Mathematics Subject Classification (2010): 35B35, 93B05, 93B07, 93B52, 93C20, 93D15, 76B75,
                    65M10, 65M12

Printed on acid-free paper

Springer is part of Springer Science+Business Media (www.springer.com)

Preface

One of the findings of the 1988 Report by the Panel on Future Directions in Control Theory, chaired by Wendell H. Fleming, was:

> Many fundamental theoretical issues, such as control of nonlinear multivariable systems, or *control of nonlinear partial differential equations*, are not yet understood.

Nowadays, more than 20 years later, we believe we can say that a lot of fundamental issues concerning the latter topic have definitely been understood, thanks to the efforts of many researchers who produced a large body of results and techniques. And yet, this process has led to an enormous amount of open questions that will need to be addressed by new generations of scientists. Surveying the most important advances of the last two decades and outlining future research directions were the main motivations that led us to organize the CIME Course on Control of Partial Differential Equations that took place in Cetraro (CS, Italy), July 19–23, 2010. We hope this volume, which is one of the outcomes of that event, will provide an ultimate formative step for those who attended the course, and will represent an authoritative reference for those who were unable to do so.

The course consisted of five series of lectures, which are now the source of the chapters of this monograph. Specifically, the following topics were covered:

- *Stabilization of evolution equations* (by Fatiha Alabau-Boussouira): these lectures discussed recent advances, as well as classical methods, for the stabilization of wave-like equations. Special attention was paid to nonlinear problems, memory-damping, and indirect stabilization of coupled PDEs. All the problems were treated by a unified methodology based on energy estimates. It was shown how the introduction of optimal-weight convexity methods leads to easy computable upper energy decay estimates, and how these results can be completed by lower energy estimates for several examples.
- *Control of the Liouville equation* (by Roger Brockett): these equations describe the evolution of an initial density of points that move according to a given differential equation, and may depend on a control which can be chosen in order to satisfy some prescribed goals. This framework also allows to overcome

limitations of the classical theory: for example, the expense required to implement control laws. Several results (e.g., on ensemble control: controlling, with a
single control, a finite but often large number of copies of a given system) as well
as open problems were presented.

- *Control in fluid mechanics* (by Olivier Glass): the lectures treated various issues
  related to the controllability of two well-known equations in fluid mechanics,
  namely the Euler equation for perfect incompressible fluids in both Eulerian and
  Lagrangian coordinates, and the one-dimensional isentropic Euler equation for
  compressible fluids in the framework of entropy solutions. Special emphasis was
  put on the aspects of the theory that are connected with the nonlinear nature of
  the problem: linearization around an equilibrium gives here no information on
  the controllability of the nonlinear system.

- *Carleman estimates for elliptic and parabolic equations, with application to
  control* (by Jérôme Le Rousseau): these are weighted energy estimates for
  solutions of partial differential equations with weights of exponential type. The
  lectures derived Carleman estimates for elliptic and parabolic operators using
  several methods: a microlocal approach where the main tool is the Gårding
  inequality, and a more computational direct approach. It was also shown how
  Carleman estimates can be used to provide unique continuation properties, as
  well as approximate and null controllability results.

- *Control and numerics for the wave equation* (by Enrique Zuazua): these lectures
  provided a self-contained presentation of the theory that has been developed
  recently for the numerical analysis of the controllability properties of wave propagation phenomena. The methodology adopted the so-called discrete approach,
  which consists in analyzing whether the semidiscrete or fully discrete dynamics
  arising when discretizing the wave equation by means of the most classical
  scheme of numerical analysis share the property of being controllable, uniformly
  with respect to the mesh-size parameters, and the corresponding controls converge to the continuous ones as the mesh size tends to zero. All the results were
  illustrated by means of several numerical experiments.

Besides the above lectures, there were three seminars, given by Karine Beauchard
(Some controllability results for the 2D Kolmogorov equation), Sylvain Ervedoza
(Regularity of HUM controls for conservative systems and convergence rates for
discrete controls), and Lionel Rosier (Control of some dispersive equations for
water waves). There were also four communications given by Ido Bright (Periodic
optimization suffices for infinite horizon planar optimal control), Khai Tien Nguyen
(The regularity of the minimum time function via nonsmooth analysis and geometric
measure theory), Camille Laurent (On stabilization and control for the critical
Klein-Gordon equation on a 3-D compact manifold), and Vincent Perrollaz (Exact
controllability of entropic solutions of scalar conservation laws with three controls).
Seminars and communications are not reproduced in these notes.

One important point, contained in the 1988 Report we mentioned above, is that
*advances in the control field are made through a combination of mathematics,
modeling, computation, and experimentation.* Hoping the reader will find the present

exposition in accord with such a basic principle, we wish to thank the lecturers and authors who designed their contributions in a detailed-yet-focussed form, for helping us realize this project. Overall, we are very grateful to all the 57 participants in the CIME course, for their enthusiasm that created a friendly and stimulating atmosphere in Cetraro. Finally, special gratitude is due to the GDRE CONEDP, for providing the essential support that allowed us to receive and accept a large number of applications, and to the C.I.M.E. Foundation, for making this event possible and for its very helpful assistance before and all along the lectures.

Rome and Paris   *Piermarco Cannarsa*
                                                                                       Jean-Michel Coron

Contents

Contributors

Fatiha Alabau-Boussouira Université Paul Verlaine-Metz, Ile du Saulcy, Metz, Cedex 1, France

Roger Brockett Division of Engineering and Applied Sciences, Harvard University, Cambridge, MA, USA

Sylvain Ervedoza CNRS; Institut de Mathématiques de Toulouse, UMR 5219, Toulouse, France, Université de Toulouse; UPS, INSA, INP, ISAE, UT1, UTM ; IMT, Toulouse, France

Olivier Glass Ceremade, Université Paris-Dauphine, Paris Cedex 16, France

Jérôme Le Rousseau Université d'Orléans, Laboratoire de Mathématiques - Analyse, Probabilités, Modélisation - Orléans (MAPMO), CNRS UMR 6628, Fédération Denis Poisson, FR CNRS 2964, Orléans cedex 2, France

Enrique Zuazua Ikerbasque – Basque Foundation for Science, Bilbao, Basque Country, Spain

Chapter 1
On Some Recent Advances on Stabilization for Hyperbolic Equations

Fatiha Alabau-Boussouira

In memory of my father Abdallah Boussouira.

Abstract The purpose of these Notes is to present some recent advances on stabilization for wave-like equations together with some well-known methods of stabilization. This course will give several references on the subject but do not pretend to exhaustivity. The spirit of these Notes is more that of a research monograph. We aim to give a simplified overview of some aspects of stabilization, on the point of view of energy methods, and insist on some of the methodological approaches developed recently. We will focus on nonlinear stabilization, memory-damping and indirect stabilization of coupled PDE's and present recent methods and results. Energy methods have the advantage to handle and deal with physical quantities and properties of the models under consideration. For nonlinear stabilization, our purpose is to present the optimal-weight convexity method introduced in (Alabau-Boussouira, Appl. Math. Optim. 51(1):61–105, 2005; Alabau-Boussouira, J. Differ. Equat. 248:1473–1517, 2010) which provides *a whole methodology* to establish easy computable energy decay rates which are optimal or quasi-optimal, and works for finite as well as infinite dimensions and allow to treat, *in a unified way* different PDE's, as well as different types of dampings: localized, boundary. Another important feature is that the upper estimates can be completed by lower energy estimates for several examples, and these lower estimates can be compared to the upper ones. Optimality is proved in finite dimensions and in particular for one-dimensional semi-discretized wave-like PDE's. These results are obtained through energy comparison principles (Alabau-Boussouira, J. Differ. Equat. 248: 1473–1517, 2010), which are, to our knowledge, new. This methodology can be extended to the infinite dimensional setting thanks to still energy comparison

F. Alabau-Boussouira (✉)
Université Paul Verlaine-Metz, Ile du Saulcy, 57045 Metz, Cedex 1, France
e-mail: alabau@univ-metz.fr

F. Alabau-Boussouira et al., *Control of Partial Differential Equations*,
Lecture Notes in Mathematics 2048, DOI 10.1007/978-3-642-27893-8_1,
© Springer-Verlag Berlin Heidelberg 2012

principles supplemented by interpolation techniques. The optimal-weight convexity method is presented with two approaches: a direct and an indirect one. The first approach is based on the multiplier method and requires the assumptions of the multiplier method on the zone of localization of the feedback. The second one is based on an indirect argument, namely that the solutions of the corresponding undamped systems satisfy an observability inequality, the observation zone corresponding to the damped zone for the damped system. The advantage is that, this observability inequality holds under the sharper optimal Geometric Control Condition of Bardos et al. (SIAM J. Contr. Optim. 30:1024–1065, 1992). The optimal-weight convexity method also extends to the case memory-damping, for which the damping effects are nonlocal, and leads to nonautonomous evolution equations. We will only state the results in this latter case. Indirect stabilization of coupled systems have received a lot of attention recently. This subject concerns stabilization questions for coupled PDE's with a reduced number of feedbacks. In practice, it is often not possible to control all the components of the vector state, either because of technological limitations or cost reasons. From the mathematical point of view, this means that some equations of the coupled system are not directly stabilized. This generates mathematical difficulties, which requires to introduce new tools to study such questions. In particular, it is important to understand how stabilization may be transferred from the damped equations to the undamped ones. We present several recent results of polynomial decay for smooth initial data. These results are based on energy methods, a nondifferential integral inequality introduced in (Alabau, Compt. Rendus Acad. Sci. Paris I 328:1015–1020, 1999) [see also (Alabau-Boussouira, SIAM J. Contr. Optim. 41(2):511–541, 2002; Alabau et al. J. Evol. Equat. 2:127–150, 2002)] and coercivity properties due to the coupling operators.

Several parts of the material presented here are extracted from published articles, in particular from the author.

1.1 Introduction

Let us start with some motivations of the material presented in these Notes. An important issue in engineering for material sciences is the stabilization of flexible structures such as beams, plates, or mechanical structures such as antennas of satellites. Oscillations or vibrations of elastic, or visco-elastic materials or structures are described by reversible PDE's. In general, for such applications, it is important to reduce vibrations by implementing feedback laws within the system. These feedbacks are built in such a way to stabilize the system, that is to reduce the oscillations of the solutions as time increases. A common way to measure this decay is to consider the natural energy associated to the system. One of the purposes of this course is to study the asymptotic behavior of the energy of the solutions of the stabilized systems i.e. determine whether convergence toward equilibrium states as times goes to infinity holds, determine its speed of convergence if necessary or study how many feedback controls are required in case of coupled systems. We will introduce some basic definitions of stabilization for PDE's and present some recent advances on this subject.

For further motivation, let us quote Dafermos [46] who considered the strong stabilization of nonlinearly locally damped wave equations. He has written in 1978 about the approach which consists in using the information that dissipative systems exhibits Lyapunov functional—indeed their energy—which is constant on the ω-limit sets, to prove strong stabilization results, the following text:

> Another advantage of this approach is that it is so simplistic that it requires only quite weak assumptions on the dissipative mechanism. The corresponding drawback is that the deduced information is also weak, never yielding, for example, decay rates of solutions.

Hence understanding to what extent and generality on the feedbacks and more precisely on

- Their growth
- The geometric properties of the domains on which they are active
- Their autonomous or non autonomous characters

and on the systems themselves: either in finite dimensions or infinite dimensions, for general classes of PDE's ... one can give sharp—even optimal—energy decay rates is a strong mathematical motivation. It has inspired a wide research on these questions and the sections on nonlinear stabilization presented in these Notes. A lot of seminal and deep results due to several authors have been derived since the early seventy's. We shall recall a part of them in the course of the sections. Stabilization of ordinary differential equations goes back to the pioneering works of Lyapunov and Lasalle. The important property is that trajectories decay along Lyapunov functions. If trajectories are relatively compact in appropriate spaces and the system is autonomous, then one can prove that trajectories converge to equilibrium asymptotically. However, the construction of Lyapunov functions is not easy in general. A somehow natural and physical Lyapunov function is the energy associated to the solutions of the system, which decays as time increases, due to the feedback action. The purpose of these Notes is to show how energy methods associated to some other tools such as convexity, differential and nondifferential integral inequalities, and comparison principles are powerful for obtaining optimal or quasi-optimal energy decay rates, linked with physical properties of dissipative systems. In this analysis, an essential question is that of optimality. Deriving decay rates is not sufficient, it is essential to know if these decay rates are optimal. We will show how to prove optimality for finite dimensional dissipative systems and how to relate techniques in finite and infinite dimensions.

There are still several hard and important open questions in the domain of stabilization. We indicate some of them at the end of these Notes. We hope that they will inspire young researchers.

In Sect. 1.1, we give a short description of Dafermos'result and Lasalle principle for strong stabilization. Linear stabilization together with the multiplier method for localized damping and the compactness–uniqueness method as in Zuazua [105] (see also [91]) for linear stabilization of semilinear waves are presented in Sect. 1.2. We present in Sect. 1.3, the optimal-weight convexity method for general nonlinear dampings for ODE's, that is for finite dimensions, together with nonlinear Gronwall inequalities and a key comparison Lemma for proving optimality. Section 1.4 recalls well-known results for polynomial stabilization for PDE's, that is for infinite dimensions. We present the optimal-weight convexity method for PDE's and further extensions and applications in Sect. 1.5. Section 1.6 is devoted first to memory-stabilization for polynomial kernels and then to memory-stabilization for general decaying kernels extending the optimal-convexity method to nonfrictional dampings. We consider indirect stabilization of coupled PDE's in Sect. 1.7. We present briefly a general approach based on an integral inequality which is not of differential nature, but relies on the underlying semigroup property. We give then examples of applications to some coupled systems of PDE's. We conclude in Sect. 1.8 by a list of some open problems in stabilization.

Let us now present shortly some of the problems which will be considered through some significant examples.

1.1.1   On Nonlinear and Memory Stabilization

We consider a range of classes of examples for which, the optimal-weight convexity method provides in a unified sharp—even optimal in the finite dimensional case— easily computable energy decay rates for arbitrary feedback growth close to the origin.

Consider first the case of a nonlinearly damped oscillator, which models the displacement of a mass subjected to the action of spring. The displacement u of the mass is described by the scalar equation

Example 1.1.1.
$$u''(t) + ku(t) + f(u) + \rho(u') = 0, \tag{1.1}$$

where $k > 0$ is a physical parameter, u' and u'' denote respectively the velocity and the acceleration, f includes some eventual nonlinear phenomenon, whereas $\rho(u')$ is the nonlinear damping. We assume that f and ρ are smooth functions and for the sake of simplicity that

$$sf(s) \geq 0 \quad \forall s \in \mathbb{R}.$$

we define

$$F(u) = \int_0^u f(s)\,ds$$

Multiplying (1.1) by u', we easily derive that

$$E'(t) = -u'\rho(u') \tag{1.2}$$

where the energy E of the solution u is defined by

$$E(t) = \left(\frac{1}{2}|u'(t)|^2 + k|u(t)|^2 + F(u(t))\right). \tag{1.3}$$

Thus, if the damping function satisfies

$$s\rho(s) \geq 0 \quad \forall s \in \mathbb{R}, \tag{1.4}$$

then the energy is decaying as time increases. The relation (1.2) is called a dissipation relation.

Example 1.1.2. Let us now consider the case of a vibrating membrane fixed on the boundary. Then the evolution of the displacement of a point x of the membrane at time t is described by the wave equation

$$\begin{cases} u_{tt} - \Delta u + \rho(., u_t) = 0 & \text{in} \quad (0, \infty) \times \Omega, \\ u = 0 & \text{on} \quad (0, \infty) \times \Gamma, \\ (u, u_t)(0) = (u^0, u^1) & \text{on} \quad \Omega, \end{cases} \tag{1.5}$$

where Ω is a bounded domain of $\mathbb{R}^N$ with sufficiently smooth boundary Γ. We assume that ρ is a smooth function satisfying (1.4). We define the energy of a solution by

$$E(t) = \frac{1}{2} \int_\Omega \left(|u'|^2 + |\nabla u|^2\right) dx \tag{1.6}$$

Formally multiplying the first equation of (1.5) by u_t, integrating over Ω and using Green's formula, we obtain for strong solutions the relation

$$E'(t) = - \int_\Omega u' \rho(u') \, dx \leq 0,$$

so that the energy of solutions is dissipated through the nonlinear damping term $\rho(u')$ in (1.5).

Example 1.1.3. Assume now that the damping occurs only through a part of the boundary Γ_1 of the membrane, then one has to consider

$$\begin{cases} u_{tt} - \Delta u = 0 \\ u = 0 \text{ in } [0, \infty) \times \Gamma_1 \\ \frac{\partial u}{\partial \nu} + \eta(.)u + \rho(., u_t) = 0 \text{ in } [0, \infty) \times \Gamma_0 \\ (u, u_t)(0, .) = (u^0, u^1) \text{ in } \Omega, \end{cases} \tag{1.7}$$

where $\{\Gamma_0, \Gamma_1\}$ is a partition of Γ and where η is a nonnegative function. The natural energy is

$$E(t) = \frac{1}{2}\Big(\int_{\Omega} |u_t(t)|^2 + |\nabla u(t)|^2 \, dx + \int_{\Gamma_0} \eta |u|^2 \, d\sigma\Big)$$

The dissipation relation (for strong solutions) is:

$$E'(t) = -\int_{\Gamma_0} u_t \rho(., u_t) \, dx \le 0, \quad t \ge 0.$$

The above three examples can be written under the abstract form

$$u'' + Au + f(u) + Bu' = 0, \tag{1.8}$$

in which u stands for the unknown, and takes values in $\mathbb{R}$ for the oscillator example, whereas it takes values in suitable subsets of the Sobolev space $H^1(\Omega)$ in the second and third examples.

Example 1.1.4. Finally as a last model example, let us consider the case of linear memory type stabilization. In this latter case, the damping is no longer frictional as for the three above models but is nonlocal in time. A model example is given by

$$\begin{cases} u_{tt} - \Delta u + k * \Delta u = 0 \\ u = 0 \text{ in } [0, \infty) \times \Gamma \\ (u, u_t)(0, .) = (u^0, u^1) \text{ in } \Omega, \end{cases} \tag{1.9}$$

where

$$(k * v)(t) = \int_0^t k(t - s)v(s) \, ds.$$

and the kernel k is positive, and decaying at infinity. In this case, the natural energy is

$$E_u(t) = \frac{1}{2}\|u_t(t)\|_{L^2(\Omega)}^2 \, dx + \frac{1}{2}\Big(1 - \int_0^t k(s) \, ds\Big)\|\nabla u(t)\|_{L^2(\Omega)}^2$$

$$+ \frac{1}{2}\int_0^t k(t - s)\|\nabla u(t) - \nabla u(s)\|_{L^2(\Omega)}^2 \, ds$$

and the dissipation relation is:

$$E_u'(t) = -\frac{1}{2}k(t)\|\nabla u(t)\|_{L^2(\Omega)}^2 + \frac{1}{2}\int_0^t k'(t - s)\|\nabla u(s) - \nabla u(t)\|_{L^2(\Omega)}^2 \, ds \le 0$$

A common feature to the previous four model examples, is that one can associate a natural energy to each of them, and that this energy is dissipated as time increases, due to the damping action, indeed the properties used by Dafermos for his strong stabilization result. We shall stress the common mathematical features between these examples and in particular between finite and infinite dimensional systems in the framework of nonlinear stabilization and linear stabilization of memory type and give a unified methodology for these different systems. In particular, we shall see that under minimal hypotheses on the feedbacks and using the optimal-weight convexity method [17], introduced in 2005 and simplified in [8], we can in a unified way give sharp, optimal or quasi-optimal energy decay rates for the four above classes of examples (see [10] for the last example).

1.1.2 On Indirect Stabilization for Coupled Systems

Stabilization of coupled systems by a reduced number of stabilizers, that is indirect stabilization have also received recently a lot of attention for parabolic equations and hyperbolic equations.

As an example, consider the following coupled wave system:

$$
\begin{cases} u_{tt} - \Delta u + a(x)u_t + \alpha v = 0 \\ v_{tt} - \Delta v + \alpha u = 0 \end{cases} \quad \text{in } (0,\infty)\times\Omega, \quad u = 0 = v \quad \text{on} \quad (0,\infty)\times\partial\Omega .
$$

where α is a constant coupling parameter and $a \geq 0$ in Ω and $a \geq a_- > 0$ on a nonempty subset ω of Ω. Here, the first equation is damped through a linear locally distributed feedback. One can remark that no feedback is applied to the second equation. Hence this second equation is conservative when $\alpha = 0$. One can define the total energy of a solution as

$$
E(t) = \frac{1}{2} \int_{\Omega} \left(|u_t|^2 + |\nabla u|^2 + |v_t|^2 + |\nabla v|^2 + 2\alpha uv \right) dx
$$

One can also show that the energy of strong solutions is dissipated, that is

$$
E'(t) = - \int_{\Omega} a(x)|u_t|^2 \, dx
$$

Is this dissipation which involves only the first component of the unknown sufficient to stabilize the full system and if so at which rate? Is it possible to study more general coupled systems of hyperbolic equations? We shall also consider these questions in this Notes.

Systems as above are said to be:

- *Strongly stable* if $E(t) \to 0$ as $t \to \infty$.
- *Polynomially stable* if $E(t) \leq C t^{-\beta}$ for all $t > 0$ and sufficiently smooth initial data.
- *(Uniformly) exponentially stable* if $E(t) \leq C e^{-\alpha t} E(0)$ for all $t \geq 0$ and some constants $\alpha > 0$ and $C \geq 0$, independent of u^0, u^1.

This course will focus on some of the above issues, such as geometrical aspects, nonlinear damping, indirect damping for coupled systems and memory damping.

1.2   Notation

We denote by $|\cdot|$ the usual Euclidean norm in $\mathbb{R}^N$. Let Ω be an open bounded domain of $\mathbb{R}^N$. We denote by $|\Omega|$ its Lebesgue measure.

We will use both notations $\frac{\partial}{\partial x_i}$ or ∂_i to denote the partial derivative with respect to x_i.

We denote by ∇u the gradient of u with respect to the space variables.

For a multi-integer $\alpha = (\alpha_1, \ldots, \alpha_N) \in \mathbb{N}^N$, we denote by $|\alpha| = \sum_{i=1}^N \alpha_i$ and $\partial^\alpha v = \frac{\partial^{|\alpha|}}{\partial_1^{\alpha_1} \ldots \partial_N^{\alpha_N}}$. Moreover, we will also use the notation $\partial_{ij}^2 u$ for $\partial_i \partial_j u$.

The time derivative of u will be denoted by u_t or u'.

In this text we will only consider real-valued functions. We denote the norm on $L^2(\Omega)$ by $\|\cdot\|$.

We denote by $L^\infty(\Omega)$ = the class of all (essentially) bounded measurable functions in Ω.

For $m \in \mathbb{N}^\star$, we denote by $H^m(\Omega)$ the Sobolev space defined by $H^m(\Omega) = \{v \in L^2(\Omega), \partial^\alpha u \in L^2(\Omega) \forall |\alpha| \le m\}$, where the derivatives are taken in the sense of distributions on Ω.

Moreover, we do not distinguish between the norms of vector valued and scalar functions, e.g. the L^2 norm of the gradient of u is still denoted by $\|\nabla u\|$ while $\|\nabla^2 u\|$ stands for $(\sum_{i,j=1}^n \|\partial_i \partial_j u\|^2)^{1/2}$.

1.3   Strong Stabilization

Abstract One important property of dissipative systems is that trajectories decay along a Lyapunov function which is in that case the energy. If trajectories are relatively compact in appropriate spaces and the system is autonomous, then one can prove that trajectories converge to equilibria asymptotically. Dafermos and Lasalle used this property to study convergence of solutions toward equilibrium and to prove strong stabilization. The purpose of this section is to recall these important results.

Definition 1.3.1. Let $T(t)_{t \ge 0}$ be a continuous semigroup on a Banach X. We recall that the ω-limit set of z_0 in X, denoted by $\omega(z_0)$, is defined by

$$\omega(z_0) = \{z \in X, \exists (t_n)_n \in [0, \infty) \text{ such that } t_n \longrightarrow \infty, z = \lim_{n \to \infty} T(t_n)z_0\}.$$

We recall the following basic result on ω-limit sets.

Proposition 1.3.2. *Let $(T(t))_{t\geq 0}$ be a continuous semigroup on a Banach space X and $z_0 \in X$ be given such that the orbit $\gamma(z_0) \equiv \cup_{t\geq 0} T(t)z_0$ is relatively compact in X. Then*

$$d(T(t)z_0, \omega(z_0)) \longrightarrow 0 \quad t \longrightarrow \infty ,$$

and the ω-limit set of z_0 is a non-empty compact, connected subset of X and it is invariant under the action of the semigroup.

Remark 1.3.3. A proof is given in [46] (see also [60]).

1.3.1 Dafermos' Strong Stabilization Result

We present in this section Dafermos' strong stabilization result and Lasalle principle.

Let Ω be a bounded open connected subset of $\mathbb{R}^N$ with a smooth boundary denoted by Γ and ω be an open subset of Ω of positive measure.

We set $H = L^2(\Omega)$ and $V = H_0^1(\Omega)$ in all what follows. We consider the following second order equation:

$$\begin{cases} u_{tt} - \Delta u + a(.)g(u_t) = 0 , & t > 0 , x \in \Omega \\ u = 0 , & t > 0 , x \in \Gamma , \\ u(0)(.) = u^0(.) , \ u'(.) = u^1(.) , & x \in \Omega , \end{cases} \tag{1.10}$$

where a is a smooth function and satisfies $a(x) \geq 0, x \in \Omega, a > 0$ in a nonempty subset ω of Ω; g is assumed to be continuously differentiable, strictly monotone with $g(0) = 0$ and g' bounded on $\mathbb{R}$ (for simplicity).

We recall the following existence proof.

Theorem 1.3.4. *Let $(u^0, u^1) \in V \times H$, the problem (1.10) has a unique solution $u \in \mathscr{C}([0, +\infty); V) \times \mathscr{C}^1([0, +\infty); H)$.*
Moreover, for all $(u^0, u^1) \in H^2(\Omega) \cap H_0^1(\Omega) \times V$, the solution is in $L^\infty([0, +\infty); H^2(\Omega) \cap H_0^1(\Omega)) \cap W^{1,\infty}([0, +\infty); V) \cap W^{2,\infty}([0, +\infty); H)$ and its energy defined by:

$$E(t) = \frac{1}{2}\left(|u'(t)|_H^2 + |\nabla u(t)|_H^2\right),$$

satisfies the following dissipation relation:

$$E'(t) = -\int_{\Omega} a(x)u_t(t)(x)g(u_t(t)(x)) \, dx \leq 0 . \tag{1.11}$$

The proof is based on semigroup theory. We define the unbounded operator $A(u, p) = (p, \Delta u - ag(p))$ with domain $\mathscr{D}(A) = (H^2(\Omega) \cap H_0^1(\Omega)) \times H_0^1(\Omega)$ in the energy space $V \times H$ equipped with the natural product norm. Then $-A$ is a maximal monotone operator and generates a continuous semigroup $(T(t))_{t\geq 0}$ on $V \times H$.

We note that the above system has a unique stationary solution, that is *a unique equilibrium state*, given by $u \equiv 0$.

Theorem 1.3.5 (Dafermos [46]). *Let* $(u^0, u^1) \in V \times H$, *then*

$$E(t) \longrightarrow 0 \quad , \quad t \longrightarrow \infty. \tag{1.12}$$

The proof of this result is based on the Lasalle invariance principle [67, 68] and on the dissipation properties of the energy of first order, that is the energy of (u_t, u_{tt}) for smoother initial data.

Proposition 1.3.6 (Lasalle invariance principle). *Let* $(T(t))_{t\geq 0}$ *be a continuous semigroup on a Banach space* X *such that the orbit* $\gamma(z_0) \equiv \cup_{t\geq 0} T(t)z_0$ *of a point* $z_0 \in X$ *is relatively compact in* X. *Assume that* $V : X \mapsto \mathbb{R}$ *is continuous and is a Lyapunov function for* $(T(t))_{t\geq 0}$, *that is* $V(T(t)z) \leq V(z)$ *for all* $z \in X$ *and all* $t \geq 0$. *Then* V *is constant on the* ω-*limit set of* z_0.

Proof. Since $(T(t))$ satisfies the semigroup property, V is a Lyapunov function, and the orbit of z_0 is relatively compact in X, we deduce that $t \mapsto V(T(t)z_0)$ is a non-increasing bounded function. Hence there exists $V_\infty \in X$ such that $V_\infty = \lim_{t\to\infty} V(T(t)z_0)$. Let $z \in \omega(z_0)$ be given. Then there exists $(t_n)_n \to \infty$, such that $z = \lim_{n\to\infty} T(t_n)z_0$. Since V is continuous on X, we have $V(z) = \lim_{n\to\infty} V(T(t_n)z_0) = V_\infty$. Hence V is constant on the ω-limit set of z_0. $\qquad\square$

Proof (of Theorem 1.3.5). Let $(u^0, u^1) \in (H^2(\Omega) \cap H_0^1(\Omega)) \times H_0^1(\Omega)$. We denote by $(T(t))_{t\geq 0}$ the continuous semigroup generated by the unbounded A defined above and $(u, v) = T(t)(u^0, u^1)$. Differentiating the first equation of (1.10) with respect to time, we obtain

$$v_{tt} = \Delta v - a(x)g'(v)v_t \tag{1.13}$$

Defining

$$E_1(t) = \frac{1}{2}\left(|v_t(t)|_H^2 + |\nabla v(t)|_H^2, \right),$$

we derive from (1.13) that

$$E_1'(t) = -\int_\Omega a(x)g'(v)v_t^2 \, dx \leq 0.$$

Hence, the sets $\{v(t,.), t \geq 0\}$ and $\{v_t(t,.), t \geq 0\}$ are bounded respectively in $H_0^1(\Omega)$ and $L^2(\Omega)$. Thus, due to Rellich theorem the set $\{v(t,.), t \geq 0\}$ is relatively compact in $L^2(\Omega)$. Using the first equation in (1.10) and the hypotheses on a and g, we deduce that the set $\{\Delta u(t,.), t \geq 0\}$ is bounded in $L^2(\Omega)$. Therefore, the set $\{u(t,.), t \geq 0\}$ is relatively compact in $H_0^1(\Omega)$.

On the other hand, thanks to (1.11), E is a Lyapunov function for $T(t)$ which is continuous on $V \times H$. Using Lasalle's invariance principle (Proposition 1.3.6), we deduce that E is constant on $\omega(u^0, u^1)$. Let $(z_0, z_1) \in \omega(u^0, u^1)$ be given and set $(z, q)(t) = T(t)(z_0, z_1)$. Since E is constant on $\omega(u^0, u^1)$ and thanks to the dissipation relation written for (z, q), we deduce that

$$\int_\Omega a(x)qg(q)\, dx = 0\,,$$

so that thanks to our assumptions on a and g, we have $a(x)g(q(t, x)) \equiv 0$ in Ω. Hence (z, q) satisfies

$$\begin{cases} z_{tt} - \Delta z = 0 & t > 0\,, x \in \Omega \\ z = 0\,, & t > 0\,, x \in \Gamma\,, \\ z(0)(.) = z_0(.)\,, \ z_t(.) = z_1(.) & \in \Omega\,, \\ z_t = 0 \text{ on } \{y \in \Omega\,, a(y) \neq 0\}\,. \end{cases}$$

Thanks to the fact that $a > 0$ on an open subset of Ω and to a well-known unique continuation result for wave equation, we deduce that $z = z_t \equiv 0$, so that $(z_0, z_1) = (0, 0)$. Hence $\omega(u^0, u^1) = \{0\}$. Since $H^2(\Omega) \cap H_0^1(\Omega) \times H_0^1(\Omega)$ is dense in $V \times H$ and thanks to the dissipation relation, we conclude that the result holds true for initial data in $V \times H$.   □

Remark 1.3.7. Dafermos' result has been extended to the case where g is assumed to be a maximal monotone graph in Haraux [59]. A presentation of Lasalle invariance principle can also be found in [65] for the case of linearly boundary damped wave equation. Many authors have contributed to strong and weak stabilization of PDE's, we refer to the reader to the above papers to find more references on this subject.

1.4   Linear Stabilization

Abstract The purpose of this section is to consider the case of linear stabilization of the wave equation by a locally distributed feedback. We recall the geometrical conditions on the damping region both from the multiplier and geometric optics conditions. We then present the piecewise multiplier method by Zuazua (Comm. Part. Differ. Equat. 15:205–235, 1990) for the case of one or several observation points and Liu (SIAM J. Contr. Optim. 35:1574–1590, 1997) with a point of view which will prepare the nonlinear stabilization case. In particular, we stress the fact that the multiplier method allows to identify a dominant energy, namely the kinetic energy. The exponential decay of the energy, provided that the damping region satisfies the piecewise multiplier condition, is proved thanks to a linear Gronwall inequality. We conclude this section by the extension of this result on the semilinear wave equation by the compactness–uniqueness method (Zuazua, Comm.

Part. Differ. Equat. 15:205–235, 1990; Rauch and Taylor, Indiana Univ. Math. J.
24:79–86, 1974).

1.4.1   Introduction

Most of the results below have been derived and extended to other PDE's by
many authors. We refer the reader to Lions [76] for a wide view on the subject of
control theory for PDE's and in particular on exact controllability and observability
questions; many of the enclosed results have been used for stabilization, in particular
the multiplier method. Boundary stabilization for wave-like equations, together with
a presentation of the multiplier method is given in Komornik [65]. Moreover a
recent book by Tucsnak and Weiss [99] considers many aspects of observability
and control questions for operators semigroup. We also refer the reader to two
recent excellent books by Coron [45] for stabilization of nonlinear control systems,
and in particular of $2 - D$ Euler equations and by Dáger and Zuazua [49] for
control questions on one-dimensional flexible multi-structures (see also [100] for
stabilization results for one-dimensional networks). For stabilization questions and
in particular the construction of feedbacks through the algebraic Riccati approach,
together with well-documented PDE's examples, we refer the reader to the book by
Lasiecka and Triggiani [73]. For a dynamical system approach, the interested reader
will find many important results in Haraux [58, 60]. All these books are also a large
source of references.

We now consider the case of a linearly locally damped wave-equation

$$\begin{cases} u_{tt} - \Delta u + a(x)u_t = 0 & \text{in} \quad (0, \infty) \times \Omega, \\ u = 0 \quad \text{on} \quad \Sigma = (0, \infty) \times \Gamma \\ (u, u_t)(0) = (u^0, u^1) \quad \text{on} \quad \Omega, \end{cases} \tag{1.14}$$

in a bounded domain $\Omega \subset \mathbb{R}^N$ with a smooth boundary Γ, $a \geq 0$ almost everywhere
in Ω. We define the energy as before. For strong solutions, it satisfies the following
dissipation relation:

$$E'(t) = - \int_{\Omega} a(x)|u_t(t, x)|^2 \, dx \leq 0. \tag{1.15}$$

1.4.2   Geometrical Aspects

The property of *finite speed of propagation* holds for the wave equation. It means
that, if the initial conditions u^0, u^1 have compact support, then the support of $u(t, \cdot)$
evolves in time at a finite speed. Thus, if the support of the set on which the damping
is active is such that some rays issued do not meet it, one can build solutions of the
damped wave equation whose energy concentrates close to these rays, so that the
energy is not decaying uniformly as time increases. This explains why, for the wave

equation, the geometry of Ω, in particular its size and localization, plays an essential role in all the issues related to control and stabilization.

We denote the support region in which the feedback is active by ω. It is taken as a subset of Ω of positive Lebesgue measure. More precisely, a is assumed to be continuous on $\overline{\Omega}$ and such that

$$a \geq 0 \quad \text{on} \quad \Omega \quad \text{and} \quad a \geq a_0 \quad \text{on} \quad \omega, \tag{1.16}$$

for some constant $a_0 > 0$. In this case, the feedback is said to be *distributed*. Moreover, it is said to be *globally* distributed if $\omega = \Omega$ and *locally* distributed if $\Omega \setminus \omega$ has positive Lebesgue measure.

The *multiplier method* and the method of geometric optics based on *microlocal analysis* are the two main methods which have been developed and used to study first exponential energy decay rates for hyperbolic reversible damped equations, and then extended to treat nonlinear dampings, source terms ... The method of geometric optics gives the sharpest geometrical results. The seminal work of Bardos et al. [27], gives geodesics sufficient conditions on the region of active control for exact controllability, and consequently of stabilization to hold. These conditions say that each ray of geometric optics should meet the control region. Burq and Gérard [36] showed that these results hold under weaker regularity assumptions on the domain and coefficients of the operators (see also [35, 38, 39]). These geodesics conditions are not explicit, in general, but they allow to get decay estimates of the energy under very general hypotheses.

The multiplier method is an explicit method. It is based on energy estimates and Gronwall inequalities. The cases of boundary control and stabilization is considered in the fundamental contributions of several authors, as Ho [42, 62], Lions [76], Lasiecka Triggiani, Komornik–Знатіа [64], and many others. Zuazua [105] gives an explicit geometric condition on ω for a semilinear wave equation subject to a damping locally distributed on ω, namely it should contain an ε-neighborhood of the whole boundary of Ω. His result is presented in Sects. 2.3 of 1.2. Using the observability estimates obtained by Fu et al. [54] combined with Zuazua's results [105], Zuazua's exponential stabilization result extend to the geometric case for which ω contains only a ε-neighborhood of the part of the boundary which is not visible from a given point x_0 of $\mathbb{R}^N$, which symbolizes an observator placed at this location. Such a condition excludes the case of a damping coefficient which vanishes at both poles of a sphere. Liu [78] piecewise multiplier method generalizes Zuazua's result (see also [84]). It allows several observation points and thus to relax the geometric assumptions as we will see below.

Another important tool is based on sharp trace regularity results, established by Lasiecka and Triggiani [72, 73]. This method which allows to estimate boundary terms in energy estimates. There also exist intermediate results between the geodesics conditions of Bardos–Lebeau–Rauch and the multiplier method, obtained by Miller [86] using differentiable escape functions. We do not address in these Notes observability estimates and exact controllability results for wave-like equations. For such questions, we recall that the method of Carleman estimates is also one of the major tool for such results. Results using Carleman estimates have been recently derived by Tebou [97].

Let us now explicit Zuazua's and K. Liu's extension. Zuazua's multiplier geometric condition can be described as follows. If a subset O of $\overline{\Omega}$ is given, one can define an ε-neighbourhood of O in $\overline{\Omega}$ as the subset of points of Ω which are at distance at most ε of O. Zuazua proved that if the set ω is such that there exists a point $x_0 \in \mathbb{R}^n$—an *observation* point—for which ω contains an ε-neighbourhood of $\Gamma(x^0) = \{x \in \partial\Omega , (x - x^0) \cdot \nu(x) \geq 0\}$, then the energy decays exponentially. In this note, we refer to this condition as (MGC).

If a vanishes for instance in a neighbourhood of the two poles of a ball Ω in $\mathbb{R}^n$, one cannot find an observation point x_0 such that (MGC) holds. Liu [78] (see also [84]) introduced a piecewise multiplier method which allows to choose several observation points, and therefore to handle the above case. Let us denote by $x^j \in \mathbb{R}^N$, $j = 1, \ldots, J$, these distinct observation points. Introduce disjoint lipschitzian domains Ω_j of Ω, $j = 1, \ldots, J$, and define

$$\gamma_j(x^j) = \{x \in \partial\Omega_j , (x - x^j) \cdot \nu_j(x) \geq 0\}$$

Here ν_j stands for the unit outward normal vector to the boundary of Ω_j. Then the piecewise multiplier geometrical condition for ω is :

(PWMGC) $\left\{ \omega \supset \mathcal{N}_\varepsilon\left(\cup_{j=1}^J \gamma_j(x^j) \cup \left(\Omega \setminus \cup_{j=1}^J \Omega_j\right) \right) \right.$

It will be denoted by (PWMGC) condition in the sequel.

With this method, one can handle the situation in which a vanishes in a neighbourhood of the two poles of a ball in $\mathbb{R}^n$ as follows. One chooses two subsets Ω_1 and Ω_2 containing, respectively, the two regions where a vanishes and apply the piecewise multiplier method with $J = 2$ and with the appropriate choices of two observation points and ε.

Let us now present in a few words, the principles of the multiplier method. It consists of integrating by parts expressions of the form

$$\int_t^T \int_O \left(u_{tt} - \Delta u + a(x)u_t\right) M u \, dx \, dt = 0 \quad 0 \leq t \leq T ,$$

where u stands for a (strong) solution of (1.14), with an appropriate choice of Mu. Multipliers have generally the form

$$Mu = \left(m(x) \cdot \nabla u + c\,u\right)\psi(x) ,$$

where m depends on the observation points and ψ is a cut-off function. Other multipliers of the form $Mu = \Delta^{-1}(\beta u)$, where β is a cut-off function and Δ^{-1} is the inverse of the Laplacian operator with homogeneous Dirichlet boundary conditions, have also been used in the literature (see [44]).

Theorem 1.4.1. *Assume that the geometric conditions (MGC) or (PWMGC) holds. Let $(u^0, u^1) \in V \times H$ be given and denote by u a solution of (1.14) and E its energy. Then E satisfies an estimate of the form*

$$\int_t^T E(s)\, ds \leq C_1 E(t) + C_2 \int_t^T \left(\int_\Omega a(x)|u_t|^2 + \int_\omega |u_t|^2 \right) ds \quad t \geq 0. \quad (1.17)$$

where C_i, $i = 1, 2$ denote generic positive constants.

Proof. We will prove that the energy of solutions satisfies an estimate of the form (1.17) using the piecewise multiplier method. We proceed as in [84]. We set just for commodity

$$\rho(x, s) = a(x)s \quad x \in \Omega, s \in \mathbb{R}. \quad (1.18)$$

Let $(u^0, u^1) \in H^2(\Omega) \cap H_0^1(\Omega) \times H_0^1(\Omega)$. Let $\varepsilon_0 < \varepsilon_1 < \varepsilon_2 < \varepsilon$ and define for $i = 0, 1, 2$:

$$Q_i = \mathcal{N}_{\varepsilon_i} \left[\cup_j \Gamma_j(x_j) \cup (\Omega \setminus \cup_j \Omega_j) \right] \quad (1.19)$$

Since $(\overline{\Omega_j} \setminus Q_1) \cap \overline{Q_0} = \emptyset$, we can construct a function $\psi_j \in \mathscr{C}_0^\infty(R^N)$ which satisfies

$$0 \leq \psi_j \leq 1, \psi_j = 1 \text{ on } \overline{\Omega_j} \setminus Q_1, \psi_j = 0 \text{ on } Q_0. \quad (1.20)$$

We define the $\mathscr{C}^1$ vector field on Ω:

$$h(x) = \begin{cases} \psi_j(x) m_j(x) \text{ if } x \in \Omega_j \\ 0 \text{ if } x \in \Omega \setminus \cup_j \Omega_j, \end{cases}$$

and the multiplier $h \cdot \nabla u$ Proceeding as in [84], we consider the expression

$$\int_S^T \int_{\Omega_j} h(x) \cdot \nabla u(u'' - \Delta u + \rho(x, u'))\, dx\, dt = 0.$$

For the sake of concision, we will omit the $dx\, dt$ in the following integrals. This gives, after appropriate integration by parts

$$\int_S^T \int_{\Gamma_j} \left(\partial_{v_j} u h \cdot \nabla u + \frac{1}{2}(h \cdot v)(u'^2 - |\nabla u|^2) \right) = \left[\int_{\Omega_j} u' h \cdot \nabla u \right]_S^T$$

$$+ \int_S^T \int_{\Omega_j} \left(\frac{1}{2} \operatorname{div} h\, (u'^2 - |\nabla u|^2) + \sum_{i,k} \frac{\partial h_k}{\partial x_i} \frac{\partial u}{\partial x_i} \frac{\partial u}{\partial x_k} + \rho(x, u')h \cdot \nabla u \right). \quad (1.21)$$

Thanks to the choice of ψ_j, only the boundary term on $(\Gamma_j \setminus \gamma_j(x_j)) \cap \Gamma$ is nonvanishing in the left hand side of (1.21). But on this part of the boundary $u = 0$, so that $u' = 0$ and $\nabla u = \partial_v u\, v = \partial_{v_j} u\, v_j$. Hence, the left hand side of (1.21) reduces to

$$\frac{1}{2} \int\limits_{S}^{T} \int\limits_{(\Gamma_j \setminus \gamma_j(x_j)) \cap \Gamma} (\partial_{\nu_j} u)^2 \psi_j (m_j \cdot \nu_j) \leq 0. \tag{1.22}$$

Therefore, since $\psi_j = 0$ on Q_0 and thanks to the above inequality used in (1.21), we deduce that

$$\left[\int\limits_{\Omega_j} u'h \cdot \nabla u \right]_S^T + \int\limits_{S}^{T} \int\limits_{\Omega_j \setminus Q_0} \left(\frac{1}{2} divh \, (u'^2 - |\nabla u|^2) \right. \tag{1.23}$$

$$\left. + \sum_{i,k} \frac{\partial h_k}{\partial x_i} \frac{\partial u}{\partial x_i} \frac{\partial u}{\partial x_k} + \rho(x, u')h \cdot \nabla u \right) \leq 0.$$

Using $\psi_j = 1$ on $\overline{\Omega}_j \setminus Q_1$ and summing the resulting inequalities on j, we obtain

$$\left[\int\limits_{\Omega} u'h \cdot \nabla u \right]_S^T + \int\limits_{S}^{T} \int\limits_{\Omega \setminus Q_1} \frac{1}{2} (Nu'^2 + (2 - N)|\nabla u|^2) + \int\limits_{S}^{T} \int\limits_{\Omega} \rho(x, u')h \cdot \nabla u \tag{1.24}$$

$$\leq -\sum_j \int\limits_{S}^{T} \left[\int\limits_{\Omega_j \cap Q_1} \frac{1}{2} div(\psi_j m_j) \, (u'^2 - |\nabla u|^2) + \sum_{i,k} \frac{\partial h_k}{\partial x_i} \frac{\partial u}{\partial x_i} \frac{\partial u}{\partial x_k} \right]$$

$$\leq C \int\limits_{S}^{T} \int\limits_{\Omega \cap Q_1} (u'^2 + |\nabla u|^2),$$

where C is a positive constant which depends only on ψ_j and m_j. We now use the second multiplier $u(N-1)/2$ and therefore evaluate the term

$$\frac{N-1}{2} \int\limits_{S}^{T} \int\limits_{\Omega} u(u'' - \Delta u + \rho(x, u')) = 0.$$

Hence, one has

$$\frac{N-1}{2} [\int\limits_{\Omega} uu']_S^T + \frac{N-1}{2} \int\limits_{S}^{T} \int\limits_{\Omega} |\nabla u|^2 - u'^2 + u\rho(x, u') = 0. \tag{1.25}$$

We set $M(u) = h \cdot \nabla u + \frac{N-1}{2} u$. Adding (1.25) to (1.24), we obtain

$$\int\limits_{S}^{T} E \, dt \leq C \int\limits_{S}^{T} \int\limits_{\Omega \cap Q_1} |\nabla u|^2 + u'^2 - [\int\limits_{\Omega} M(u)u']_S^T \tag{1.26}$$

$$- \int\limits_{S}^{T} \int\limits_{\Omega} M(u)\rho(x, u').$$

We estimate the terms on the right hand side of (1.26) as follows. First, since E is nonincreasing, we deduce that

$$\left| \left[\int_{\Omega} M(u)u' \right]_S^T \right| \leq CE(S) . \tag{1.27}$$

We estimate the last term of (1.26) as follows

$$\left| \int_S^T \int_{\Omega} M(u)\rho(x, u') \right| \leq \frac{C}{\delta} \int_S^T \int_{\Omega} |\rho(x, u')|^2 + \delta \int_S^T E \, dt \quad \forall \delta > 0 . \tag{1.28}$$

The difficulty is now to estimate the first term on the right hand side of (1.26). We just follow here the usual technique for the wave equation as developed in [84]. We give the steps for the sake of the completeness. Since $\overline{\mathbb{R}^N \setminus Q_2} \cap \overline{Q_1} = \emptyset$, there exists a function $\xi \in \mathscr{C}_0^{\infty}(\mathbb{R}^N)$ such that

$$0 \leq \xi \leq 1 , \xi = 1 \text{ on } Q_1 , \xi = 0 \text{ on } R^N \setminus Q_2 . \tag{1.29}$$

Multiplying the first equation of (1.14) by ξu and integrating the resulting equation on $[S, T] \times \Omega$, we obtain after several integration by parts:

$$\int_S^T \int_{\Omega} \xi |\nabla u|^2 = \int_S^T \int_{\Omega} \xi |u'|^2 + \frac{1}{2} \Delta \xi u^2 \tag{1.30}$$

$$- \left[\int_{\Omega} \xi u u' \right]_S^T - \int_S^T \int_{\Omega} \xi u \rho(x, u') .$$

Hence, we have

$$\int_S^T \int_{\Omega \cap Q_1} |\nabla u|^2 \leq C \int_S^T \int_{\Omega \cap Q_2} |u'|^2 + u^2 + |\rho(x, u')|^2 + CE(S) . \tag{1.31}$$

Since $\overline{\mathbb{R}^N \setminus \omega} \cap \overline{Q_2} = \emptyset$, there exists a function $\beta \in \mathscr{C}_0^{\infty}(\mathbb{R}^N)$ such that

$$0 \leq \beta \leq 1 , \beta = 1 \text{ on } Q_2 , \beta = 0 \text{ on } R^N \setminus \omega . \tag{1.32}$$

We fix t and consider the solution z of the following elliptic problem:

$$\Delta z = \beta(x)u , \ x \in \Omega , \tag{1.33}$$

$$z = 0 \text{ on } \Gamma . \tag{1.34}$$

Hence, z and z' satisfy the following estimates:

$$|z|_{L^2(\Omega)} \leq C |u|_{L^2(\Omega)} , \tag{1.35}$$

$$|z'|_{L^2(\Omega)}^2 \leq C \int_{\Omega} \beta |u'|^2 . \tag{1.36}$$

Multiplying the first equation of (1.14) by z and integrating the resulting equation on $[S, T] \times \Omega$, we obtain after integration by parts:

$$\int_S^T \int_\Omega \beta u^2 = \left[\int_\Omega z u' \right]_S^T + \int_S^T \int_\Omega -z'u' + z\rho(x, u') . \tag{1.37}$$

Hence, using the estimates (1.35) and (1.36) in the above relation, we have

$$\int_S^T \int_{\Omega \cap Q_2} |u|^2 \le \frac{C}{\eta} \int_S^T \int_\omega |u'|^2 + \frac{C}{\eta} \int_S^T \int_\Omega |\rho(x, u')|^2 + \eta \int_S^T E$$

$$+ CE(S) \quad \forall\, \eta > 0 . \tag{1.38}$$

We now use the estimates (1.27), (1.28), (1.31) and (1.38) in (1.26). This gives

$$\int_S^T E \le CE(S) + C\delta \int_S^T E + \frac{C}{\delta} \int_S^T \left[\int_\omega |u'|^2 + \int_\Omega |\rho(x, u')|^2 \right] \quad \forall\, \delta > 0 . \tag{1.39}$$

Choosing δ sufficiently small, we obtain finally

$$\int_S^T E \le CE(S) + C \int_S^T \left[\int_\omega |u'|^2 + \int_\Omega |\rho(x, u')|^2 \right] . \tag{1.40}$$

Hence, we proved that E satisfies an estimate of the form (1.17).   □

Once this estimate is proved, one can use the dissipation relation to prove that the energy satisfies integral inequalities of Gronwall type. This is easy in the linear case and is the subject of the next section.

1.4.3 Exponential Decay for Linear Feedbacks

Exponential decay will be deduced from (1.17) using the following linear Gronwall inequality (see also [56]):

Theorem 1.4.2 (Komornik [65], Theorem 8.1). *Let* $E : [0, \infty) \mapsto [0, \infty)$ *be a non-increasing function satisfying, for some constant* $T > 0$, *the linear Gronwall inequality*

$$\int_t^\infty E(s)\, ds \le TE(t), \quad \forall\, t \ge 0 . \tag{1.41}$$

Then, E satisfies

$$E(t) \le E(0)e^{1-t/T}, \quad t \ge 0 . \tag{1.42}$$

Proof. Define

$$f(t) = \int_t^\infty E(s)\, ds\,, \quad t \geq 0\,.$$

Thanks to (1.41), f satisfies

$$Tf'(t) + f(t) \leq 0\,, \quad \forall t \geq 0\,,$$

so that

$$f(t)e^{t/T} \leq f(0) = \int_0^\infty E(s)\, ds \leq TE(0)\,.$$

Hence, we have

$$\int_t^\infty E(s)\, ds \leq TE(0)e^{-t/T}\,, \quad t \geq 0\,.$$

Since E is a nonnegative and nonincreasing function

$$TE(t) \leq \int_{t-T}^t E(s)\, ds \leq \int_{t-T}^\infty E(s)\, ds \leq TE(0)e^{-(t-T)/T}\,,$$

so that (1.42) is proved.   □

Remark 1.4.3. One can remark that for $t \leq T$, $E(t) \leq E(0)e^{1-t/T}$.

We can establish:

Theorem 1.4.4. *Assume the hypotheses of Theorem 1.4.1. Let $(u^0, u^1) \in V \times H$ be given and denote by u a solution of* (1.14) *and E its energy. Then E satisfies*

$$E(t) \leq C\, E(0)e^{-\gamma t} \quad \forall t \geq 0 \tag{1.43}$$

Proof. Using the dissipation relation (1.15), one has

$$\int_t^T \int_\Omega a|u_t|^2\, dx\, ds \leq \int_t^T -E'(s)\, ds \leq E(t)\,, 0 \leq t \leq T\,.$$

On the other hand, thanks to assumption (1.16) on the coefficient a, we have

$$\int_t^T \int_\omega u_t^2\, dx\, ds \leq \frac{1}{a_0} \int_t^T \int_\Omega a|u_t|^2\, dx\, ds \leq \frac{1}{a_0} E(t)\,, 0 \leq t \leq T\,.$$

By the above inequalities and (1.17), E satisfies

$$\int_t^T E(s)\, ds \leq cE(t)\,, \quad \forall\, 0 \leq t \leq T\,. \tag{1.44}$$

Thanks to Theorem 1.4.2, E decays exponentially at infinity. Extension to initial data only in the energy space is easy by using density of $(H^2(\Omega) \cap H_0^1(\Omega)) \times H_0^1(\Omega)$ in the energy space and on the dissipativity property. $\qquad\square$

Remark 1.4.5. An alternative method is to introduce a modified (or perturbed) energy E_ε which is equivalent to the natural one for small values of the parameter ε as in Komornik and Zuazua [64]. Then one shows that this modified energy satisfies a differential inequality so that it decays exponentially at infinity. The exponential decay of the natural energy follows then at once. In this case, the modified energy is indeed a strict Lyapunov function for the PDE. The natural energy cannot be in general such a strict Lyapunov function due to the finite speed of propagation (consider initial data which have compact support compactly embedded in $\Omega \setminus \omega$). There are also very interesting approaches using the frequency domain approach, or spectral analysis such as developed by Liu [78], Liu and Zheng [79].

In the sequel, we concentrate on the integral inequality method. This method has been generalized in several directions and we present in this course some results concerning extensions to

- Nonlinear feedback
- Indirect or single feedback for coupled system
- Memory type feedbacks

1.4.4 The Compactness–Uniqueness Method

We shall present in this part, the compactness–uniqueness method as in Zuazua [105] (see [91] for the first introduction of compactness arguments), to prove exponential decay of the energy for solutions of linearly locally damped semilinear wave equations. This method is used for extension of the previous results to the semilinear wave equation, that is

$$\begin{cases} u_{tt} - \Delta u + f(u) + a(x)u_t = 0 & \text{in} \quad (0, \infty) \times \Omega, \\ u = 0 & \text{on} \quad (0, \infty) \times \Gamma \\ (u, u_t)(0) = (u^0, u^1) & \text{on} \quad \Omega, \end{cases} \tag{1.45}$$

in a bounded connected domain $\Omega \subset \mathbb{R}^N$ with a smooth boundary Γ and where $a \in L^\infty(\Omega)$ satisfies (1.16) with $a_0 > 0$. We assume that $f \in \mathscr{C}^1(\mathbb{R})$ satisfies

$$f(s)s \geq 0 \quad \forall s \in \mathbb{R}. \tag{1.46}$$

together with the growth condition

$$\begin{cases} \exists\, M > 0, \text{ and } p > 1 \text{ such that } (N-2)p \leq N \\ |f(x) - f(y)| \leq M\left(1 + |x|^{p-1} + |y|^{p-1}\right)|x - y|, \quad \forall\, x, y \in \mathbb{R} \end{cases} \tag{1.47}$$

Under these conditions, the above problem is well-posed in the energy space $H_0^1(\Omega) \times L^2(\Omega)$. We define the energy of solutions by

$$E(t) = \frac{1}{2} \int_\Omega \left(|u'(t)|^2 + |\nabla u(t)|^2 \right) dx + \int_\Omega F(u)\, dx, \qquad (1.48)$$

where

$$F(s) = \int_0^s f(y)\, dy. \qquad (1.49)$$

For strong solutions, the energy satisfies the following dissipation relation:

$$E'(t) = -\int_\Omega a(x)|u_t(t,x)|^2\, dx \leq 0. \qquad (1.50)$$

For the sake of concision, we shall present Zuazua's result in the superlinear case (see [105] for the globally Lipschitz case), that is when f satisfies in addition

$$\exists \delta > 0, \text{ such that } f(s)s \geq (2 + \delta)F(s) \quad \forall s \in \mathbb{R}. \qquad (1.51)$$

For an arbitrary point $x_0 \in \mathbb{R}^N$, we set $m(x) = x - x_0$ and define

$$\Gamma(x_0) = \{x \in \Gamma, m(x) \cdot v(x) > 0\}, \qquad (1.52)$$

where v is the unit outward normal at Γ.

Theorem 1.4.6 (Zuazua [105], Theorem 2.1). *Assume that $a \in L^\infty(\Omega)$ satisfies (1.16) with $a_0 > 0$, where ω is a neighbourbood of Γ. Assume moreover that $f \in \mathscr{C}^1(\mathbb{R})$ satisfies (1.46)–(1.47) and (1.51). Then there exists $\gamma > 0$ and $C > 0$ such that for any initial data $(u^0, u^1) \in H_0^1(\Omega) \times L^2(\Omega)$, the energy satisfies*

$$E(t) \leq CE(0)e^{-\gamma t} \quad \forall\, t \geq 0. \qquad (1.53)$$

Proof. Let $T > 0$ be fixed sufficiently large. Let $\alpha \in R$ be a positive constant that will be suitably chosen later on. We use the multiplier $Mu = (x - x_0) \cdot \nabla u + \alpha u$ and evaluate by integration by parts the left hand side of

$$\int_0^T \int_\Omega (u_{tt} - \Delta u + f(u) + au_t)Mu\, dx\, dt = 0$$

This gives

$$\left[\int_\Omega (u_t Mu + \alpha a u^2/2) \right]_0^T + (N/2 - \alpha)\int_0^T \int_\Omega u_t^2 + (1 + \alpha - N/2)\int_0^T \int_\Omega |\nabla u|^2$$

$$+\alpha \int_0^T \int_\Omega u f(u) - N \int_0^T \int_\Omega F(u) + \int_0^T \int_\Omega au_t m \cdot \nabla u$$

$$= \int\limits_0^T 1/2 \int\limits_\Gamma m \cdot v \left|\frac{\partial u}{\partial v}\right|^2 \leq \int\limits_0^T 1/2 \int\limits_{\Gamma(x_0)} m \cdot v \left|\frac{\partial u}{\partial v}\right|^2$$

Thanks to (1.51), there exists $\alpha \in ((N-2)/2, N/2)$ if $N > 1$ and $\alpha \in (0, 1/2)$ if $N = 1$ such that

$$f(s)s \geq \frac{N+\theta}{\alpha} F(s) \quad \forall\, s \in \mathbb{R}.$$

for a certain $\theta > 0$. Using this property together with some standard estimates, we obtain

$$\int\limits_0^T E(t)\,dt \leq C \left(\int\limits_0^T \int\limits_{\Gamma(x_0)} m \cdot v \left|\frac{\partial u}{\partial v}\right|^2 + \int\limits_0^T \int\limits_\Omega a|u_t|^2 + \left|\left[\int\limits_\Omega (u_t M u + \alpha a u^2/2) \right]_0^T \right| \right) \tag{1.54}$$

Let $\widetilde{\omega}$ be a neighbourhood of $\Gamma(x_0)$ compactly included in ω. Then there exists a vector field $h \in (W^{1,\infty}(\Omega))^N$ such that $h = v$ on $\Gamma(x_0)$, $h \cdot v \geq 0$ a.e. on Γ and $h = 0$ in $\Omega \backslash \widetilde{\omega}$ (see [76] for a proof). Moreover there exists a function $\eta \in W^{1,\infty}(\Omega)$, such that $0 \leq \eta \leq 1$ a.e. in Ω, $\eta = 1$ a.e. in $\widetilde{\omega}.$, $\eta = 0$ a.e. in $\Omega \backslash \omega$, and $|\nabla \eta|^2/\eta \in L^\infty(\omega)$ (see [76] for a proof). Using successively the multipliers $h \cdot \nabla u$ and ηu, one can control the L^2 norm in $L^2([0, T] \times \Gamma(x_0))$ of the trace of the normal derivative of u and the L^2 norm in $L^2([0, T] \times \Omega)$ of $|\nabla u|$. Combining this with (1.54), we have

$$\int\limits_0^T E(t)\,dt \leq C \left(\int\limits_0^T \int\limits_\Omega (a|u_t|^2 + u^2) + \left|\left[\int\limits_\Omega (u_t M u + \alpha a u^2/2) \right]_0^T \right| + \left|\left[\int\limits_\Omega \eta u(u_t + au/2) \right]_0^T \right| \right.$$

$$\left. + \left|\left[\int\limits_\Omega u_t h \cdot \nabla u \right]_0^T \right| \right) \leq C_1 E(T) + C \int\limits_0^T \int\limits_\Omega a|u_t|^2 + u^2$$

Since E is nonincreasing, we can bound below the left hand side by $TE(T)$. Hence for T sufficiently large, we deduce that

$$E(T) \leq C_0(\delta) \left(\int\limits_0^T \int\limits_\Omega (a|u_t|^2 + u^2) \right) \tag{1.55}$$

where the above constant $C_0(\delta)$ depends on f only through the constant $\delta > 0$ of assumption (1.51).

The last term on the right hand side involves the L^2 norm of the solution. Using a contradiction argument based on compactness arguments, we will see that this term can be absorbed by the first term on the right hand side, that is we will prove that there exists a positive constant C such that

$$\int\limits_0^T \int\limits_\Omega u^2 \leq C \int\limits_0^T \int\limits_\Omega a|u_t|^2 \tag{1.56}$$

for all solutions. This result is proved in Zuazua [105]. Argue by contradiction that (1.56) does not hold. Then, one can build a sequence $(u_n)_n$ of nonvanishing solutions of (1.45) such that setting $\lambda_n = ||u_n||_{L^2((0,T)\times\Omega)}$ and $v_n = u_n/\lambda_n$, v_n is solution of

$$\begin{cases} (v_n)_{tt} - \Delta v_n + f_n(v_n) + a(x)(v_n)_t = 0 & \text{in} \quad (0,\infty) \times \Omega, \\ v_n = 0 & \text{on} \quad (0,\infty) \times \Gamma \end{cases} \tag{1.57}$$

where

$$f_n(s) = \frac{1}{\lambda_n} f(\lambda_n s) \quad \forall\, s \in \mathbb{R}, \forall\, n \in \mathbb{N}. \tag{1.58}$$

Then we can remark that the functions f_n satisfies an hypothesis similar to (1.51) where F is replaced by F_n the primitive of f_n which vanishes at 0, and with the same constant δ than in (1.51). Moreover v_n satisfies

$$||v_n||_{L^2((0,T)\times\Omega)} = 1, \tag{1.59}$$

and

$$\int_0^T \int_\Omega a|(v_n)_t|^2 \longrightarrow 0 \tag{1.60}$$

Remarking now that v_n satisfies an equation similar to (1.45) with f replaced by f_n, the energy of v_n satisfies the estimate (1.55) with a constant C_0 which still only depends on δ. Thus, from (1.55) applied to the sequences v_n and its energy, we deduce that $(v_n)_n$ is bounded in $H^1((0,T) \times \Omega)$. Hence, we can extract a subsequence, still denoted by $(v_n)_n$ such that

$$\begin{cases} v_n \longrightarrow v \text{ weakly in } H^1((0,T) \times \Omega) \\ v_n \longrightarrow v \text{ strongly in } L^2((0,T) \times \Omega) \\ v_n \longrightarrow v \text{ a.e. in } (0,T) \times \Omega \end{cases} \tag{1.61}$$

Moreover, thanks to (1.59)–(1.60), we have

$$||v||_{L^2((0,T)\times\Omega)} = 1, \tag{1.62}$$

and

$$v_t = 0 \text{ a.e. in } (0,T) \times \{x \in \Omega, a(x) > 0\}. \tag{1.63}$$

We prove by contradiction that the sequence $(\lambda_n)_n$ is bounded. So assume that, up to subsequence, $\lambda_n \longrightarrow \infty$. Using (1.51), we remark that

$$F(s) \geq \min(F(1), F(-1))|s|^{2+\delta} \quad \forall\, |s| \geq 1.$$

On the other hand, thanks to (1.55) written for v_n and its energy and thanks to the definition of energy, we deduce that the sequence $(F_n(v_n))_n$ is bounded in

$L^1((0, T) \times \Omega)$ by a constant $C_2 > 0$. Using these two properties, we obtain

$$\lambda_n^\delta \int_0^T \int_{\{|v_n| \geq \lambda_n^{-1}\}} |v_n|^{2+\delta} + \int_0^T \int_{\{|v_n| \leq \lambda_n^{-1}\}} F(\lambda_n v_n) \leq C_2$$

This implies that

$$\int_0^T \int_\Omega |v_n|^{2+\delta} \longrightarrow 0$$

which contradicts (1.62). Hence $(\lambda_n)_n$ is bounded, so that up to a subsequence we can assume that either that $\lambda_n \longrightarrow \lambda > 0$ or $\lambda_n \longrightarrow 0$.

Let us first assume that $\lambda_n \longrightarrow \lambda > 0$. Since $(v_n)_n$ is bounded in $L^\infty((0, T); H_0^1(\Omega)) \cap W^{1,\infty}((0, T); L^2(\Omega))$, it is relatively compact in $L^\infty((0, T); H^{1-\varepsilon}(\Omega))$ for every $\varepsilon \in (0, 1)$. Since $(N - 2)p \leq N$, we deduce that the sequence $(f_n(v_n))_n$ converges strongly to $q = f(\lambda v)/\lambda$ in $L^\infty((0, T); L^r(\Omega))$ for every $r \geq 1$. Then, one can pass to the limit in (1.57), and deduce that v satisfies

$$\begin{cases} v_{tt} - \Delta v + q(t,x)v = 0 & \text{in} \quad (0, T) \times \Omega, \\ v = 0 & \text{on} \quad (0, T) \times \Gamma \end{cases}$$

We set $w = v_t$. Then w satisfies

$$w_{tt} - \Delta w + f'(\lambda v)w = 0. \tag{1.64}$$

and $w \equiv 0$ a.e. in $(0, T) \times \omega$.

Let us now assume that $\lambda_n \longrightarrow 0$. In this case, $w = v_t$ satisfies (1.64) with $q(t, x) = f'(0)$. Moreover $w \equiv 0$ a.e. in $(0, T) \times \omega$.

Hence for both cases, $w \in L^2((0, T) \times \Omega)$ satisfies an equation of the form

$$w_{tt} - \Delta w + b(t,x)w = 0. \tag{1.65}$$

where $b \geq 0$ a.e. in Ω and $b \in L^\infty(0, T, L^N(\Omega))$ and $w \equiv 0$ a.e. in $(0, T) \times \omega$. Applying a unique continuation of Ruiz [94], we deduce that if $T \geq$ diameter of Ω, then $w \equiv 0$ in $(0, T) \times \Omega$. Thus $v = v(x) \in H_0^1(\Omega)$ and solves an elliptic equation of the form

$$-\Delta v + q(t,x)v = 0 \text{ in } \Omega, \forall t \geq 0$$

where $q \geq 0$. Thus $v \equiv 0$ which contradicts (1.62). Hence we prove that (1.56) holds for all solutions. Using this in (1.55). We deduce that

$$E(T) \leq C_1 \int_0^T \int_\Omega a|u_t|^2 = C_1(E(0) - E(T)),$$

so that

$$E(T) \leq \sigma E(0)$$

where

$$\sigma = \frac{C_1}{C_1 + 1} \in (0, 1).$$

This together with the semigroup property implies that

$$E(t) \le CE(0)e^{-\gamma t} \quad \forall t \ge 0,$$

with $C = 1 + 1/C_1$ and $\gamma = \frac{1}{T}\ln(C)$. □

Remark 1.4.7. In his paper Zuazua's result is presented in a larger manner, so that if a unique continuation result is available for the linear wave equation with nonnegative and time and space dependent potential essentially bounded, then his compactness–uniqueness method can be used to derive an exponential decay of the semilinear wave equation. Indeed unique continuation and even more observability estimates, have been obtained in this latter case by Fu et al. in [54] (see also the references therein). Hence Zuazua's result presented above also holds when ω is a neighbourhood in Ω of the set $\Gamma(x_0)$.

1.5 Nonlinear Stabilization in Finite Dimensions

Abstract We present here the optimal-weight convexity method, introduced in the infinite dimensional case in (Alabau-Boussouira, Appl. Math. Optim. 51(1):61–105, 2005), adapted to the finite dimensional case in (Alabau-Boussouira, J. Differ. Equat. 248:1473–1517, 2010) with optimality results in this latter case. Hence, we consider in this section the case of nonlinear stabilization for ordinary differential equations. The aim is to give a complete characterization (*optimal*) of the energy decay rates for general damping functions with applications to the semi-discretization of PDE's. We will give general tools based on nonlinear Gronwall inequalities, convexity properties and a key comparison lemma (Alabau-Boussouira, J. Differ. Equat. 248:1473–1517, 2010) to establish this characterization. This approach is based on the *convexity properties close to 0 of a function H linked to the feedback [see (Alabau-Boussouira, Appl. Math. Optim. 51(1):61–105, 2005)]* and on a *new criteria to classify the feedbacks' behavior based on the behavior at 0 of a function Λ_H introduced for the first time, as far as we know, in (Alabau-Boussouira, J. Differ. Equat. 248:1473–1517, 2010)*. We combine these new mathematical tools to establish optimal upper energy decay rates and energy comparison principles. These tools will also be used in the infinite dimensional case combined with the multiplier method to handle geometrical aspects (see Sect. 1.5).

1.5.1 Nonlinear Gronwall Inequalities

We have already seen in Sect. 1.2 how a linear Gronwall inequality lead to exponential decay of solution of linearly damped wave equations. For nonlinear feedbacks, nonlinear Gronwall inequalities will be useful. The main difficulty is to identify a suitable weight function for proving weighted nonlinear Gronwall inequalities, the weight being known only in the peculiar case of polynomially growing feedbacks close to the origin (see below and Sect. 1.4). Let us first recall the following polynomial nonlinear Gronwall inequality (see also the references in [65]).

Theorem 1.5.1 (Komornik [65], Theorem 9.1). *Let* $E : [0, \infty) \mapsto [0, \infty)$ *be a non-increasing function satisfying, for some constants* $r > 0$ *and* $T > 0$, *the nonlinear Gronwall inequality*

$$\int\limits_{t}^{\infty} E^{r+1}(s)\, ds \leq T E^r(0) E(t), \quad \forall\, t \geq 0. \tag{1.66}$$

Then, E satisfies

$$E(t) \leq E(0)\left(\frac{T + r t}{T + r T}\right)^{-1/r} \tag{1.67}$$

Proof. The result clearly holds for the case $E(0) = 0$. Assume that $E(0) \neq 0$. Replacing E by $E/E(0)$, we can assume without loss of generality that $E(0) = 1$. We define a function $F : [0, \infty) \mapsto [0, \infty)$ by

$$F(t) = \int\limits_{t}^{\infty} E^{r+1}(s)\, ds\,.$$

Differentiating F and using (1.66), we have

$$-F' \geq T^{-r-1} F^{r+1} \quad \text{a.e. on } (0, \infty)$$

Set $T^+ = \sup\{t \geq 0,\, E(t) > 0\}$. Then, using the inequality

$$F(0) \leq T E(0)^{r+1},$$

we deduce that

$$F(s) \leq T^{(r+1)/r}(T + rs)^{-1/r} \quad \forall\, s \in [0, T^+). \tag{1.68}$$

Moreover, since $F \equiv 0$ on $[T^+, \infty)$, this last inequality holds for every $s \geq 0$. Thanks to the fact that E is nonnegative and nonincreasing, we have

$$F(s) \geq \int\limits_{s}^{T+(r+1)s} E^{r+1}\, dt \geq (T + rs) E^{r+1}(T + (r+1)s)$$

Using (1.68) in this last inequality and setting $t = T + (r+1)s$, we obtain (1.67). □

The weight function in the above nonlinear Gronwall type inequality is given by $w(E)$ where $w(s) = s^r$ for $s \geq 0$. To study energy decay rates for general feedbacks, and not only linear or polynomial growing feedbacks, one needs to prove more general nonlinear Gronwall inequalities, which generalizes the case of polynomial growing feedbacks. To state the results, we need to introduce some notation. Let $\eta > 0$ and $T_0 > 0$ be fixed given real numbers and L be a strictly increasing function from $[0, +\infty)$ on $[0, \eta)$, with $L(0) = 0$ and $\lim_{y \to +\infty} L(y) = \eta$. We define as in Alabau-Boussouira [17, 18] and for any $r \in (0, \eta)$, a function K_r from $(0, r]$ on $[0, +\infty)$ as follows

$$K_r(\tau) = \int_\tau^r \frac{dy}{yL^{-1}(y)} , \tag{1.69}$$

and ψ_r which is a strictly increasing onto function defined from $[\frac{1}{L^{-1}(r)}, +\infty)$ on $[\frac{1}{L^{-1}(r)}, +\infty)$ by:

$$\psi_r(z) = z + K_r(L(\frac{1}{z})) \geq z , \quad \forall z \geq \frac{1}{L^{-1}(r)} , \tag{1.70}$$

so that $\lim_{s \to \infty} \psi_r^{-1}(s) = \infty$. Then we prove the following result.

Theorem 1.5.2 (Alabau-Boussouira [17]). *We assume that E is a nonincreasing, absolutely continuous function from $[0, +\infty)$ on $[0, +\infty)$, satisfying $0 < E(0) < \eta$ and the inequality*

$$\int_S^T E(t)L^{-1}(E(t)) \, dt \leq T_0 E(S), \quad \forall 0 \leq S \leq T . \tag{1.71}$$

Then E satisfies the following estimate:

$$E(t) \leq L\left(\frac{1}{\psi_r^{-1}(\frac{t}{T_0})}\right), \quad \forall t \geq \frac{T_0}{L^{-1}(r)} , \tag{1.72}$$

where r is any real such that

$$\frac{1}{T_0} \int_0^{+\infty} E(\tau)L^{-1}(E(\tau)) \, d\tau \leq r \leq \eta .$$

Thus, we have $\lim_{t \to +\infty} E(t) = 0$, the decay rate being given by the estimate (1.72).

Proof. We define a function k and a function M by

$$k(t) = \int_t^{+\infty} M(E(\tau)) \, d\tau , \quad \forall t \geq 0, \tag{1.73}$$

where M is given by

$$M(y) = yL^{-1}(y), \quad \forall\, y \geq 0. \tag{1.74}$$

Then, thanks to (1.71), we have

$$k(t) \leq T_0 E(t), \quad \forall\, t \geq 0. \tag{1.75}$$

Moreover, since L^{-1} is a strictly nonnegative function, M is an increasing nonnegative function. Thus, differentiating (1.73), and using (1.75), we deduce that

$$-k'(s) = M(E(s)) \geq M\left(\frac{k(s)}{T_0}\right), \quad \forall\, s \geq 0,$$

Integrating this last inequality between 0 and t and making the change of variable

$$y = \frac{k(t)}{T_0}$$

in the above integral, we obtain

$$\int_{\frac{k(t)}{T_0}}^{B} \frac{dy}{M(y)} \geq \frac{t}{T_0}, \quad \forall\, t \geq 0,$$

where B is defined by

$$0 < B = \frac{1}{T_0} \int_0^{+\infty} E(\tau) L^{-1}(E(\tau))\, d\tau \leq E(0) < \eta.$$

Since M is positive on $(0, \eta]$, we deduce that for all $r \in [B, \eta]$, we have

$$\int_{\frac{k(t)}{T_0}}^{r} \frac{dy}{M(y)} \geq \frac{t}{T_0}, \quad \forall\, t \geq 0, \tag{1.76}$$

Thanks to the definition of K_r and since L^{-1} is strictly increasing on $[0, \eta)$, we deduce that for all $r \in [B, \eta)$ and all $\tau \in (0, r]$, we have

$$\frac{1}{L^{-1}(r)}\left(\ln r - \ln \tau\right) \leq K_r(\tau), \quad \forall\, 0 < \tau \leq r.$$

Hence, $\lim_{\tau \to 0^+} K_r(\tau) = +\infty$ holds. Thus, K_r is a strictly decreasing function from $(0, r]$ onto $[0, +\infty)$. This, together with (1.76), give the estimate

$$k(t) \leq T_0 K_r^{-1}(\frac{t}{T_0}) , \quad \forall \, t \geq 0 .$$

In particular, since M is increasing and nonnegative on $[0, \eta)$, whereas E is nonincreasing, we deduce that

$$\theta M \Big(E(t + \theta) \Big) \leq \int_t^{t+\theta} M(E(\tau)) \, d\tau \leq k(t) \leq T_0 K_r^{-1}(\frac{t}{T_0}) , \quad \forall \, t \geq 0 , \forall \, \theta > 0 . \tag{1.77}$$

Hence, we have the following estimate

$$E(t) \leq M^{-1} \Big(\min_{\theta \in (0,t]} (T_0 \gamma_t(\theta)) \Big) , \quad \forall \, t > 0 , \tag{1.78}$$

where we set

$$\gamma_t(\theta) = \frac{1}{\theta} K_r^{-1} \Big(\frac{t - \theta}{T_0} \Big) , \quad \forall \, \theta \in (0, t] . \tag{1.79}$$

Let now $t > 0$ be fixed for the moment. Thus, $\theta^\star$ is a critical point of γ_t, if and only if it satisfies the relation:

$$K_r^{-1} \Big(\frac{t - \theta^\star}{T_0} \Big) + \frac{\theta^\star}{T_0 K_r' \Big(K_r^{-1} (\frac{t - \theta^\star}{T_0}) \Big)} = 0 .$$

Hence, using the definition of M and of ψ_r, $\theta^\star$ is a critical point of γ_t, if and only if it satisfies

$$\psi_r \Big(\frac{\theta^\star}{T_0} \Big) = \frac{t}{T_0} ,$$

Noticing that ψ_r is strictly increasing and onto from $[\frac{1}{L^{-1}(r)}, +\infty)$ on $[\frac{1}{L^{-1}(r)}, +\infty)$, we deduce that for all $t \geq T_0 \frac{1}{L^{-1}(r)}$, γ_t has a unique critical point $\theta(t)$ at which it attains a minimum. This point $\theta(t)$ is given by:

$$\theta(t) = T_0 \psi_r^{-1}(\frac{t}{T_0}) . \tag{1.80}$$

Moreover, thanks to the definition of $\theta(t)$, we remark that

$$M^{-1}(T_0 \phi(\theta(t))) = K_r^{-1} \Big(\frac{t - \theta(t)}{T_0} \Big) = L \Big(\frac{T_0}{\theta(t)} \Big) .$$

Thus, using these identities in (1.78), together with (1.80), we obtain (1.72). Remarking now that, $\psi_r^{-1}(\tau) \to +\infty$ as $\tau \to +\infty$ and since L is continuous at

0 with $L(0) = 0$, we deduce that $\lim\limits_{t \to +\infty} L\left(\dfrac{1}{\psi_r^{-1}(\frac{t}{T_0})}\right) = 0$. So that the upper bound
in (1.72) goes to 0 as time goes to ∞. $\square$

The function L above will be chosen later on in an optimal way using convexity arguments. We recall some definitions and introduce some notation for convex functions. We recall that if ϕ is a proper convex function from $\mathbb{R}$ on $\mathbb{R} \cup \{+\infty\}$, then its convex conjugate ϕ^* is defined as:

$$\phi^*(y) = \sup_{x \in \mathbb{R}}\{xy - \phi(x)\}.$$

Most of the properties given below, are established in [17] but in a somehow different context. We prefer therefore to reformulate the requested results for clarity in the next proposition, which proof is left to the reader.

Theorem 1.5.3. *Let H be a given strictly convex $\mathscr{C}^1$ function from $[0, r_0^2]$ to $\mathbb{R}$ such that $H(0) = H'(0) = 0$, where $r_0 > 0$ is sufficiently small. We define*

$$\widehat{H}(x) = \begin{cases} H(x)\,, & \text{if } x \in [0, r_0^2]\,, \\ +\infty\,, & \text{if } x \in \mathbb{R} - [0, r_0^2]\,, \end{cases} \tag{1.81}$$

and

$$L(y) = \begin{cases} \dfrac{\widehat{H}^*(y)}{y} & , \text{if } y \in (0, +\infty)\,, \\ 0 & , \text{if } y = 0\,, \end{cases} \tag{1.82}$$

where $\widehat{H}^*$ stands for the convex conjugate function of $\widehat{H}$. We also define a function Λ_H on $(0, r_0^2]$ by

$$\Lambda_H(x) = \dfrac{H(x)}{x H'(x)}\,. \tag{1.83}$$

Then the following properties hold:

- *L is the strictly increasing continuous onto function from $[0, +\infty)$ on $[0, r_0^2)$ given by:*

$$L(y) = \begin{cases} (H')^{-1}(y) - \dfrac{H((H')^{-1}(y))}{y} & , \text{if } y \in [0, H'(r_0^2)]\,, \\ r_0^2 - \dfrac{H(r_0^2)}{y} & , \text{if } y \in [H'(r_0^2), +\infty)\,. \end{cases} \tag{1.84}$$

- *L is differentiable on $(0, H'(r_0^2))$ and*

$$L'(v) = \dfrac{H((H')^{-1}(v))}{v^2}\,, \qquad v \in (0, H'(r_0^2)) \tag{1.85}$$

- *L satisfies also*

$$L(v) = (H')^{-1}(v)\Big(1 - \Lambda_H\big((H')^{-1}(v)\big)\Big),\qquad(1.86)$$

and in particular

$$L(H'(r_0^2)) \le r_0^2.\qquad(1.87)$$

- $\Lambda_H(x) \in [0, 1]$ *for all* $x \in [0, r_0^2]$
- *Moreover all the above results are still valid in the case $r_0 = \infty$ replacing the closed right intervals by open right intervals of extremity $+\infty$. In particular replacing $H'(r_0^2)$ by ∞.*

Remark: The assumptions on H can be relaxed. In particular, the above definitions still make sense if H is only assumed to be convex in a neighborhood of 0, to vanish at 0 and to be nonnegative in a neighborhood of 0. But in this latter case, no explicit decay rates will be found. We prefer here to give sufficient conditions which lead to explicit and simple optimal energy decay rates under still very general assumptions on the feedbacks. Depending on the behavior of the function Λ_H at the origin, we can establish a simple explicit upper estimate of the energy decay rates. We refer to [8] for more results.

Theorem 1.5.4 (Alabau-Boussouira [8]). *Let H be a given strictly convex $\mathscr{C}^1$ function from $[0, r_0^2]$ to $\mathbb{R}$ such that $H(0) = H'(0) = 0$, where $r_0 > 0$ is sufficiently small. We define $\widehat{H}$ by (1.81), L by (1.82) and Λ_H by (1.83). Let E be a given nonincreasing, absolutely continuous, nonnegative real function defined on $[0, +\infty)$, $T_0 > 0$ be a fixed real number and $\beta > 0$ a given real number such that E satisfies the nonlinear Gronwall inequality*

$$\int_S^T E(t) L^{-1}\Big(\frac{E(t)}{2\beta}\Big)\, dt \le T_0 E(S),\qquad \forall\, 0 \le S \le T.\qquad(1.88)$$

under the condition

$$0 < \frac{E(0)}{2L(H'(r_0^2))} \le \beta,\qquad(1.89)$$

Then, if $\limsup_{x\to 0+} \Lambda_H(x) < 1$, E decays at infinity as follows:

$$E(t) \le 2\beta (H')^{-1}\Big(\frac{DT_0}{t}\Big),\qquad(1.90)$$

for t sufficiently large and where D is a positive constant which does not depend on $E(0)$. Otherwise, we have the following general decay rate

$$E(t) \le 2\beta L\Big(\frac{1}{\psi_r^{-1}(\frac{t}{T_0})}\Big),\qquad \forall\, t \ge \frac{T_0}{H'(r_0^2)},\qquad(1.91)$$

where

$$\psi_r(x) = \frac{1}{H'(r_0^2)} + \int\limits_{\frac{1}{x}}^{H'(r_0^2)} \frac{1}{v^2\left(1 - \Lambda_H\left((H')^{-1}(v)\right)\right)} \, dv, \quad x \ge \frac{1}{H'(r_0^2)}. \quad (1.92)$$

Finally all the above results are still valid in the case $r_0 = +\infty$ *removing the condition* (1.89) *and replacing respectively* $H'(r_0^2)$ *by* ∞ *and* $1/H'(r_0^2)$ *by* 0 *in* (1.92).

Proof. We set

$$\widehat{E}(t) = \frac{E(t)}{2\beta} \, .$$

Using (1.88), $\widehat{E}$ satisfies (1.71). Then, since E is nonincreasing, and thanks to Proposition 1.5.3 applied to $\widehat{E}$, we deduce that

$$\widehat{E}(t) \le \widehat{E}(0) \le L(H'(r_0^2)) = r_0^2 - \frac{H(r_0^2)}{H'(r_0^2)} < r_0^2. \quad (1.93)$$

We set $\eta = r_0^2$ and define L as in Proposition 1.5.3. Then, L is a strictly increasing onto function from $[0, +\infty)$ on $[0, \eta)$. We define B by

$$0 < B = \frac{1}{T_0} \int\limits_0^{+\infty} \widehat{E}(\tau) L^{-1}(\widehat{E}(\tau)) \, d\tau \le \widehat{E}(0) < \eta. \quad (1.94)$$

We also set $r = L(H'(r_0^2))$. Then, thanks to (1.93) and (1.94), $r \in [B, \eta)$. Thus, $\widehat{E}$ satisfies the hypotheses of Theorem 1.5.2 with r and B defined as above, so that the following estimate holds

$$\widehat{E}(t) \le L\left(\frac{1}{\psi_r^{-1}(\frac{t}{T_0})}\right), \quad \forall t \ge \frac{T_0}{H'(r_0^2)}, \quad (1.95)$$

where ψ_r is defined in (1.70) and K_r is defined by (1.69). Making the change of variable $v = L^{-1}(y)$ in (1.69) and using the formulas (1.85) and (1.86), we obtain

$$x \le \psi_r(x) = x + K_r(L(\frac{1}{x})) = x + \int\limits_{\frac{1}{x}}^{H'(r_0^2)} \frac{\Lambda_H\left((H')^{-1}(v)\right)}{v^2\left(1 - \Lambda_H\left((H')^{-1}(v)\right)\right)} \, dv$$

$$= \frac{1}{H'(r_0^2)} + \int\limits_{\frac{1}{x}}^{H'(r_0^2)} \frac{1}{v^2\left(1 - \Lambda_H\left((H')^{-1}(v)\right)\right)} \, dv, \quad x \ge \frac{1}{H'(r_0^2)}$$

$$(1.96)$$

We deduce easily estimate (1.91) in the general case.

We now assume that $\limsup_{x\to 0+} \Lambda_H(x) < 1$. Thus, there exists $0 < 2\varepsilon_0 < 1 - \limsup_{x\to 0+} \Lambda_H(x)$ and there exists $\delta > 0$ such that

$$\frac{1}{1 - \Lambda_H(x)} < \frac{1}{\varepsilon_0}, \quad \forall\, x \in (0, \delta].$$

Using this upper bound with $x = (H')^{-1}(v)$ in (1.96), we deduce that

$$\psi_r(x) \leq \frac{1}{H'(r_0^2)} + \frac{1}{\varepsilon_0} \int_{\frac{1}{x}}^{H'(\delta)} \frac{1}{v^2}\, dv + \int_{H'(\delta)}^{H'(r_0^2)} \frac{1}{v^2 \left(1 - \Lambda_H\left((H')^{-1}(v)\right)\right)}\, dv.$$

$$\leq Dx, \quad \text{for } x \text{ sufficiently large}, \tag{1.97}$$

where D is a positive constant which depends on r_0 and δ. Since ψ_r is strictly increasing, we deduce that

$$\frac{1}{\psi_r^{-1}(\frac{t}{T_0})} \leq \frac{DT_0}{t}, \quad \text{for } t \text{ sufficiently large}.$$

Thanks to the definition of $\widehat{E}$, estimate (1.95), formula (1.86) and since L is an increasing function and Λ_H is a nonnegative function, we obtain the desired estimate (1.90). $\qquad\square$

Remark 1.5.5. Note also that for general types of dampings, the required weight function L^{-1} is not defined on all $\mathbb{R}^+$. This is not surprising in view of this degree of generality. This makes the exposition more technical. A clear exposition of the optimal results and previous results in the literature in the linear and polynomial cases is given in [65]. Convexity properties for general feedbacks have been first introduced in the context of damped hyperbolic PDE's by Lasiecka and Tataru [70], who used an ODE approach rather than integral inequalities. Liu and Zuazua [80] and Martinez [85], and Eller Lagnese and Nicaise for Maxwell equations used convexity properties in different ways. A more general nonlinear Gronwall inequality than stated above is proved in [17] (see also [18]).

1.5.2 A Comparison Lemma

The following result will be determinant for optimality results and energy comparison principles as introduced in [7, 8].

Theorem 1.5.6 (Alabau-Boussouira [8]). *Let H be a given strictly convex $\mathcal{C}^1$ function from $[0, r_0^2]$ to $\mathbb{R}$ such that $H(0) = H'(0) = 0$, where $r_0 > 0$ is sufficiently small and define Λ_H as in (1.83). Let z be the solution of the ordinary differential equation:*

$$z'(t) + \kappa\, H(z(t)) = 0,\, z(0) = z^0 \quad t \geq 0, \tag{1.98}$$

where $z_0 > 0$ and $\kappa > 0$ are given. Then $z(t)$ is defined for every $t \geq 0$ and decays to 0 at infinity. Moreover assume that either

$$0 < \liminf_{x \to 0} \Lambda_H(x) \leq \limsup_{x \to 0} \Lambda_H(x) < 1 , \qquad (1.99)$$

or that there exists $\mu > 0$ such that

$$0 < \liminf_{x \to 0} \left(\frac{H(\mu x)}{\mu x} \int_x^{z_1} \frac{1}{H(y)} \, dy \right), \text{ and } \limsup_{x \to 0} \Lambda_H(x) < 1 , \qquad (1.100)$$

for a certain $z_1 \in (0, z_0]$ (arbitrary). Then there exists $T_1 > 0$ such that for all $R > 0$ there exists a constant $C > 0$ such that

$$(H')^{-1}\left(\frac{R}{t}\right) \leq C^2 z(t) , \quad \forall \, t \geq T_1 . \qquad (1.101)$$

Remark 1.5.7. This lemma allows us to compare time-pointwise estimates to energy types estimates. The upper estimates obtained by the optimal-weight convexity method are time-pointwise estimates, whereas the lower estimates derived by an energy comparison principle are in an energy formulation. To get optimality results, it is essential to be able to compare these two kind of estimates. The above comparison Lemma is the key to perform this comparison. It will also be useful to get lower estimates for infinite dimensional systems which can be compared to the upper estimates.

The proof of this result relies on the two next propositions. Their proofs is given in Alabau-Boussouira [8].

Proposition 1.5.8. *Let H be a given strictly convex $\mathscr{C}^1$ function from $[0, r_0^2]$ to $\mathbb{R}$ such that $H(0) = H'(0) = 0$, where $r_0 > 0$ is sufficiently small and define Λ_H as in (1.83). Then if $\lambda_+ = \limsup_{x \to 0} \Lambda_H(x) < 1$, there exist $\eta_+ \in (\lambda_+, 1)$ and $\delta > 0$ such that, the function*

$$x \mapsto x^{-1/\eta_+} H(x) \text{ is increasing on } (0, \delta] . \qquad (1.102)$$

Proposition 1.5.9. *Let H be a given strictly convex $\mathscr{C}^1$ function from $[0, r_0^2]$ to $\mathbb{R}$ such that $H(0) = H'(0) = 0$, where $r_0 > 0$ is sufficiently small and define Λ_H as in (1.83). We define the application σ by*

$$\sigma(x) = \frac{H(x)}{x} \int_x^{z_0} \frac{1}{H(y)} \, dy , \quad x \in (0, z_0] , \qquad (1.103)$$

where $z_0 > 0$ is an arbitrary real number. Then if

$$\liminf_{x \to 0} \Lambda_H(x) > 0 , \qquad (1.104)$$

there exists $\delta > 0$ and $\eta^- > 0$ such that σ satisfies the following estimate

$$\eta^- \leq \sigma(x), \quad x \in (0, \delta]. \tag{1.105}$$

Proof (of Theorem 1.5.6). Thanks to (1.98), z satisfies

$$t = \frac{1}{\kappa} \int_{z(t)}^{z(0)} \frac{1}{H(y)} \, dy \tag{1.106}$$

Moreover, thanks to our hypotheses on H, z is a positive, decreasing function from $[0, \infty)$ onto $(0, z(0)]$. In particular z decays to 0 as t goes to ∞. Thanks to (1.98), we have the relation

$$\sigma(z(t)) = \kappa \, t \, \frac{H(z(t))}{(z(t))}, \quad \forall \, t \geq 0. \tag{1.107}$$

Since, z decays to 0 as t goes to ∞, for all $\delta > 0$, there exists $T_1 > 0$ such that $0 < z(t) \leq \delta$ for all $t \geq T_1$.

Assume first that (1.99) *holds*: Then (1.105) of Proposition 1.5.9 holds for a certain sufficiently small $\delta > 0$. Combining this with the above property of z, we deduce that

$$\eta^- \frac{1}{\kappa \, t} \leq \frac{H(z(t))}{(z(t))}, \quad \forall \, t \geq T_1. \tag{1.108}$$

Thanks to the last inequality of (1.99), there exists $\eta_+ > 0$ such that (1.102) holds. We choose a constant C such that

$$C^2 \geq \max(1, \left(\frac{\kappa \, R}{\eta^-}\right)^{\frac{\eta_+}{1-\eta_+}}) \tag{1.109}$$

Thanks to Proposition 1.5.8 and to $C \geq 1$, and using (1.108), we obtain

$$\eta^- \frac{1}{\kappa \, t} \leq \frac{H(z(t))}{(z(t))} \leq \frac{H(C^2 z(t))}{(C^2 z(t))} C^{2(1-1/\eta_+)}, \quad \forall \, t \geq T_1.$$

Thanks to our choice of C in (1.109), and to the convexity of H, we obtain

$$R \leq t \, \frac{H(C^2 z(t))}{(C^2 z(t))} \leq t \, H'(C^2 z(t)), \quad \forall \, t \geq T_1.$$

Since H' is an increasing function, we have

$$\left(H'\right)^{-1}\left(\frac{R}{t}\right) \leq C^2 z(t), \quad \forall \, t \geq T_1.$$

Thus our claim is proved in the case where (1.99) holds.

Assume now that (1.100) *holds*: Then there exists $\mu > 0$, $\eta_0 > 0$ and $\delta > 0$ such that

$$\eta_0 < \left(\frac{H(\mu x)}{\mu x} \int_x^{z(0)} \frac{1}{H(y)} \, dy \right), \quad \forall \, x \in (0, \delta]. \tag{1.110}$$

Since, z decays to 0 as t goes to ∞, there exists as before, $T_1 > 0$ such that $0 < z(t) \leq \delta$ for all $t \geq T_1$. Combining this with the above inequality and using the relation (1.106), we deduce that

$$\frac{\eta_0}{\kappa t} \leq \frac{H(\mu z(t))}{\mu z(t)}, \quad \forall \, t \geq T_1. \tag{1.111}$$

Thanks to the last inequality of (1.100), there exists $\eta_+ > 0$ such that (1.102) holds. We choose a constant C such that

$$C^2 \geq \max \left(\mu, \mu \left(\frac{R\kappa}{\eta_0} \right)^{\frac{\eta_+}{1-\eta_+}} \right) \tag{1.112}$$

thanks to Proposition 1.5.8 and since $C \geq \mu$, we have

$$\left(\frac{C^2}{\mu} \right)^{(1-\eta_+)/\eta_+} \frac{\eta_0}{\kappa t} \leq \frac{H(C^2 z(t))}{(C^2 z(t))}, \quad \forall \, t \geq T_1.$$

Thanks to our choice of C in (1.112), and to the convexity of H, we obtain the desired estimate (1.101) as before.   □

1.5.3 Energy Decay Rates Characterization: The Scalar Case

We consider the scalar case:

$$u'' + v\,u + f(u) + \rho(u') = 0. \tag{1.113}$$

where $v > 0$ and u is a scalar unknown. We set

$$F(s) = \int_0^s f(\sigma) \, d\sigma, \tag{1.114}$$

and assume that

$$\exists \, \tilde{\mu} > 0 \text{ such that } 0 \leq F(s) \leq \tilde{\mu} s f(s), \quad \forall \, s \in \mathbb{R}. \tag{1.115}$$

We define the energy of a solution u as:

$$E(t) = \frac{1}{2} \left(|u'(t)|^2 + v \, |u(t)|^2 \right) + F(u). \tag{1.116}$$

Thanks to our assumption (1.115), the energy is nonnegative on the maximal interval of existence of solutions. We assume that the feedback satisfies the assumption

$$
(HS1) \begin{cases}
\rho \in \mathscr{C}(\mathbb{R}) \,, \text{ is monotone increasing}\,, \rho(0) = 0 \,, \\
\exists \text{ a strictly increasing odd function } g \text{ such that} \\
c\,|s| \le |\rho(s)| \le C\,|s| \,, \quad \forall |s| \ge 1 \,, \\
c\,g(|s|) \le |\rho(s)| \le C\,g^{-1}(|s|) \,, \quad \forall\,|s| \le 1 \,, \\
\exists r_0 > 0 \text{ such that } g \in \mathscr{C}^1([0, r_0]), g(0) = g'(0) = 0 \,,
\end{cases}
$$

where g^{-1} denotes the inverse function of g and where c, C are positive constants.

Remark 1.5.10. The assumption of linear growth at infinity is made for the sake of simplicity. It can be removed using Dafermos' strong stabilization [46] (see Sect. 1.1).

Define v as the solution of the ordinary differential equation:

$$
v'(t) + g(v(t)) = 0 \,, v(0) = \sqrt{2E(0)}, t \ge 0 \,. \tag{1.117}
$$

Note that this ODE is also considered in [101] in the case of a one-dimensional nonlinearly boundary damped wave equation, for which the authors can establish some comparison properties for peculiar initial data and for certain sequences of time going to infinity. The characterization below is new and more general. Note also that the characterization below is known in the case of polynomial feedbacks and is due to Haraux [60]. In the general case, we prove

Theorem 1.5.11 (Alabau-Boussouira [8]). *Assume that f is a continuous and locally Lipschitz function on $\mathbb{R}$ which satisfies (1.115), and that $\rho = g$ satisfies $(HS1)$. We define a function H by*

$$
H(x) = \sqrt{x}\, g(\sqrt{x}) \,, x \in [0, r_0^2] \tag{1.118}
$$

and assume that H is strictly convex on $[0, r_0^2]$. We also define $\widehat{H}$ by (1.81), L by (1.82) and Λ_H by (1.83). Let $(u^0, u^1) \in \mathbb{R}^2$, satisfying $0 < |u^1| + |u^0|$ be given, u be the solution of the Cauchy problem (1.113) corresponding to this initial data, and E be its energy. Assume that either

$$
0 < \liminf_{x \to 0} \Lambda_H(x) \le \limsup_{x \to 0} \Lambda_H(x) < 1 \,,
$$

or that there exists $\mu > 0$ such that

$$
0 < \liminf_{x \to 0} \left(\frac{H(\mu x)}{\mu x} \int_x^{z_1} \frac{1}{H(y)}\, dy \right), \text{ and } \limsup_{x \to 0} \Lambda_H(x) < 1 \,,
$$

for a certain $z_1 \in (0, z_0]$ (arbitrary). Then the energy of solution of satisfies the estimate

$$E(t) = O(v^2(t)) = O\left((H')^{-1}(\frac{1}{t})\right), \quad \textit{uniformly for large time}$$

Remark 1.5.12. We refer to [8] for more general results.

For the sake of clearness, the proof of this theorem is divided in three lemmas. The first Lemma below establishes an energy comparison principle and allows us to give a lower estimate of the energy of the solutions.

Lemma 1.5.13 (An energy comparison principle). *(Alabau-Boussouira [8]) Assume that f is a continuous and locally Lipschitz function on $\mathbb{R}$ which satisfies (1.115), and that $\rho = g$ satisfies $(HS1)$. Moreover assume that H is increasing and $H(0) = 0$. Let u be a solution of (1.113) and E be its energy. Then the following lower estimate holds*

$$\frac{1}{2}v^2(t) \leq E(t), \quad \forall\, t \geq 0, \tag{1.119}$$

where v is the solution of (1.117).

Remark 1.5.14. The above lower estimate does not require the hypotheses on the behavior of Λ_H in a neighbourhood of 0.

Proof. Thanks to the dissipation relation (1.122), and to our assumptions on g, we have

$$-E'(t) \leq u'(t)g(u'(t)) = H\left((u')^2\right), \quad \forall\, t \geq 0.$$

Hence, thanks to (1.117), we have

$$\left(\frac{v^2}{2} - E\right)'(t) = H\left((u')^2\right) - H\left((v(t))^2\right) \leq H(2E(t)) - H\left((v(t))^2\right), \quad \forall\, t \geq 0,$$

and

$$v^2(0) = 2E(0).$$

Since H is strictly increasing on $\mathbb{R}$, we deduce easily by comparison principles for ODE's that (1.119) holds. $\qquad\square$

Lemma 1.5.15. *Assume that f is a continuous and locally Lipschitz function on $\mathbb{R}$ which satisfies (1.115), and that ρ satisfies $(HS1)$. We assume that H defined by (1.118) is strictly convex on $[0, r_0^2]$. We also define $\widehat{H}$ by (1.81), L by (1.82) and Λ_H by (1.83). Let $(u^0, u^1) \in \mathbb{R}^2$, satisfying $0 < |u^1| + |u^0|$ be given, u be the solution of the Cauchy problem (1.113) corresponding to this initial data, and E be its energy. Assume that $\limsup_{x \to 0+} \Lambda_H(x) < 1$, then E decays at infinity as*

$$E(t) \leq 2\beta(H')^{-1}(\frac{D}{t}) \tag{1.120}$$

for t sufficiently large, where $\beta = \beta(\tilde{\mu}, v, r_0, E(0)) = \beta_{E(0)}$ is an explicit constant (see [8] for the explicit expression).

Proof. Set

$$\theta = \min\left(1, \frac{1}{2\tilde{\mu}}\right). \tag{1.121}$$

$$E'(t) = -u'(t)\,\rho(u'(t)) \leq 0. \tag{1.122}$$

Thanks to this dissipation, we can note that the solutions of (1.113) are defined for all $t \geq 0$. We set

$$w(s) = L^{-1}\left(\frac{s}{2\beta}\right) \quad \forall\, s \in [0, 2\beta r_0^2). \tag{1.123}$$

Then, w is a nonnegative $\mathscr{C}^1$ and strictly increasing function defined from $[0, r_0^2)$ onto $[0, +\infty)$ where L is given by (1.84) and where $\beta = \beta_{E(0)}$ is an explicit constant depending (see [8]). We multiply (1.113) by $w(E(t))u(t)$ and integrate the resulting equation on $[S, T]$. Since E is nonincreasing, w is nondecreasing and thanks to our assumption (1.115) on f, this gives

$$\theta \int_S^T Ew(E)\,dt \leq \int_S^T w(E)|u'|^2\,dt - \frac{1}{2}\int_S^T w(E)\rho(u')u$$

$$+\frac{1}{2}\int_S^T w'(E)\,E'u'u\,dt - \frac{1}{2}\Big[w(E)u'u\Big]_S^T$$

$$\leq \int_S^T w(E)|u'|^2\,dt + \frac{1}{4v\theta}\int_S^T w(E)|\rho(u')|^2\,dt$$

$$+\frac{v\theta}{4}\int_S^T w(E)|u|^2\,dt + \frac{1}{\sqrt{v}}E(S)w(E(S)), \quad \forall\, 0 \leq S \leq T.$$

where θ is defined by (1.121). Thus, we have

$$\int_S^T Ew(E)\,dt \leq \frac{2}{\theta}\int_S^T w(E)|u'|^2\,dt + \frac{1}{2v\theta^2}\int_S^T w(E)|\rho(u')|^2\,dt \tag{1.124}$$

$$+\frac{2}{\sqrt{v\theta}}E(S)w(E(S)), \quad \forall\, 0 \leq S \leq T.$$

We first assume that $r_0 < \infty$. We proceed as in [17]. We first remark that thanks to $(HS1)$, we have (up to the positive constants c and C, which may change)

$$\begin{cases} c\,|s| \leq |\rho(s)| \leq C\,|s|, & \forall |s| \geq r_0, \\ c\,g(|s|) \leq |\rho(s)| \leq C\,g^{-1}(|s|), & \forall\, |s| \leq r_0, \end{cases}$$

Step 1: *Estimate of the linear kinetic energy*

For $|u'| \leq r_0$, we have

$$H(|u'|^2) = |u'|g(|u'|) \leq \frac{1}{c}u'\rho(u').$$

This, together with Young's inequality imply

$$w(E)|u'|^2 \leq \widehat{H}^{\star}(w(E)) + H(|u'|^2) \leq \widehat{H}^{\star}(w(E)) + \frac{1}{c}u'\rho(u'), \quad \text{for } |u'| \leq r_0.$$

On the other hand, we have

$$w(E)|u'|^2 \leq \frac{1}{c}w(E)u'\rho(u'), \quad \text{for } |u'| \geq r_0.$$

Combining the above two inequalities and the dissipation relation, we obtain

$$\int_S^T w(E)|u'|^2 \, dt \leq \int_S^T \widehat{H}^{\star}(w(E)) \, dt + \frac{1}{c}[1 + w(E(S))]E(S), \quad \forall \, 0 \leq S \leq T.$$

$$(1.125)$$

Step 2: *Estimate of the nonlinear kinetic energy*

For $|u'| \leq r_0$, we have, thanks to Young's inequality

$$w(E)\frac{|\rho(u')|^2}{C^2} \leq \widehat{H}^{\star}(w(E)) + H\left(\left|\frac{\rho(u')}{C}\right|^2\right)$$

$$\leq \widehat{H}^{\star}(w(E)) + \frac{1}{C}u'\rho(u'), \quad \text{for } |u'| \leq r_0.$$

On the other hand, we have

$$w(E)|\rho(u')|^2 \leq Cw(E)u'\rho(u'), \quad \text{for } |u'| \geq r_0.$$

Combining the above two inequalities and the dissipation relations, as above, we have

$$\int_S^T w(E)|\rho(u')|^2 \, dt \leq C^2 \int_S^T \widehat{H}^{\star}(w(E)) \, dt \qquad (1.126)$$

$$+C[1 + w(E(S))]E(S), \quad \forall \, 0 \leq S \leq T.$$

Using (1.125)–(1.126) in (1.124), we obtain the estimate

$$\int_S^T Ew(E) \, dt \leq \beta \int_S^T \widehat{H}^{\star}(w(E)) \, dt + \left(\frac{2}{\sqrt{v\theta}} + \frac{2}{\theta c} + \frac{C}{2v\theta}\right) \quad (1.127)$$

$$[1 + w(E(S))]E(S), \quad \forall \, 0 \leq S \leq T,$$

where β can be easily computed. Moreover, one can check that

$$\frac{E(0)}{2\beta} \le L(H'(r_0^2)) = r_0^2(1 - \Lambda_H(r_0^2)),$$

so that

$$w(E(S)) \le L^{-1}(\frac{E(0)}{2\beta}) \le H'(r_0^2), \quad \forall \, 0 \le S. \tag{1.128}$$

Thanks to the definition of our weight function w, we have

$$\beta \widehat{H}^{\star}(w(E)) = \frac{Ew(E)}{2}.$$

Using this last relation in (1.127), together with (1.128), we deduce that

$$\int_S^T EL^{-1}(\frac{E}{2\beta}) \, dt \le T_0 \, E(S), \quad \forall \, 0 \le S \le T, \tag{1.129}$$

where $T_0 = T_0(\tilde{\mu}, \nu, r_0)$ is an explicit constant which can easily be computed (see [8] for the explicit expression). We conclude the proof by applying Theorem 1.5.4 in the case $r_0 < \infty$. The proof for $r_0 = \infty$ can be found in [8]. $\qquad \square$

Lemma 1.5.16. *Assume the hypotheses of Theorem 1.5.11. Then, we have*

$$E(t) \le C_2 v^2(t) \quad \text{for } t \text{ sufficiently large}, \tag{1.130}$$

where C_2 is a positive constant.

Proof. For both cases (1.99) or (1.100), the assumption $\limsup_{x \to 0} \Lambda_H(x) < 1$ holds, so that we can apply Lemma 1.5.15. Thus, (1.90) holds for $t \ge T_2$ for a sufficiently large T_2. On the other hand, setting $z(t) = v^2(t)$ for all $t \ge 0$, then z satisfies the ODE (1.117) with $\kappa = 2$ and $z_0 = 2E(0)$. Since we assume that either (1.99) or (1.100) holds, we can apply Lemma 1.5.6. Thus the upper estimate of (1.130) is proved with $C_2 = 2\beta C^2$. $\qquad \square$

Proof (of Theorem 1.5.11). Thanks to Lemma 1.5.13, Lemma 1.5.15 and Lemma 1.5.16, we have

$$\frac{1}{2}v^2(t) \le E(t) \le 2\beta(H')^{-1}(\frac{D}{t}) \le C_2 v^2(t) \quad \text{for } t \text{ sufficiently large},$$

which gives the desired result. $\qquad \square$

Remark 1.5.17. The lower estimate in Lemma 1.5.13 [8] establishes an energy comparison principle between the solution of the second order nonlinear oscillator type equation (1.113) and the first order nonlinear differential equation (1.117). The inequality (1.130) states that the energy of the nonlinear oscillator is bounded below by the energy of the first order ODE (1.117). The notion of this energy principle is new, as far as we know. It consists in comparing the dissipation of energies for these two ODE's.

1.5.4 The Vectorial Case and Semi-discretized PDE's

The extension of the former results to a vectorial case is motivated by discretization of hyperbolic dissipative systems. The proofs of the results given below are given in [8]. Let us consider a frictional dissipative wave equation in the one-dimensional space domain $(0, 1)$.

$$\begin{cases} u_{tt}(t,x) - u_{xx}(t,x) + f(x, u(t,x)) + \rho(x, u_t(t,x)) = 0, & 0 < t, 0 < x < 1, \\ u(t,x) = 0, & \text{for } x = 0, x = 1, 0 < t, \\ u(0,x) = u^0(x), \ u_t(0,x) = u^1(x), & 0 < x < 1. \end{cases}$$

$$(1.131)$$

We assume that this system is dissipative, thus ρ is monotone nondecreasing with respect to the second variable and $\rho(.,0) = 0$. A semi-discretization of the above equation in space, with for instance a uniform mesh $x_i = i\,h$ for $i = 0, \dots, n+1$ with a parameter of discretization $h = 1/(n+1)$, gives the finite dimensional system

$$\begin{cases} u_i'' - \dfrac{u_{i+1} + u_{i-1} - 2u_i}{h^2} + f_i(u_i) + \rho_i(u_i') = 0, & 0 < t, i = 1, \dots, n, \\ u_0(t) = u_{n+1}(t) = 0, & 0 < t, \\ u_i(0) = u_{i,0}, \ u_i'(0) = u_{i,1}, & i = 1, \dots, n, \end{cases}$$

$$(1.132)$$

where u_i is a function of t which stands for an approximation of the solution u at point x_i and where $f_i(s) = f(x_i, s)$ and $\rho_i(s) = \rho(x_i, s)$ for all $s \in \mathbb{R}$. Thanks to our assumption that the system (1.131) is dissipative, we check easily that its discretized version (1.132) is also dissipative. Several questions raise. In the case of linear dissipation, the energy of the above semi-discretized wave equation decays exponentially. Zuazua [96] proved that the above discretization does not lead to uniform decay rates with respect to the parameter h of discretization. Hence, the exponential decay rate of the discretized energy is not uniform with respect to h. These results are strongly related to observability estimates for the undamped equation, which are not uniform with respect to h. This goes back to the pioneering work of Glowinski Li and Lions [55] (see also [63]). This phenomenon is due to high frequency numerical spurious oscillations. Tebou and Zuazua [96] (see the review by Zuazua in [107]) proved that if one adds a suitable numerical viscosity which vanishes as the discretization parameter goes to 0, then the corresponding energy of the solution of the new scheme decays exponentially uniformly with respect to h. Recently, adapted numerical schemes have been proposed by Zuazua–Ervedoza

[52], Münch–Pazoto [87] (see also the references therein). The semi-discretization of the nonlinear dissipative case has not been considered as far as we know. A first step is to obtain sharp, if possible optimal, upper energy decay rates for fixed discretization parameters. Hence, our motivation is to extend the optimal energy decay rates for nonlinear dissipation to the discretized system. The system (1.132) can be written in the abstract finite dimensional form given by

$$u'' + Au + f(u) + \rho(u') = 0 . \tag{1.133}$$

where the unknown $u \in \mathbb{R}^n$, f and ρ_i are defined as above and where the matrix A is given by

$$A = h^{-2} \begin{pmatrix} 2 & -1 & 0 & \dots & 0 \\ -1 & 2 & -1 & \dots & 0 \\ \dots & \dots & \dots & \dots & \dots \\ 0 & \dots & 0 & -1 & 2 \end{pmatrix}$$

Remark 1.5.18. A similar formulation holds for other hyperbolic equations, such as Petrowsky equation for instance (see [8]).

We denote respectively by $\langle , \rangle$ and $| \cdot |$ the euclidian scalar product and norm in $\mathbb{R}^n$. We also denote by $v = (v_1, \dots, v_n)$ vectors in $\mathbb{R}^n$. Let A be a real symmetric definite positive matrix of order n. Thus, there exists $\alpha_0 > 0$ such that

$$|A^{1/2}\xi|^2 \geq \alpha_0 |\xi|^2 , \forall \; \xi \in \mathbb{R}^n . \tag{1.134}$$

We assume that the feedback has the form:

$$\rho : v = (v_1, \dots, v_n) \in \mathbb{R}^n \mapsto \big(g_1(v_1), \dots, g_n(v_n)\big) .$$

We also assume that f is a vectorial function of the form:

$$f(u) = (f_1(u_1), \dots, f_n(u_n)) , \quad \forall \; u = (u_1, \dots, u_n) \in \mathbb{R}^n .$$

We define the energy of a solution of (1.133) by

$$E(t) = \frac{1}{2}\big(|u'(t)|^2 + |A^{1/2}u(t)|^2\big) + F(u) . \tag{1.135}$$

where

$$F(u) = \sum_{i=1}^{n} F_i(u_i) ,$$

with

$$F_i(u_i) = \int_0^{u_i} f_i(\sigma) \, d\sigma .$$

We assume that the semilinearity satisfies

$$\exists \; \tilde{\mu} > 0 \text{ such that } 0 \leq F(\xi) \leq \tilde{\mu}\langle \xi , f(\xi) \rangle , \quad \forall \; \xi \in \mathbb{R}^n . \tag{1.136}$$

Theorem 1.5.19 (Alabau-Boussouira [8]). *Assume that f is continuous and locally Lipschitz on $\mathbb{R}^n$ and satisfies (1.136). Assume also that the functions g_i satisfy*

$$(HOV1) \begin{cases} g_i \in \mathscr{C}(\mathbb{R}) \text{ and is monotone increasing} \\ s\, g_i(s) \geq 0, \quad \forall\, s \in \mathbb{R}, \forall\, i = 1,\ldots,n \\ \exists \text{ a strictly increasing odd function } g \text{ such that} \\ c|s| \leq |g_i(s)| \leq C\,|s|, \quad \forall |s| \geq 1, s \in \mathbb{R}, \\ cg(|s|) \leq |g_i(s)| \leq k_1\, g(|s|), \quad \forall\, s \in \mathbb{R}, \forall\, i = 1,\ldots,n, \\ |g(s)| \leq k_2 g^{-1}(|s|), \quad \forall\, |s| \leq 1, s \in \mathbb{R}, \forall\, i = 1,\ldots,n, \\ \exists\, r_0 > 0 \text{ such that } g \in \mathscr{C}^1([0,r_0]), g(0) = g'(0) = 0, \end{cases}$$

where g^{-1} denotes the inverse function of g and where k_i are positive constants for $i = 1, 2$. Moreover, we assume that the function defined by (1.118) is strictly convex on $[0, r_0^2]$ and we define Λ_H by (1.83). Assume that either

$$0 < \liminf_{x \to 0} \Lambda_H(x) \leq \limsup_{x \to 0} \Lambda_H(x) < 1,$$

or that there exists $\mu > 0$ such that

$$0 < \liminf_{x \to 0} \left(\frac{H(\mu x)}{\mu x} \int_x^{z_1} \frac{1}{H(y)}\, dy \right), \text{ and } \limsup_{x \to 0} \Lambda_H(x) < 1,$$

for a certain $z_1 \in (0, z_0]$ (arbitrary). Let u be a solution of (1.133), and E be its energy defined by (1.135). Then the energy of solution of satisfies

$$E(t) = O(v^2(t)) = O\left((H')^{-1}(\frac{D}{t})\right) \quad \text{uniformly with respect to } t, \quad (1.137)$$

where D is a positive constant and where v is the solution of the ordinary differential equation:

$$v'(t) + n\, k_1\, g(v(t)) = 0, \, v(0) = \sqrt{2E(0)}, t \geq 0. \quad (1.138)$$

From the above theorem, we deduce easily the following corollary.

Theorem 1.5.20. *Assume that the vectorial functions defined by $f_i(.) = f(x_i, .)$ and $g_i(.) = \rho(x_i, .)$ for $i = 1,\ldots n$ satisfy the hypotheses of Theorem 1.5.19, where the points x_i, $i = 1,\ldots,n$ denotes the discretization points. Furthermore, we assume that g involved in $(HOV1)$ is such that (1.99) holds or that there exists $\mu > 0$ and $z_1 \in (0, E(0)]$ such that (1.100) holds, then the energy of the solutions of (1.132) satisfy (1.137). Hence, the energy decay rate given in Theorem 1.5.19 is optimal for the semi-discretized system (1.132).*

Remark 1.5.21. The decay rate given in the above theorem depends on n and thus on the discretization parameter h.

1.5.5 Examples of Feedbacks and Optimality

Theorem 1.5.22. *Let u be a solution of* (1.113) *and E be its energy. Then we have the following results.*

 Example 1: *let g be given by* $g(x) = x^p$ *where* $p > 1$ *on* $(0, r_0]$.
 Then

$$E(t) = O\left(t^{\frac{-2}{p-1}}\right),\qquad(1.139)$$

uniformly for t sufficiently large and for all (u^0, u^1) *in* $\mathbb{R}^2$.
 Example 2: *let g be given by* $g(x) = x^p (\ln(\frac{1}{x}))^q$ *where* $p > 1$ *and* $q > 1$ *on* $(0, r_0]$.
 Then

$$E(t) = O\left(t^{-2/(p-1)}(\ln(t))^{-2q/(p-1)}\right)\qquad(1.140)$$

uniformly for large t.
 Example 3: *let g be given by* $g(x) = e^{-\frac{1}{x^2}}$ *on* $(0, r_0]$.
 Then

$$E(t) = O\left((\ln(t))^{-1}\right),$$

uniformly for large t.
 Example 4: *let g be given by* $g(x) = e^{-(\ln(\frac{1}{x}))^p}$, $1 < p < 2$, $x \in [0, r_0]$.
 Then

$$E(t) = O\left(e^{-2(\ln(t))^{1/p}}\right)$$

uniformly for large t.
 Example 5: *let g be given by* $g(x) = x(\ln(\frac{1}{x}))^{-p}$ *on* $(0, r_0]$, *where* $p > 0$.

Then

$$E(t) \leq C \,\beta_{E(0)} e^{-2(\frac{pt}{DT_0})^{1/(p+1)}} t^{-1/(p+1)}\qquad(1.141)$$

for t sufficiently large. Optimality cannot be asserted for this latter example.

Proof. For Examples 1 up 5, H satisfies the hypotheses of convexity of Theorem 1.5.11. For examples 1 and 2, Λ_H has a finite limit in 0, which is in $(0, 1)$ so that the first alternative of Theorem 1.5.11 is satisfied. Applying this theorem we obtain the announced characterization of the asymptotic behavior of the energy at infinity for these two examples. For Examples 3 and 4, Λ_H tends to 0 at 0. One can prove that the second alternative of Theorem 1.5.11 is satisfied. Applying this theorem we obtain the announced characterization of the asymptotic behavior of the energy at infinity for these two examples. For Example 5, now Λ_H tends to 1 at 0. We have no characterization of the asymptotic behavior but still a sharp upper estimate given by (1.91). □

Remark 1.5.23. Let us comment Example 5 for which optimality cannot be asserted. Indeed if one considers that $f \equiv 0$ and ρ is linear in (1.113), then we can give easily examples of parameters such that there exist two branches of solutions with a different asymptotic behavior (see [8] for more details). If Λ_H

tends to 1 at 0, g is close to a linear function close to the origin, which gives some hints that this case is close to what happens in the linear situation. If the feedback is not close to linear growth around the origin, then our characterization tells that all solutions (except the vanishing one) have the same asymptotic behavior at infinity.

1.6 Polynomial Feedbacks in Infinite Dimensions

Abstract The purpose of this section is to present the first results obtained chronogically for specific nonlinear feedbacks, that is the case of polynomial feedbacks. Many authors have obtained results in this direction: Lagnese [66] (J. Differ. Equat. 50:163–182, 1983), Haraux (Systèmes Dynamiques Dissipatifs et Applications, Masson, Paris, Milan, Barcelone, 1991), Zuazua [106] (SIAM J. Contr. Optim. 28:265–268, 1989), Komornik (Exact Controllability and Stabilization. The Multiplier Method, Wiley, Masson, Paris, 1994), Nakao [89] (Math. Ann. 305:403–417, 1996), Conrad and Rao (Asymptot. Anal. 7:159–177, 1993) and the references therein. These results concern localized and boundary damped wave-like equations and use either the method of perturbed energy combined with nonlinear differential inequalities or polynomial Gronwall inequalities for the natural energy together with the multiplier method. We shall present the approach based on polynomial Gronwall inequalities combined with the multiplier method. In this context, we will as for the linear feedback case, separate the different steps: identification of dominant energies and treatment of the nonlinearity of the feedbacks.

For the sake of clarity of exposition, we will only consider the case of a polynomially locally damped wave-equation, that is

$$\begin{cases} u_{tt} - \Delta u + \rho(.,u_t) = 0 & \text{in} \quad (0,\infty) \times \Omega, \\ u = 0 & \text{on} \quad \Sigma = (0,\infty) \times \Gamma \\ (u,u_t)(0) = (u^0,u^1) & \text{on} \quad \Omega, \end{cases} \tag{1.142}$$

where in the sequel g will be assume to have a polynomial growth close to the origin. We set $H = L^2(\Omega)$ and $V = H_0^1(\Omega)$. We consider feedbacks ρ satisfying

$$(HF) \begin{cases} \rho \in \mathscr{C}(\overline{\Omega} \times \mathbb{R}) \text{ and is monotone increasing with respect to the} \\ \text{second variable and } \exists\, a \in \mathscr{C}(\overline{\Omega})\,, a \geq 0 \text{ on } \Omega \text{ such that} \\ a(x)|v| \leq |\rho(x,v)| \leq C\,a(x)|v|\,, \quad \forall x \in \Omega\,, \text{ if } |v| \geq 1\,, \\ a(x)g(|v|) \leq |\rho(x,v)| \leq C\,a(x)g^{-1}(|v|)\,, \quad \forall x \in \Omega\,, \text{ if } |v| \leq 1\,, \\ a(x) \geq a_- > 0\,, \quad \forall\, x \in \omega\,, \end{cases}$$

where g^{-1} denotes the inverse function of g and where C is a positive constant.

We recall the following classical existence and regularity result (see e.g. [58, 69] for the proof) using the theory of maximal nonlinear monotone operator:

Theorem 1.6.1. *Assume hypothesis* (HF). *Then for all* $(u^0, u^1) \in V \times H$, *the problem* (1.142) *has a unique solution* $u \in \mathscr{C}([0, +\infty); V) \times \mathscr{C}^1([0, +\infty); H)$. *Moreover, for all* $(u^0, u^1) \in D(A) \times V$, *the solution of is in* $L^\infty([0, +\infty); D(A)) \times W^{1,\infty}([0, +\infty); V) \times W^{2,\infty}([0, +\infty); H)$ *and its energy defined by:*

$$E(t) = \frac{1}{2}\left(|u'(t)|_H^2 + |\nabla u(t)|_H^2\right),\tag{1.143}$$

satisfies the following dissipation relation:

$$E'(t) = -\int_\Omega u'(t)(x)\rho(x, u'(t))(x)\,dx \leq 0.\tag{1.144}$$

Using the multipliers $E^{(p-1)/2}Ku$ where Ku stands for the different multipliers used for the linear case in Sect. 1.2, one can prove in a similar way to the linear case

Theorem 1.6.2. *We assume that* (HF) *holds with* $g(s) = |s|^{p-1}s$ *with* $p > 1$ *for s close to 0 (polynomial case). We also assume that* ω *satisfies the geometric conditions (MGC) or (PWMGC). Let* $(u^0, u^1) \in V \times H$ *be given and denote by u a solution of* (1.142) *and E its energy defined as in* (1.143). *Then, E satisfies*

$$\int_t^T E(s)E^{(p-1)/2}(s)\,ds \leq \delta_1 E^{(p+1)/2}(t) + \delta_2 \int_t^T E^{(p-1)/2}(s)\int_\Omega |\rho(., u_t)|^2$$

$$+\delta_3 \int_t^T E^{(p-1)/2}(s)\int_\omega |u_t|^2\,ds \quad \forall 0 \leq t \leq T.\tag{1.145}$$

where δ_i, $i = 1, \ldots, 3$ *are positive constants.*

Theorem 1.6.3. *We make the assumptions of Theorem 1.6.2. Let* $(u^0, u^1) \in V \times H$ *be given and denote by u a solution of* (1.142) *and E its energy defined as in* (1.143). *Then E satisfies*

$$\int_S^T E^{(p+1)/2}(s)\,ds \leq T_0 E(S) \quad \forall\, 0 \leq S \leq T.\tag{1.146}$$

Proof. We set $\Omega_t = \{x \in \Omega, |u_t| \leq 1\}$ and $\omega_t = \{x \in \omega, |u_t| \leq 1\}$. Thanks to (HF), we have

$$|\rho(., u_t)|^2 \leq C\left(u_t\rho(., u_t)\right)^{2/(p+1)} \quad \text{on the set } x \in \Omega_t.$$

In a similar way, we have

$$|u_t|^2 \leq C\left(u_t\rho(., u_t)\right)^{2/(p+1)} \quad \text{on the set } x \in \omega_t.$$

Hence we have

$$\int_S^T E^{(p-1)/2} \int_{\Omega_t} |\rho(.,u_t)|^2 \le C \int_S^T E^{(p-1)/2}(-E')^{2/(p+1)}\, dt$$

and

$$\int_S^T E^{(p-1)/2} \int_{\omega_t} |u_t|^2 \le C \int_S^T E^{(p-1)/2}(-E')^{2/(p+1)}\, dt$$

Using Young's inequality, we have for any $\varepsilon > 0$

$$\int_S^T E^{(p-1)/2}\left(\int_{\Omega_t} |\rho(.,u_t)|^2 + \int_{\omega_t} |u_t|^2\right) \le \varepsilon \int_S^T E^{(p+1)/2}$$

$$+C_\varepsilon E(S) \quad \forall\, 0 \le S \le T.$$

On the other hand, thanks to (HF), we have

$$\int_S^T E^{(p-1)/2}\left(\int_{\Omega\setminus\Omega_t} |\rho(.,u_t)|^2 + \int_{\omega\setminus\omega_t} |u_t|^2\right) \le C\, E(S) \quad \forall\, 0 \le S \le T.$$

Combining these different estimates, we obtain (1.146). □

Remark 1.6.4. The idea of splitting the set Ω in two subsets, one with velocities close to zero and its complementary goes back to an original idea of Zuazua.

Theorem 1.6.5. *We make the assumptions of Theorem 1.6.2. Let $(u^0,u^1) \in V \times H$ be given and denote by u a solution of (1.142) and E its energy defined as in (1.143). Then E satisfies*

$$E(t) \le C(E(0))\, t^{\frac{-2}{p-1}}, \tag{1.147}$$

for t sufficiently large and for all (u^0,u^1) in $H_0^1(\Omega) \times L^2(\Omega)$.

Proof. If $E(0) = 0$ then the result holds easily. Assume now that $E(0) \ne 0$. Thanks to (1.146) and using Theorem 9.1 [65], that is Theorem 1.5.1 in the present Notes, with $r = (p-1)/2$, we conclude. □

1.7 The Optimal-Weight Convexity Method

1.7.1 Introduction and Scope

Abstract The main purpose of this section is to present the optimal-weight convexity method based on the construction of an optimal-weight function for general Gronwall inequalities, determined as an unknown of an explicit equation, thanks to convexity properties of a suitable feedback-dependent function H. We introduced the method in (Alabau-Boussouira, Appl. Math. Optim. 51(1): 61–105, 2005) for the infinite dimensional case and in (Alabau-Boussouira, J. Differ. Equat. 248: 1473–1517, 2010) for its version in the finite dimensional case, already presented in Sect. 1.3, and its simplification. The extension of this method to nonfrictional dampings such as linear memory-damping with general decaying kernels, is presented in [10]. This approach is build through several new results: optimal decay of solutions of general Gronwall inequalities, feedback classification through a new function Λ_H (introduced in (Alabau-Boussouira, J. Differ. Equat. 248: 1473–1517, 2010)), constructive and simplified decay rates, comparison arguments. This approach provides *a whole and complete methodology* to establish easy explicit computable general optimal and quasi-optimal energy decay rates with applications in finite and infinite dimensions settings, for localized, boundary and memory-dampings, as well as for many PDE's [see (Alabau-Boussouira, Appl. Math. Optim. 51(1): 61–105, 2005; Alabau-Boussouira, J. Differ. Equat. 248: 1473–1517, 2010) for various examples (see also [16])].

Convexity properties were already used by several authors for wave-like equations in case of frictional dampings. Energy decay rates for feedbacks with general growth were first considered, as far as we know, by Lasiecka and Tataru [70] in 1993. In this seminal paper, the authors consider the case of a semilinear wave equation subjected to nonlinear frictional damping and used some convexity properties to give upper energy decay rates by means of a nonlinear differential ordinary differential equation with a nonlinearity given as the solution of several successive implicit relations. Explicit computable decay estimates were given for the case of linear or polynomial feedbacks as applications. A similar approach was given by Liu and Zuazua in [80] including in addition feedbacks with nonlinear growth at infinity. The authors used sharper properties of convex functions and in particular their convex conjugates. Simple explicit and easily computable formula for upper energy decay estimates for general feedbacks is given in Martinez in [84, 85]. His method also involves convexity properties but in a different way than in [17, 70, 80]. It does not always give, at least in one shoot, the expected optimal energy decay rates as for instance the polynomial, polynomial-logarithmic or close to exponential as shown in [17]. Indeed his upper estimates are sharp in the case of very fast decaying feedbacks at the origin. Upper estimates thanks to convexity properties and the use of integral inequalities have been derived for Maxwell equations by Eller Lagnese and Nicaise [51]. Other results have been derived later on in [43] (see also [98] and references therein).

Our purpose here is to give a method which provides, in one shoot, easy computable energy decay rates which are optimal or quasi-optimal, and works for finite as well as infinite dimensions and allow to treat different PDE's, and different types of dampings: localized, boundary and of memory-type. Another important aspect is also to present these results in a unified way, so that thanks to additional geometric hypotheses in the infinite dimensional case, the same hypotheses on the feedback than in the corresponding finite dimensional case, lead to the same sharp upper estimates. It can also been extended in a nontrivial way to linear memory-damping, which involves nonlocal damping operators (see [10]). Another feature is the one concerned with optimality of the energy decay rates. In Sect. 1.3, we already use this method and prove optimality of our estimates. In [17], we use the method of Martinez and Vancostenoble [101] to give examples of optimality of our estimates in infinite dimensions.

In general, in infinite dimensions, the optimality problem is largely open. The difficulty is to obtain lower estimates. Indeed, we will see that in the PDE case, the energy lower estimate obtained in Sect. 1.3 are no longer available under the same general form. Haraux [57] derives, as far as we know, the first result in this direction. He proves a "weak" lower energy estimate, obtained through a "weak" lower velocity estimate for a one-dimensional wave equation, locally polynomially damped with very smooth solutions and using a 1-D interpolation result. We show in Sect. 1.5.5, how to prove "strong" lower energy estimates and "weak" lower velocity estimates for less regular solutions, multi-dimensional domains— in particular domains of annulus type, boundary as well as localized dampings and general feedback growths. These results are obtained combining interpolation theory together with a comparison Lemma established for the first time, as far as we know, in [8] (see Lemma 1.5.6 in Sect. 1.3). Thus, we can still derive lower energy estimates, but which are not equivalent as time goes to infinity to our sharp upper estimates. Also the required smoothness of solutions for general multidimensional domains is an open question. Vancostenoble and Martinez [101] (see also [102]) consider optimality estimates for decaying feedbacks at the origin for a one-dimensional (or in an annulus in $\mathbb{R}^3$ for radial solutions) nonlinearly boundary damped wave equation for specific initial data, namely with zero velocity in Ω. Their proof is based on an explicit formula of the energy at specific times through a sequence. We also mention Carpio [40], who takes another strategy based on differential inequalities and Lyapunov approach to track the sharp dependence with respect to initial data of the constant involved in upper estimates in case of polynomial feedbacks. She also proves optimality of the constant with respect to the initial data in the upper estimates in some situations.

Let us present the main lines of the optimal-weight convexity method. It is based on the following steps:

- **Step I:** *Dominant kinetic energy estimates*

    Use multipliers of the form $K(u)w(E)$ to prove that the energy satisfies an estimate of the form

$$\int_S^T w(E(t))E(t)\,dt \leq \delta_1 E(S)w(E(S)) + \delta_2 w(E)\int_S^T \int_{O_2} |\rho(.,u_t)|^2 \quad (1.148)$$

$$+\delta_3 w(E)\int_S^T \int_\omega |u_t|^2 \quad \forall\, 0 \leq S \leq T$$

where the $\delta_i > 0$ are constants for $i = 1, \ldots 3$ and where

1. K stands for various multipliers which may depend on u, ∇u, and $(-\Delta)^{-1}(\beta u)$, β being a cut-off function. These multipliers introduced by several authors are by now well-known.
2. w at this stage is a smooth, nonnegative and nondecaying weight function.
3. $O_2 = \Omega$ in case of locally distributed feedbacks and $O_2 = \partial\Omega$ in case of boundary feedbacks, whereas ω stands for the subdomain of Ω (respectively $\partial\Omega$) where the feedback is effective in case of locally distributed (resp. boundary) feedbacks.

- **Step II:** *Determination of optimal weight function*

  The optimal weight function $w(.) = L^{-1}(./2\beta)$, which has been introduced for the first time by Alabau-Boussouira [17], is determined implicitly thanks to convexity properties of a function H, directly expressed in terms of the feedback close to the origin. Due to our choices of H, and of this optimal choice of the weight function w, we prove that E satisfies the same general nonlinear Gronwall inequality of Sect. 1.3.1 presented in the context of finite dimensional stabilization, that is

  $$\int_S^T E(t)L^{-1}(E(t)/2\beta)\,dt \le T_0 E(S), \quad \forall\, 0 \le S \le T\,.$$

Remark 1.7.1. Note that some authors use our above strategy, but replace the optimal-weight function $L^{-1}(./2\beta)$ by a weaker one as follows. We define L by

$$L(y) = \widehat{H}^\star(y) \text{ for } y > 0\,, L(0) = 0\,.$$

It is easy to check that

$$L(y) \le L_1(y)$$

where

$$L_1(.) = (H')^{-1}(.)$$

In particular, applying our convexity method, the weight-function w becomes

$$w_1(.) = L_1^{-1}(\varepsilon_0\,.) = H'(\varepsilon_0\,.)$$

where $\varepsilon_0 = 1/2\beta$. Then, they follow all the methodology of the optimal-weight convexity method, that of Sect. 1.5.3 below, replacing w by w_1 and follow our results on generalized nonlinear Gronwall weighted integral inequalities as introduced originally in [17]. The resulting estimates are weaker and the methodology is not original.

- **Step III:** *Optimal decay rates for solutions of nonlinear Gronwall inequalities*

  Applying the results of Sect. 1.3.1, we deduce that E satisfies the semi-explicit estimate:

  $$E(t) \le 2\beta\, L\left(\frac{1}{\psi_r^{-1}(\frac{t}{T_0})}\right), \quad \forall\, t \ge \frac{T_0}{L^{-1}(r)}\,,$$

where r is any real such that

$$\frac{1}{T_0} \int_0^{+\infty} E(\tau) L^{-1}(E(\tau)/2\beta) \, d\tau \le r \le \eta.$$

- **Step IV:** *Simplification of decay rates and energy comparison principle*
  We then use our simplification method introduced in [8] to prove a sharp upper simple, explicit and easily computable estimate, that is

1. Use the new function Λ_H defined in [8] which behavior at the origin measures in some way the behavior of the feedback close to the origin.
2. If $\limsup_{x\to0+} \Lambda_H(x) < 1$, we apply our results of Sect. 1.3.1 for stabilization in finite dimensions and prove that E decays at infinity as follows:

$$E(t) \le 2\beta(H')^{-1}\left(\frac{DT_0}{t}\right),$$

   for t sufficiently large and where D is a positive constant which does not depend on $E(0)$.
3. Under further hypotheses on the behavior of Λ_H close to 0, we give an upper energy estimate based on the energy comparison principle introduced in our paper [8], that is

$$E(t) \le C_2 v^2(t) \quad \text{for } t \text{ sufficiently large},$$

   where C_2 is a positive constant and v is the solution of the ordinary differential equation

$$v'(t) + g(v(t)) = 0, \, v(0) = \sqrt{2E(0)}, t \ge 0.$$

We will complete the above strategy by a methodology to derive lower energy estimates (see Sect. 5.7).

1.7.2 Dominant Kinetic Energy Estimates

This part already exists in the linear stabilization problem, even though it is not presented under this form in general. It becomes important to distinguish this step to understand where and how convexity properties are useful. Let Ω be a bounded open subset of $\mathbb{R}^N$ with a smooth boundary denoted by Γ and ω be an open subset of Ω of positive measure. We assume that the feedback ρ satisfies (HF) with now a general function g.

Theorem 1.7.2. *Assume that (HF) holds where ω satisfies the geometric conditions (MGC) or $(PWMGC)$. We respectively define H, $\widehat{H}$, and L by (1.118), (1.81) and (1.82). Let $(u^0, u^1) \in V \times H$ be given and denote by u a solution of (1.142) and E its energy defined as in (1.6). We assume that $E(0) > 0$. Then for any* *nonnegative, nondecaying $\mathscr{C}^1$ function w defined on $[0, E(0)]$, E satisfies*

$$\int\limits_t^T E(s)w(E(s))\,ds \leq \delta_1 E(t)w(E(t)) + \delta_2 \int\limits_t^T w(E(s)) \int\limits_\Omega |\rho(.,u_t)|^2$$

$$+\delta_3 \int\limits_t^T w(E(s)) \int\limits_\omega |u_t|^2\,ds\,\forall\,0 \leq t \leq T. \qquad (1.149)$$

where δ_i, $i = 1,\ldots,3$ are positive constants.

Remark 1.7.3. The above estimate shows that the nonlinear kinetic energy and the localized linear kinetic energy dominates the behavior of the total energy. A similar estimate holds for the boundary damping case (see [8, 17]). This estimate holds also for other equations such as Petrowsky equation, coupled systems such as Timoshenko beams . . . For the sake of presentation and clearness we do not include these results here. They can be found in [8, 17] in an abstract form with several PDE examples of applications.

Proof. Let w be a nonnegative, nondecaying $\mathscr{C}^1$ function defined on $[0, E(0)]$. The proof follows that of the linear stabilization in Sect. 1.2, except that one has to choose multipliers of the form $w(E)Ku$. Let $(u^0, u^1) \in H^2(\Omega) \cap H_0^1(\Omega) \times H_0^1(\Omega)$. Let $\varepsilon_0 < \varepsilon_1 < \varepsilon_2 < \varepsilon$ and define for $i = 0, 1, 2$ the subsets Q_i as in (1.19) and the functions $\psi_j \in \mathscr{C}_0^\infty(R^N)$ which satisfies (1.20). We also define h as the $\mathscr{C}^1$ vector field on Ω:

$$h(x) = \begin{cases} \psi_j(x)m_j(x) \text{ if } x \in \Omega_j \\ 0 \text{ if } x \in \Omega \backslash \cup_j \Omega_j, \end{cases}$$

and consider the multiplier $w(E)Mu$ where $Mu = h\cdot\nabla u + u(N-1)/2$. We consider the expression

$$\int\limits_S^T \int\limits_{\Omega_j} w(E)Mu\,(u'' - \Delta u + \rho(x,u'))\,dx\,dt = 0.$$

This gives, after appropriate integration by parts and proceeding as in Sect. 1.2

$$\int\limits_S^T Ew(E)\,dt \leq C \int\limits_S^T w(E) \int\limits_{\Omega\cap Q_1} u'^2 + |\nabla u|^2 - [w(E)\int\limits_\Omega M(u)u']_S^T \quad (1.150)$$

$$+ \int\limits_S^T E'w(E) \int\limits_\Omega M(u)u' - \int\limits_S^T w(E) \int\limits_\Omega M(u)\rho(x,u').$$

We need to estimate the terms on the right hand side of (1.150). Due to the weight $w(E)$ which depends only on time there is an additional term

$$\int\limits_S^T E'w'(E) \int\limits_\Omega M(u)u'$$

which can be easily bounded above by

$$|\int_S^T E'w'(E)\int_\Omega M(u)u'| \le c\int_S^T (-E')w'(E)E\,dt$$

$$\le CE(S)w(E(S)) \quad \forall\, 0 \le S \le T.$$

In a similar way, we have

$$|[w(E)\int_\Omega M(u)u']_S^T| \le CE(S)w(E(S)) \quad \forall\, 0 \le S \le T.$$

As for the linear case, the only difficulty is to estimate the first term on the right hand side of (1.150). The proof follows that of the linear case up to the multiplicative additional weight $w(E)$ which depends only on time. One can show that additional terms due to integration by parts can be bounded above. More precisely, we proceed as follows. We define $\xi \in \mathscr{C}_0^\infty(\mathbb{R}^N)$ as in (1.29). Multiplying the first equation of (1.142) by $w(E)\xi u$ and integrating the resulting equation on $[S,T]\times\Omega$, we obtain after some integration by parts:

$$\int_S^T w(E)\int_\Omega \xi|\nabla u|^2 = \int_S^T w(E)\int_\Omega \xi|u'|^2 + \frac{1}{2}\Delta\xi u^2 + \int_S^T E'w'(E)\int_\Omega \xi uu'$$

$$- \left[w(E)\int_\Omega \xi uu'\right]_S^T - \int_S^T w(E)\int_\Omega \xi u\rho(x,u'). \tag{1.151}$$

Hence, we have

$$\int_S^T w(E)\int_{\Omega\cap Q_1} |\nabla u|^2 \le C\int_S^T w(E)\int_{\Omega\cap Q_2} |u'|^2 + u^2 + |\rho(x,u')|^2$$

$$+CE(S)w(E(S)). \tag{1.152}$$

We define $\beta \in \mathscr{C}_0^\infty(\mathbb{R}^N)$ such that (1.32) holds. We fix t and consider the solution z of the elliptic problem (1.33) and (1.34). Multiplying the first equation of (1.142) by $w(E)z$ and integrating the resulting equation on $[S,T]\times\Omega$,

$$\int_S^T w(E)\int_{\Omega\cap Q_2} |u|^2 \le \frac{C}{\eta}\int_S^T w(E)\int_\omega |u'|^2 + \frac{C}{\eta}\int_S^T w(E)\int_\Omega |\rho(x,u')|^2$$

$$+\eta\int_S^T Ew(E) + CE(S)w(E(S)) \quad \forall\, \eta > 0. \tag{1.153}$$

We use the above estimates and proceed in a similar way than for the linear case, this gives

$$\int\limits_{S}^{T} Ew(E) \le CE(S)w(E(S)) + C\delta \int\limits_{S}^{T} Ew(E) \qquad (1.154)$$

$$+\frac{C}{\delta} \int\limits_{S}^{T} w(E)\left[\int\limits_{\omega} |u'|^2 + \int\limits_{\Omega} |\rho(x, u')|^2\right] \quad \forall \, \delta > 0 \,.$$

Choosing δ sufficiently small, we obtain finally

$$\int\limits_{S}^{T} Ef(E) \le CE(S)w(E(S)) + C \int\limits_{S}^{T} w(E)\left[\int\limits_{\omega} |u'|^2 + \int\limits_{\Omega} |\rho(x, u')|^2\right]. \quad (1.155)$$

Hence, we proved that E satisfies an estimate of the form (1.149). $\square$

1.7.3 Weight Function As an Optimal Unknown

In general, the usual ways used to prove energy decay rates are either to prove that a perturbed energy, equivalent to the original one, is a Lyapunov function satisfying a nonlinear differential inequality or to prove that the original energy directly satisfies a nonlinear Gronwall inequality. In this latter case (also in the former one), a crucial point to be sure to loose no information on the speed of decay of the energy is to find mathematical arguments to build an optimal weight. This is an important novelty of our work in [17] and later on, of its simplification in [8]. This work is based on an approach for which, using convexity arguments and in particular Young's inequality, we determine the optimal weight w as the solution of an implicit relation, that is

$$\beta \widehat{H}^{\star}(w(E(t))) = \frac{1}{2}E(t)w(E(t)) \quad \forall \, t \ge 0 \,.$$

to establish that the energy E satisfies a suitable nonlinear Gronwall inequality. The above implicit relation has a unique solution which involves the inverse function of L defined by (1.82).

Theorem 1.7.4. *Assume that ω satisfies the geometric conditions (MGC) or (PWMGC). We respectively define H, $\widehat{H}$, and L by (1.118), (1.81), (1.82). Let $(u^0, u^1) \in V \times H$ be given and denote by u a solution of (1.142) and E its energy defined as in (1.6). We assume that $E(0) > 0$ and define $\beta > 0$ as the explicit constant depending on $E(0)$, $|\omega|$, $|\Omega|$ defined in (1.170). We define w as in (1.123), then E satisfies*

$$\int\limits_{S}^{T} E(s)w(E(s)) \, ds \le T_0 E(S) \quad \forall \, 0 \le S \le T \,. \qquad (1.156)$$

Proof. We set $\varepsilon_0 = g(r_0)$. Assuming that r_0 is small enough, we can assume that $0 < \varepsilon_0 < 1$.

From (HF), we can easily deduce that ρ satisfies the following inequalities:

$$c_1 a(x)|v| \leq |\rho(x,v)| \leq c_2 a(x)|v| , \quad \forall\, x \in \Omega , \forall\, \varepsilon_0 \leq |v| , \qquad (1.157)$$

and

$$c_1 a(x) g(|v|) \leq |\rho(x,v)| \leq c_2 a(x) g^{-1}(|v|) , \quad \forall\, x \in \Omega , \forall\, |v| \leq \varepsilon_0 . \qquad (1.158)$$

for certain positive constants c_1, c_2. We set for all fixed $t \geq 0$, $\Omega_1^t = \{x \in \Omega ,$ $|u'(t,x)| \leq \varepsilon_0\}$. We also set

$$c_g = \frac{1}{c_2\|a\|_\infty} . \qquad (1.159)$$

Thus, by definition of c_g and thanks to (1.158), we have:

$$c_g^2 |\rho(x, u'(t)(x))|^2 \leq r_0^2 , \quad \forall x \in \Omega_1^t .$$

Using Jensen's inequality, we have

$$H\!\left(\frac{1}{|\Omega_1^t|} \int_{\Omega_1^t} c_g^2 |\rho(x, u'(t)(x))|^2 \, dx\right) \leq \frac{1}{|\Omega_1^t|} \int_{\Omega_1^t} H(c_g^2 |\rho(x, u'(t)(x))|^2) \, dx$$

$$\leq \frac{1}{|\Omega_1^t|} \int_{\Omega_1^t} c_g |\rho(x, u'(t)(x)| g(c_g |\rho(x, u'(t)(x)))| \, dx .$$

$$(1.160)$$

Using (1.158) and the fact that H is increasing, we deduce that

$$\int_S^T w(E(t)) \int_{\Omega_1^t} |\rho(x, u')|^2 \, dx\, dt \leq \int_S^T \frac{|\Omega_1^t|}{c_g^2} w(E(t)) H^{-1} \qquad (1.161)$$

$$\left(\frac{1}{|\Omega_1^t|} \int_{\Omega_1^t} c_g u' \rho(x, u') \, dx\right) dt$$

Since $\widehat{H}$ is a convex and proper function, we can apply Young's inequality (see [93, 103]), so that

$$\int_S^T w(E(t)) \int_{\Omega_1^t} |\rho(x, u')|^2 \, dx\, dt \leq \frac{|\Omega|}{c_g^2} \int_S^T \widehat{H}^\star(w(E(t))) \, dt \qquad (1.162)$$

$$+ \frac{1}{c_g} E(S) , \quad \forall\, 0 \leq S \leq T .$$

On $\Omega \backslash \Omega_1^t$, we have $|u'(t)| \geq \varepsilon_0$. Hence, thanks to (1.157), we have

$$|\rho(x, u')|^2 \leq \frac{1}{c_g} u' \rho(x, u') , \quad \forall \, x \in \Omega \setminus \Omega_1^t .$$ (1.163)

This implies

$$\int_S^T w(E(t)) \int_{\Omega \setminus \Omega_1^t} |\rho(x, u')|^2 \, dx \, dt \leq \frac{1}{c_g} \int_S^T (-E'(t)) w(E(t)) \, dt .$$ (1.164)

We now turn to the localized kinetic energy, that is we want to estimate the term $\int_S^T w(E(t)) \int_\omega |u'(t)|^2 \, dx \, dt$. We set as in [17]

$$r_1^2 = H^{-1}(c_1 a_- c_g H(r_0^2)) ,$$ (1.165)

and

$$\varepsilon_1 = \min (r_0, g(r_1)) .$$ (1.166)

One can remark that $\varepsilon_1 \leq \varepsilon_0$. We define, for fixed $t \geq 0$, the set $\omega_1^t = \{x \in \omega , |u'(t)(x)| \leq \varepsilon_1\}$. Thanks to (1.158), our choice of ε_1 and to Jensen's inequality, we obtain

$$H\left(\frac{1}{|\omega_1^t|} \int_{\omega_1^t} |u'(t)(x)|^2 \, dx\right) \leq \frac{1}{|\omega_1^t|} \int_{\omega_1^t} H(|u'(t)(x)|^2) \, dx$$

$$\leq \frac{1}{|\omega_1^t|} \int_{\omega_1^t} u'(t)(x) g(u'(t)(x)) \, dx$$

$$\leq \frac{1}{|\omega_1^t| c_1 a_-} \int_{\omega_1^t} u'(t)(x) \rho(x, u'(t)(x)) \, dx .$$

Using once again Young's inequality, we derive

$$\int_S^T w(E(t)) \int_{\omega_1^t} |u'(t)(x)|^2 \, dx \, dt \leq |\omega| \int_S^T \widehat{H}^\star(w(E(t))) \, dt$$

$$+ \frac{1}{c_1 a_-} E(S), \quad \forall 0 \leq S \leq T.$$ (1.167)

On $\omega \setminus \omega_1^t$, we have $|u'(t)| \geq \varepsilon_1$. For $|u'(t)| \geq \varepsilon_0$, (1.157) holds. We can easily check that using (1.158), similar inequalities holds for $\varepsilon_1 \leq |u'(t)| \leq \varepsilon_0$. This implies

$$\int_S^T w(E(t)) \int_{\omega \setminus \omega_1^t} |u'|^2 \, dx \, dt \leq \frac{1}{c_1 a_-} \int_S^T (-E'(t)) w(E(t)) \, dt .$$ (1.168)

Inserting now (1.162), (1.164), (1.167) and (1.168) in (1.149), we obtain

$$\int_S^T w(E(t))E(t)\,dt \le \delta_1\, E(S)f(E(S)) + \left(\frac{\delta_2}{c_g} + \frac{\delta_3}{c_1 a_-}\right)E(S)$$

$$+\left(\frac{\delta_2}{c_g} + \frac{\delta_3}{\tilde{c}_1 a_-}\right)\int_S^T (-E'(t))w(E(t))$$

$$+\left(\delta_2\frac{|\Omega|}{c_g^2} + \delta_3|\omega|\right)\int_S^T \widehat{H}^\star(w(E(t)))\,dt\,. \quad (1.169)$$

We recall that L, defined by (1.82) is a strictly increasing function from $[0, +\infty)$ onto $[0, r_0^2)$. We choose a real number $\beta = \beta_{E(0)} > 0$ as follows:

$$\beta = \max\left(\delta_3\,|\omega| + \frac{\delta_2\,|\Omega|}{c_g^2}, \frac{E(0)}{2\,L(H'(r_0^2))}\right) \quad (1.170)$$

Now, we choose the weight function w as announced, that is:

$$w(s) = L^{-1}\left(\frac{s}{2\beta}\right) \quad \forall\, s \in [0, 2\beta r_0^2)\,. \quad (1.171)$$

Then w is a strictly increasing function from $[0, 2\beta r_0^2)$ onto $[0, +\infty)$. Moreover, by construction, w satisfies the relation:

$$\beta\widehat{H}^\star(w(s)) = \frac{1}{2}sw(s) \quad \forall\, s \in [0, 2\beta r_0^2)\,.$$

Since E is nonincreasing, we have

$$E(t) \le E(0) < E(0)\frac{r_0^2}{L(H'(r_0^2))} \le 2\,\beta r_0^2 \quad \forall\, t \ge 0\,.$$

Hence, one has in particular:

$$\beta\widehat{H}^\star(w(E(t))) = \frac{1}{2}E(t)w(E(t)) \quad \forall\, t \ge 0\,. \quad (1.172)$$

With this choice of β and w, the last term on the right hand side of (1.169) is bounded above by

$$\frac{1}{2}\int_S^T E(t)(w(E(t)))\,dt\,. \quad (1.173)$$

On the other hand, we recall that $-E'$ is nonnegative on $[0, +\infty)$, E is non-negative and nonincreasing on $[0, +\infty)$ whereas w is nonnegative and increasing on $[0, 2\beta r_0^2)$. Thus, the third term on the right hand side of (1.169) is bounded above by

$$\left(\frac{\delta_2}{c_g} + \frac{\delta_3}{\tilde{c}_1 a_-}\right) \int_S^T (-E'(t)) w(E(t)) \leq \left(\frac{\delta_2}{c_g} + \frac{\delta_3}{\tilde{c}_1 a_-}\right) E(S) L^{-1}\left(\frac{E(S)}{2\beta}\right).$$

(1.174)

We insert the estimates (1.173) and (1.174) in (1.169). This gives

$$\int_S^T E(t) L^{-1}\left(\frac{E(t)}{2\beta}\right) dt \leq 2\left(\delta_1 + \frac{\delta_2}{c_g} + \frac{\delta_3}{\tilde{c}_1 a_-}\right) E(S) L^{-1}\left(\frac{E(S)}{2\beta}\right)$$
$$+ 2\left(\frac{\delta_2}{c_g} + \frac{\delta_3}{c_1 a_-}\right) E(S).$$

Hence, the energy E satisfies the estimate

$$\int_S^T E(t) L^{-1}\left(\frac{E(t)}{2\beta}\right) dt \leq T_0 E(S) \quad \forall\, 0 \leq S \leq T,$$

(1.175)

where T_0 is independent of $E(0)$ and, with our choice of β is given by

$$T_0 = 2\left(\frac{\delta_2}{c_g} + \frac{\delta_3}{c_1 a_-} + (\delta_1 + \frac{\delta_2}{c_g} + \frac{\delta_3}{\tilde{c}_1 a_-}) H'(r_0^2)\right)$$

(1.176)

Thus, the functions g, H, E and β satisfy the hypotheses of Theorem 1.5.4. Applying the conclusions of this theorem, we deduce that E satisfies the desired estimate, which concludes the proof.   □

1.7.4   Simplification of the Energy Decay Rates

Corollary 1.7.5. *Assume the hypotheses of Theorem 1.7.4. We define H by (1.118) and assume that H is strictly convex on $[0, r_0^2]$. We also define $\hat{H}$ by (1.81), L by (1.82) and Λ_H by (1.83). Let $(u^0, u^1) \in V \times H$ be given and denote by u a solution of (1.142) and E its energy defined as in (1.6). Assume that $\limsup_{x\to 0} \Lambda_H(x) < 1$, then E satisfies*

$$E(t) \leq 2\beta (H')^{-1}\left(\frac{DT_0}{t}\right),$$

for t sufficiently large and where D is a positive constant which does not depend on $E(0)$ and β is given in (1.170). Moreover, assume that either

$$0 < \liminf_{x\to 0} \Lambda_H(x) \leq \limsup_{x\to 0} \Lambda_H(x) < 1,$$

or that there exists $\mu > 0$ such that

$$0 < \liminf_{x \to 0} \left(\frac{H(\mu x)}{\mu x} \int_x^{z_1} \frac{1}{H(y)} \, dy \right), \text{ and } \limsup_{x \to 0} \Lambda_H(x) < 1,$$

for a certain $z_1 \in (0, z_0]$ (arbitrary). Then the energy of solution of satisfies the estimate

$$E(t) \leq C(E(0)) v^2(t) \quad \text{for sufficiently large } t,$$

where v is the solution of the ordinary differential equation

$$v'(t) + g(v(t)) = 0, \, v(0) = \sqrt{2E(0)}, t \geq 0.$$

Remark 1.7.6. Without the additional above assumptions on Λ_H, we have the general energy decay rate provided in (1.91). In the PDE case, we cannot derive a lower estimate saying that E is bounded below by a constant time the energy of v in the above ordinary differential equation, as in Lemma 1.5.13 of Sect. 1.3. We will see in Sect. 1.5.5 that we can establish lower weaker energy estimates using interpolation.

1.7.5 Generalization to Optic Geometric Conditions: The Indirect Optimal-Weight Convexity Method

We recalled in Sect. 1.2 two methods giving geometrical conditions on the set ω on which the feedback is active, for exponential stabilization to hold in case of linear stabilization. These geometric conditions are still required in the nonlinear stabilization case. The results we presented in the above sections of Sect. 1.5 require the multiplier geometric conditions, which are known to be nonoptimal. We show in Alabau-Boussouira and Ammari [9] (see also the important work by Daoulatli et al. [50]) that it is possible to combine the optimal-weight convexity method with the quasi-optimal optic geometric conditions of Bardos et al. [27] and of Lebeau [74] on the set ω. We now present this generalization of the results on nonlinear stabilization—presented in the above section of this section—to the optic geometric conditions of Bardos et al. [27] and of Lebeau [74]. This generalization is partly based on the approach introduced by Haraux in [61] for linear stabilization for the case of bounded linear stabilization operators and extended by Ammari and Tucsnak [24] (see also [25]) for the case of linear unbounded stabilization operators. It consists in deriving stabilization for an observability estimate for the undamped equation.

Let us consider the following abstract second order equation

$$\begin{cases} \ddot{w}(t) + Aw(t) + a(.)\rho(., \dot{w}) = 0, & t \in (0, \infty), x \in \Omega \\ w(0) = w_0, \dot{w}(0) = w_1. \end{cases} \quad (1.177)$$

where Ω is a bounded open set in $\mathbb{R}^N$, with a boundary Γ of class $\mathscr{C}^2$, $X = L^2(\Omega)$, with its usual scalar product denoted by $\langle \cdot, \cdot \rangle_X$ and the associated norm $| \cdot |_X$ and where $A : D(A) \subset X \to X$ is a densely defined self-adjoint linear operator satisfying

$$\langle Au, u \rangle_X \geq \alpha |u|_X^2 \qquad \forall u \in D(A) \tag{1.178}$$

for some $\alpha > 0$. We also set $V = D(A^{1/2})$ and $|u|_V = |A^{1/2}u|_X$. We make the following assumptions on the feedback ρ and on a:

Assumption (A1): $\rho \in \mathscr{C}(\Omega \times \mathbb{R}; \mathbb{R})$ is a continuous monotone nondecreasing function with respect to the second variable on Ω such that $\rho(., 0) = 0$ on Ω and there exists a continuous strictly increasing odd function $g \in \mathscr{C}([-1, 1]; \mathbb{R})$, differentiable in a neighbourhood of 0 and satisfying $g(0) = g'(0) = 0$, with

$$\begin{cases} c_1 g(|v|) \leq |\rho(., v)| \leq c_2 g^{-1}(|v|), & |v| \leq 1, \text{ a.e. on } \Omega, \\ c_1 |v| \leq |\rho(., v)| \leq c_2 |v|, & |v| \geq 1, \text{ a.e. on } \Omega, \end{cases} \tag{1.179}$$

where $c_i > 0$ for $i = 1, 2$. Moreover $a \in \mathscr{C}(\overline{\Omega})$, with $a \geq 0$ on Ω and

$$\exists\, a_- > 0 \text{ such that } a \geq a_- \text{ on } \omega. \tag{1.180}$$

The energy of solutions is defined by

$$E_w(t) = \frac{1}{2}\left(|\dot{w}|_X^2 + |w|_V^2\right) \tag{1.181}$$

We recall that the energy of mild solutions of (1.177) satisfies the dissipation relation

$$\int_0^T \langle a(\;)\rho(., \dot{w}), \dot{w} \rangle_X = E_w(0) - E_w(T) \tag{1.182}$$

One important question is at which rate the energy of solutions of the damped equation (1.177) goes to 0 as time goes to ∞. We already give above a direct method to derive sharp energy decay rates for system (1.177) provided that the set ω satisfies the hypotheses of the piecewise multiplier method [78] (see also [105]). Here we are interested to extend these results to the Geometric Control Condition of Bardos et al. [27]. For this, we will use an indirect method based on an observability inequality for the linear undamped system

$$\begin{cases} \ddot{\phi}(t) + A\phi(t) = 0, & t \in (0, T), x \in \Omega \\ \phi(0) = \phi_0, \dot{\phi}(0) = \phi_1. \end{cases} \tag{1.183}$$

We define the energy of a solution ϕ of (1.183) by

$$E_\phi(t) = \frac{1}{2}\left(|\dot{\phi}|_X^2 + |\phi|_V^2\right) \tag{1.184}$$

Then we prove.

Theorem 1.7.7 (Alabau-Boussouira–Ammari [6]). *Assume that ρ and a satisfy the assumption $(A1)$ and that there exists $r_0 > 0$ sufficiently small so that the function H defined by (1.118) is strictly convex on $[0, r_0^2]$ with $H(0) = H'(0) = 0$. We also define $\widehat{H}$ by (1.81), L by (1.82) and Λ_H by (1.83). Assume also either that*

$$0 < \liminf_{x\to 0^+} \Lambda_H(x) \tag{1.185}$$

or

$$\lim_{x\to 0^+} \frac{H'(x)}{\Lambda_H(x)} = 0 \tag{1.186}$$

Moreover let $T > 0$ be such that there exists $c_T > 0$ the following observability inequality is satisfied for the linear damped conservative system (1.183)

$$c_T E_\phi(0) \leq \int_0^T |\sqrt{a}\dot\phi|_X^2 \, dt, \quad \forall (\phi_0, \phi_1) \in V \times X. \tag{1.187}$$

Then, the energy of the solution of (1.177) satisfies

$$E_w(t) \leq 2\beta L\left(\frac{1}{\psi_r^{-1}(\frac{t-T}{T_0})}\right), \quad \text{for } t \text{ sufficiently large,} \tag{1.188}$$

where $L\left(\frac{1}{\psi_r^{-1}(\frac{t-T}{T_0})}\right) \to 0$ as $t \to \infty$, where β is a positive constant which depends on $E_w(0)$, c_T, and T, and where ψ_r is defined in (1.92). If in addition, $\limsup_{x\to 0} \Lambda_H(x) < 1$, then E satisfies

$$E(t) \leq 2\beta(H')^{-1}\left(\frac{DT_0}{t-T}\right), \tag{1.189}$$

for t sufficiently large and where D is a positive constant which does not depend on $E(0)$ and β depends on $E_w(0)$, c_T, and T.

The proof of this result is given in [6] (see also [9] for an announcement of these results). A key point of the proof is to use the optimal-weight convexity method to prove that the energy satisfies nonlinear discrete inequalities, that is

Theorem 1.7.8 (Alabau-Boussouira–Ammari [6]). *Under the above hypotheses and setting $\widehat{E}_w = E_w/\beta$, we have*

$$\widehat{E}((k+1)T) \leq \widehat{E}(kT)\left(1 - \rho_T L^{-1}(\widehat{E}(kT))\right), \quad \forall k \in \mathbb{N}. \tag{1.190}$$

for a certain ρ_T depending on the observability constant, and the time T.

The proof of our main result relies on a result important in itself, since it allows to compare discrete energy inequalities to continuous ones, that is

Theorem 1.7.9 (Alabau-Boussouira–Ammari [6]). *Assume that the above assumption holds and let $T > 0$ and $\rho_T > 0$ be given. Let $\delta > 0$ be such that the function defined by $x \mapsto x - \rho_T M(x)$ is strictly increasing on $[0, \delta]$. Assume that $\widehat{E}$ is a nonnegative, nonincreasing function defined on $[0, \infty)$ with $\widehat{E}(0) < \delta$ and satisfying (1.190). Then $\widehat{E}$ satisfies the upper estimate*

$$\widehat{E}(t) \leq TL\left(\frac{1}{\psi_r^{-1}\left(\frac{(t-T)\rho_T}{T}\right)}\right), \quad \text{for } t \text{ sufficiently large,}$$

If moreover $\limsup_{x \to 0+} \Lambda_H(x) < 1$, *then we have the simplified decay rate*

$$\widehat{E}(t) \leq T(H')^{-1}\left(\frac{DT}{\rho_T(t-T)}\right),$$

for t sufficiently large and where D is a positive constant independent of $\widehat{E}(0)$ and of T.

1.7.5.1   Examples of Application

We first consider the geometrical situation considered by Lebeau [27, 74]. More precisely, (Ω, g) is assumed to be a $\mathscr{C}^\infty$ Riemannian compact and connex manifold, with a boundary of class ∞, whereas $-A$ is the Laplacian on Ω for the metrics g, and $a \in \mathscr{C}^\infty(\overline{\Omega}; [0, \infty))$, and where ρ and a satisfies assumptions $(A1)$. Thanks to Theorem 0 in [74] and to [27], and applying Theorem 1.7.7, we deduce the following result

Corollary 1.7.10. *Assume that ρ and a satisfy the assumption $(A1)$ and that there exists $r_0 > 0$ sufficiently small so that the function H defined by (1.118) is strictly convex on $[0, r_0^2]$. Assume also either that (1.185) or (1.186) holds. Assume moreover that the geodesics of $\overline{\Omega}$ have no contact of infinite order with Γ and that there exists a time $T_- > 0$ such that every generalized geodesics of Ω of length larger than T_- meets w. Then, the energy of the solution satiof (1.177) satisfies (1.188). If in addition, $\limsup_{x \to 0} \Lambda_H(x) < 1$, then E satisfies (1.189).*

Remark 1.7.11. Other examples are given in [6]. The above results hold true in particular for the wave equation. We generalize the above results to the case of arbitrary nonlinear boundary dampings (see [9]). We can apply our simplification for energy decay rates and give example of applications to several feedback growth (see Sect. 5.6). Under additional hypotheses, we can also compare the above decay rate to the energy decay rate of an ODE as in Sect. 1.5.4.

1.7.6   Examples of Feedbacks and Sharp Upper Estimates

We will now illustrate through five examples of feedback growth the sharp upper energy estimates that Theorem 1.7.5 gives. For the examples below, g is only given close to 0. One can notice that the upper estimates are similar to that of the finite dimensional case.

Theorem 1.7.12. *Let* $(u^0, u^1) \in V \times H$ *be given and denote by* u *a solution of* (1.142) *and* E *its energy defined as in* (1.6). *We assume that* $E(0) > 0$ *and define* $\beta > 0$ *as the explicit constant depending on* $E(0), |\omega|, |\Omega|$ *defined in* (1.170). *Then, we have*

> **Example 1:** *let* g *be given by* $g(x) = x^p$ *where* $p > 1$ *on* $(0, r_0]$.
> *Then*

$$E(t) \leq C\beta_{E(0)} \, t^{\frac{-2}{p-1}}, \tag{1.191}$$

for t *sufficiently large and for all* (u^0, u^1) *in* $H_0^1(\Omega) \times L^2(\Omega)$.
> **Example 2:** *let* g *be given by* $g(x) = x^p (ln(\frac{1}{x}))^q$ *where* $p > 1$ *and* $q > 1$ *on* $(0, r_0]$.
> *Then*

$$E(t) \leq C\beta_{E(0)} \, t^{-2/(p-1)} (\ln(t))^{-2q/(p-1)} \tag{1.192}$$

for t *sufficiently large and for all* (u^0, u^1) *in* $H_0^1(\Omega) \times L^2(\Omega)$.
> **Example 3:** *let* g *be given by* $g(x) = e^{-\frac{1}{x^2}}$ *on* $(0, r_0]$.
> *Then*

$$E(t) \leq C\beta_{E(0)} (\ln(t))^{-1},$$

for t *sufficiently large and for all* (u^0, u^1) *in* $H_0^1(\Omega) \times L^2(\Omega)$.
> **Example 4:** *let* g *be given by* $g(x) = e^{-(ln(\frac{1}{x}))^p}$, $1 < p < 2, x \in [0, r_0]$.
> *Then*

$$E(t) \leq C\beta_{E(0)} e^{-2(\ln(t))^{1/p}}$$

for t *sufficiently large and for all* (u^0, u^1) *in* $H_0^1(\Omega) \times L^2(\Omega)$.
> **Example 5:** *let* g *be given by* $g(x) = x(ln(\frac{1}{x}))^{-p}$ *where* $p > 0$.

Then

$$E(t) \leq C\beta_{E(0)} e^{-2(\frac{pt}{DT_0})^{1/(p+1)}} t^{-1/(p+1)} \tag{1.193}$$

for t *sufficiently large and for all* (u^0, u^1) *in* $H_0^1(\Omega) \times L^2(\Omega)$.

Remark 1.7.13. We can notice that the above upper estimates are similar to that of Theorem 1.5.22 in the finite dimensional case, but we cannot derive the lower estimates as for the ODE case. Also, the upper comparison with Kv^2 where v is the solution of the ODE (1.117) still holds true in the PDE case under the hypotheses $(H2)$ on H, as in the finite dimensional case (see Sect. 1.3).

We can give many other examples of feedback growth. These five examples were chosen thanks to their distinct behavior close to the origin.

The reader can also find in [8] various examples of applications to other PDE's than the wave equation, namely Petrowsky equations, Timoshenko beams ...

Remark 1.7.14. The results presented in Sect. 1.5.5 extend to the nonlinear boundary damped equations (see [9]).

The above results also generalize to nonlinearly damped semilinear wave equations (see [8] for a result in this direction).

1.7.7 Lower Energy Estimates

We saw in the previous sections how optimality can be proved in the finite dimensional case. The situation is different in the infinite dimensional framework. The difficulty is to obtain lower energy estimates. Moreover, even when lower energy estimates are known, these estimates are not of the same order than the upper ones, so that optimality is still a largely open question. The purpose of this section is to give some lower energy estimates recently derived in [7, 8]. As far as we know, in the infinite dimensional setting, very few lower estimates are available. Haraux [57] in 1995 considered the 1-D wave, globally damped wave equation with a nonlinear feedback with polynomial growth such as x^p close to the origin. He proved a weak lower velocity estimate in uniform norm, from which he deduced the following weak lower energy estimate

$$\limsup_{t \to \infty} (t^{3/(p-1)} E(t)) > 0 .$$

For initial data in $W^{2,\infty}(0, 1) \times W^{1,\infty}(0, 1)$. The proof requires the smoothness of the solutions. This regularity is proved using a clever Lyapunov function. It does not give a direct estimate of the energy. Also the proof does not work for the boundary damped one-dimensional wave equation neither for systems nor for multidimensional examples. It is important to note that Haraux' method requires both the dissipation relation and an upper energy estimate. So, if one thinks to generalization to multi-dimensional domains, the region on which the feedback is effective will have to satisfy the Geometric Control Condition, whereas the methodology, we shall present, introduced in Alabau-Boussouira [7, 8] is only based on the dissipation relation and interpolation properties. In particular, our method does not require the Geometric Control Condition, nor an upper energy estimate. Concerning optimality results, such results have been obtained for the one-dimensional wave equation with boundary feedbacks and specific initial data by Vancostenoble [102] and Martinez–Vancostenoble [101] using the explicit form of the solution through D'Alembert's formula. They consider the nonlinearly boundary damped wave equation in $\Omega = (0, 1)$ as follows

$$\begin{cases} u_{tt} - u_{xx} = 0 \text{ on } (0, +\infty) \times \Omega , \\ u = 0 \text{ on } (0, +\infty) \times \{0\} , \\ u_x + v \, \rho(u_t) = 0 \text{ on } (0, +\infty) \times \{1\} , \\ u(0, .) = u_0(.) , \ u_t(0, .) = u_1(.) \text{ on } \Omega , \end{cases}$$

For this equation, one has the following important optimality result.

Theorem 1.7.15 (Vancostenoble and Martinez [101]). *Assume that g is a strictly increasing and odd $\mathscr{C}^1$ function on $\mathbb{R}$ such that $g(0) = g'(0) = 0$. Assume that ρ is a continuous nondecreasing function on $\mathbb{R}$ such that $\rho =$ or $\rho = g^{-1}$ in a neighbourhood of 0. Then for all initial data of the form $(u_0, u_1)(x) = (2A_0x, 0)$, with $A_0 \neq 0$, the solution of the above boundary damped wave equation satisfies*

$$\exists\, n_0 \in \mathbb{N}, \forall\, n \geq n_0\,, E(2n) = \frac{1}{2}V^2(2t_n)$$

where $(t_n)_{n \geq n_0}$ is a real positive increasing sequence such that $t_n \sim n$ as $n \to \infty$, and where $V : [0, \infty) \mapsto [0, \infty)$ is the solution of the differential equation, we have seen before, that is

$$V' + g(V) = 0\,, V(0) = 2\sqrt{E(2n_0)/2}\,.$$

Vancostenoble and Martinez have also shown that explicit lower bounds for certain sequences of time can be obtained. In particular, assuming for instance the above hypotheses and that $|\rho(s)| \leq |g(s)|$ in a neighbourhood of 0 and that the function $s \mapsto s((\frac{1}{2}g^{-1})'(s) - 1)$ is increasing in a neighbourhood of 0, then for all initial data of the above form, the solution u of the $1 - D$ wave equation satisfies

$$\exists\, n_0\,, n_1 \in \mathbb{N}, \forall\, n \geq n_0\,, E(2n) \geq \frac{1}{2}\Big[(g')^{-1}\Big(\frac{1}{2(n+n_1)}\Big)\Big]^2$$

They give other examples of lower bounds, extend the results to radial solutions in annulus type domains. The starting point of their method is to use the D'Alembert formula to write the solution of the wave equation

$$u_{tt} - u_{xx} = 0\,, t > 0\,, x \in (0,1)\,,$$

that is

$$u(t,x) = f(t+x) - f(t-x)\,, t > 0\,, x \in (0,1)\,.$$

and to write the conditions that f must satisfy so that the boundary and initial conditions hold. Introducing A_n such that $E(2n) = 2A_n^2$, they prove some nice properties on the asymptotic behavior of the sequence A_n for initial data of the specific form given before. We apply their results in [17] to derive the optimality of our general estimates for some examples of dampings.

Theorem 1.7.16 (Alabau-Boussouira [17]). *There exist initial data of the form $(u^0(x), u^1(x)) = (2A_0\,x, 0)$ for all $x \in \Omega$, such that the upper energy estimates of the corresponding solution are optimal for Examples 1 to 4 of Sect. 5.6. Optimality cannot be asserted for Example 5.*

Hence optimality results have been only obtained for peculiar cases. The question is open in the general case. As we already said, this is due to two difficulties: obtaining lower energy estimates and sharpening these lower estimates so that they are of the same order than the upper ones. Even if one restricts his or her objectives to the first difficulty, in the infinite dimensional setting very few lower estimates are available. We shall present several results which allow us to generalize lower energy estimates to general dampings, less regular solutions, boundary damping cases, multi-dimensional situations, other PDE's such as Petrowsky equations or systems such as Timoshenko beams.

Let us start by a simple example which connects the infinite dimensional case to the finite one and allows us to give a first small improvement of Haraux' result by a more direct and simpler argument. We consider the equation

$$\begin{cases} \partial_t^2 u(t,x) - \partial_x^2 u(t,x) + +\rho(\partial_t u(t,x)) = 0, & 0 < t, 0 < x < 1, \\ u(t,x) = 0, & \text{for } x = 0, x = 1, 0 < t, \\ u(0,x) = u_0(x), \; \partial_t u(0,x) = u_1(x), & 0 < x < 1. \end{cases} \tag{1.194}$$

The energy of a solution is defined as usual and is denoted by E. We assume that ρ is a smooth, odd and nondecreasing function on $\mathbb{R}$, with linear growth at infinity and such that

$$\rho(s) = |s|^{p-1} s, \quad \forall \, |s| \le 1.$$

where $p > 1$.

Theorem 1.7.17 (Alabau-Boussouira [8]). *Assume the above hypotheses on ρ. Assume moreover that $(u_0, u_1) \in W^{2,\infty}(\Omega) \times W^{1,\infty}(\Omega)$. Then there exists a constant $C > 0$, which depends only on p, $E(0)$ and the norm of u_t in $L^\infty([0,\infty); W^{1,\infty}(\Omega))$ such that*

$$E(t) \ge C(t+1)^{-3/(p-1)}, \quad \forall \, t \ge 0$$

This result can be generalized. We will start by weakening the smoothness assumptions on the solutions. Let us consider

$$\begin{cases} u_{tt}(t,x) - u_{xx}(t,x) + a(x)g(u_t(t,x)) = 0, & 0 < t, x \in \Omega, \\ u(t,c) = u(t,d) = 0, & \text{for } 0 < t, \\ u(0,x) = u_0(x), \; u_t(0,x) = u_1(x), & x \in \Omega, \end{cases}$$

where $\Omega = (c,d) \subset \mathbb{R}$, with $-\infty < c < d < \infty$, $a \in L^\infty(\Omega)$ and $a \ge 0$ a.e. on Ω with $a > 0$ on an open subset ω of Ω. The energy of a solution is defined by

$$E(t) = \frac{1}{2} \int_\Omega (u_t^2 + u_x^2) \, dx.$$

Well-posedness holds for initial data in the energy space $\mathscr{H} = H_0^1(\Omega) \times L^2(\Omega)$, i.e. for all $(u_0, u_1) \in \mathscr{H}$, the above equation has a unique solution u in $\mathscr{C}(\mathbb{R}_+; H_0^1(\Omega)) \cap \mathscr{C}^1(\mathbb{R}_+; L^2(\Omega))$. Moreover if (u_0, u_1) is in $(H_0^1(\Omega) \cap H^2(\Omega)) \times H_0^1(\Omega)$ then the solution is smoother and the energy of order 1 defined as

$$E_1(t) = \frac{1}{2} \int_\Omega (u_{tt}^2 + u_{xt}^2) \, dx$$

is well-defined, and nonincreasing. In this latter case, the natural energy E satisfies the dissipation relation

$$-E'(t) = \int_\Omega a(x)u_t(t,x)g(u_t(t,x))\,dx \quad t \geq 0.$$

Assume that g satisfies

$$(H1) \begin{cases} g : \mathbb{R} \mapsto \mathbb{R} \text{ is assumed to be an odd, increasing} \\ \text{continuously differentiable function} \\ g \text{ has a linear growth at infinity}, \\ sg(s) > 0 \quad \forall s \in \mathbb{R}^*, \\ g(0) = g'(0) = 0. \end{cases}$$

and further that

$$(H2) \begin{cases} \exists r_0 > 0 \text{ such that the function } H : [0, r_0^2] \mapsto \mathbb{R} \text{ defined by} \\ H(x) = \sqrt{x}g(\sqrt{x}), \text{ is strictly convex on } [0, r_0^2], \\ \text{and either } 0 < \liminf_{x\to 0} \Lambda(x) \leq \limsup_{x\to 0} \Lambda(x) < 1, \\ \text{or there exists } \mu > 0 \text{ such that} \\ 0 < \liminf_{x\to 0} \left(\frac{H(\mu x)}{\mu x} \int_x^{z_1} \frac{1}{H(y)}\,dy \right), \\ \text{and } \limsup_{x\to 0} \Lambda(x) < 1, \text{ for some } z_1 \in (0, z_0]. \end{cases}$$

Theorem 1.7.18 (Alabau-Boussouira [4]). *Assume that* $(u_0, u_1) \in (H_0^1(\Omega) \cap H^2(\Omega)) \times H_0^1(\Omega)$ *and that g satisfies* $(H1)$. *We assume that* $(H2)$ *holds. Then the energy satisfies the lower estimate*

$$C(E_1(0))\left((H')^{-1}\left(\frac{1}{t - T_0} \right) \right)^2 \leq E(t), \quad \forall t \geq T_1 + T_0,$$

where $C(E_1(0))$ *depends explicitly on* $E_1(0)$.

Remark 1.7.19. A weaker lower estimate has first been derived in [7].

Proof. We prove that

$$u_t^2(t,x) \leq C(E_1(0))\sqrt{E(t)} \quad \forall t \geq 0, \forall x \in \Omega.$$

Thanks to the dissipation relation and defining $\widetilde{H}(x) = H(x)/x$, we have

$$-E'(t) = \int_\Omega a(x)u_t^2\widetilde{H}(u_t^2(t,x))\,dx.$$

Moreover, using Dafermos' strong stabilization result that is that $\lim_{t\to\infty} E(t) = 0$, we deduce that there exists $T_0 \geq 0$ such that u_t^2 varies in an interval in which $\widetilde{H}$ is increasing. We thus have

$$\widetilde{H}\left(|u_t(t,.)|^2\right) \le \widetilde{H}(C(E_1(0))\sqrt{E(t)}), \quad t \ge T_0, \text{ in } \Omega .$$

Using this inequality together with the dissipation relation we deduce that

$$K\left(C(E_1(0))\sqrt{E(t)}\right) \le \alpha_a\,(t - T_0), \quad \forall\, t \ge T_0 .$$

where $\alpha_a > 0$ is an explicit constant depending on a and where

$$K(\tau) = \int_\tau^{\sqrt{E(T_0)}} \frac{1}{H(y)}\, dy, \quad \tau \in (0, C(E_1(0))\sqrt{E(T_0)}],$$

Since K is nonincreasing, we deduce that

$$\left(\frac{1}{C(E_1(0)}K^{-1}\big(c\,(t - T_0)\big)\right)^2 \le E(t), \quad \forall\, t \ge T_0,$$

for a certain constant $c > 0$. We then use the key comparison Lemma 1.5.6 of Sect. 3.3 [8] we already used in the finite dimensional case. We use this Lemma with

$$z(t - T_0) = K^{-1}\big(c\,(t - T_0)\big)$$

to obtain the announced lower estimate and conclude the proof.   □

Remark 1.7.20. The lower bound has the form

$$C(E_1(0))\left((H')^{-1}\left(\frac{1}{t - T_0}\right)\right)^2$$

So if $g(s) = |s|^{p-1}s$ in a neighbourhood of 0, then $H(x) = x^{(p+1)/2}$, so that the resulting lower bound is

$$E(t) \ge C(E_1(0))(t - T_0)^{-4/(p-1)},$$

which is strictly weaker than the lower bound we stated before for polynomial dampings and smoother solutions.

Let us now state the general result for the $1 - D$ locally damped wave equation for general dampings and more regular solutions. We recall the regularity result proved by Haraux [57]. If $(u_0, u_1) \in W^{2,\infty}(\Omega) \times W^{1,\infty}(\Omega)$, the solution of the $1 - D$ locally damped wave equation is such that

$$u_t \in L^\infty([0, \infty); W^{1,\infty}(\Omega)).$$

Haraux' proof is stated for functions g with power-like growth but it can easily be checked that Haraux' proofs for regularity are still valid for general dampings g. Thanks to this regularity result, we prove

Theorem 1.7.21 (Alabau-Boussouira [8]). *Assume the above hypotheses on g and H. Assume that $(u_0, u_1) \in W^{2,\infty}(\Omega) \times W^{1,\infty}(\Omega)$ and $E(0) > 0$. Then, the energy satisfies the lower estimate*

$$C_1\Big((H')^{-1}(\frac{1}{t - T_0})\Big)^{3/2} \leq E(t), \quad \text{sufficiently large } t,$$

where $C_1 > 0$ depends on $\|u_{tx}\|_{L^\infty((0,\infty)\times\Omega)}$.

Remark 1.7.22. In the peculiar case $g(s) = |s|^{p-1}s$ close to the origin, $H(x) = x^{(p+1)/2}$, so that the resulting lower estimate we get is

$$E(t) \geq C_1(t - T_0)^{-3/(p-1)}, \quad \text{for sufficiently large } t.$$

We can remark that this estimate is strictly better than the one obtained for strong solutions. In the above two cases, there is a gap between the lower and upper estimate so that contrarily to what happens to the finite dimensional case, we cannot deduce optimality.

We now consider the boundary damping case. On one side, Vancostenoble and Martinez' proof relies on the fact that they can use the D'Alembert formula, since they considered $1 - D$ boundary damped wave equation. On the other hand, the regularity proof of Haraux does not work for the $1 - D$ boundary wave equation, nor on systems or Petrowsky equations for instance, so that one has to find other ways to establish lower estimates and in particular as before for strong solutions. Let us show that the methodology we introduced before still works for the $1 - D$ boundary damped wave equation. For this we consider

$$\begin{cases} u_{tt}(t, x) - u_{xx}(t, x) = 0, & 0 < t, x \in \Omega, \\ u_x(t, d) + \sigma u(t, d) + g(u_t(t, d)) = 0 \\ u(0, x) = u_0(x), u_t(0, x) = u_1(x), & x \in \Omega. \end{cases} \tag{1.195}$$

where $\Omega = (c, d) \subset \mathbb{R}$, with $-\infty < c < d < \infty$, $\sigma \geq 0$ and where we consider one of the following boundary conditions at $x = c$

$$\text{either } u(t, c) = 0, \quad \text{for } 0 < t, \tag{1.196}$$

or

$$u_x(t, c) + lu(t, c) = 0, \tag{1.197}$$

where $l \geq 0$. The energy of a solution is defined by

$$E(t) = \frac{1}{2}\Big(\int_\Omega (u_t^2 + u_x^2)\, dx + \sigma u^2(t, d) + lu^2(t, c)\Big). \tag{1.198}$$

We set $V = \{v \in H^1(\Omega), v(c) = 0\}$ if the first boundary condition holds, and $V = H^1(\Omega)$ if the second one holds.

Theorem 1.7.23 (Alabau-Boussouira [7]). *Assume that $(u_0, u_1) \in (V \cap H^3(\Omega)) \times V \cap H^2(\Omega))$ and also satisfies the following compatibility conditions:*

$$
\begin{cases}
u_{0xx}(c) = 0 \text{ if } (1.196) \text{ holds}, \\
u_{0x}(c) + l u_0(c) = 0, u_{1x}(c) + l u_1(c) = 0 \text{ if } (1.197) \text{ holds} \\
u_{0x}(d) + \sigma u_0(d) + g(u_1(d)) = 0, \\
u_{1x}(d) + \sigma u_1(d) + g'(u_1(d)) u_{0xx}(d) = 0, .
\end{cases}
\tag{1.199}
$$

Moreover assume that g and H satisfies the above assumptions. We set $\widetilde{H}(x) = H(x)/x$. Then the energy satisfies the lower estimate

$$
C_1 \left((\widetilde{H}')^{-1} \left(\frac{1}{t} \right) \right)^2 \leq E(t), \quad \text{sufficiently large } t,
\tag{1.200}
$$

where C_1 depends on $E_1(0)$.

The proof relies on some regularity properties, an elementary interpolation property and on the key comparison Lemma. All the above results for $1 - D$ locally as well as boundary have been generalized to radial solutions of the wave equation in annulus type domains in dimension 2 and 3. We would like to generalize these results to more general domains and PDE's.

Let us start with a lower estimate for a fourth order space PDE, that is Petrowsky equation. We first consider Petrowsky equation in a bounded open subset Ω of $\mathbb{R}^N$ with a boundary of class $\mathscr{C}^4$ denoted by Γ, where $N = 2$ or $N = 3$. We assume that ω is an open subset of Ω of positive measure. We consider the following Petrowsky equation with nonlinear damping

$$
\begin{cases}
\partial_{tt} u + \Delta^2 u + a(.) g(u_t) = 0 \quad \text{in} \quad (0, \infty) \times \Omega \\
u = 0 = \dfrac{\partial u}{\partial \nu} \quad \text{on} \quad \Sigma = \Gamma \times \mathbb{R}, \\
(u, \partial_t u)(0) = (u_0, u_1) \quad \text{on} \quad \Omega,
\end{cases}
$$

where $a \in L^\infty(\Omega)$ and $a \geq 0$ a.e. on Ω with $a > 0$ on an open subset ω of Ω. The energy of a solution is defined by

$$
E(t) = \frac{1}{2} (\int_\Omega |u_t|^2 + |\Delta u|^2) \, dx.
$$

We set $V = H_0^2(\Omega)$, $H = L^2(\Omega)$ and $Au = \Delta^2 u$ for $u \in D(A)$, where $D(A)$ is defined by $D(A) = H^4(\Omega) \cap H_0^2(\Omega)$. Well-posedness in the energy space $\mathscr{H} = H_0^2(\Omega) \times H$ holds thanks to the results on maximal monotone operators. Moreover if the initial data are smoother, that is $(u_0, u_1) \in H^4(\Omega) \cap H_0^2(\Omega) \times H_0^2(\Omega)$, then the solutions are smoother and the energy of order 1, that is

$$E_1(t) = \frac{1}{2} (\smallint_\Omega |u_{tt}|^2 + |\Delta u_t|^2)\, dx\,,$$

is nonincreasing. Assume that the following unique continuation property holds:
The subset ω is such all weak solutions of

$$\begin{cases} \partial_{tt} u + \Delta^2 u = 0 & \text{in} \quad (0, \infty) \times \Omega \\ u = 0 = \dfrac{\partial u}{\partial \nu} & \text{on} \quad \Sigma = \Gamma \times \mathbb{R}, \end{cases}$$

with

$$u_t \equiv 0 \text{ on } \omega$$

satisfy

$$u \equiv 0 \text{ in } [0, \infty) \times \Omega\,.$$

Applying Dafermos' technique, we deduce that under the above unique continuation property, strong stabilization holds, that is

$$\lim_{t \longrightarrow \infty} E(t) = 0\,.$$

We can establish a lower bound of the energy using comparison principles seen before and in particular the key Lemma. Let us define for any r real in (N, q_N), where $q_N = \infty$ if $N = 2$, whereas $q_N = 6$ if $N = 3$:

$$\theta_r = \frac{1}{1 + 2/N - 2/r}\,.$$

We recall the following Gagliardo–Nirenberg inequality (see [34])

$$\|v\|_{L^\infty(\Omega)} \le C_r \|v\|_{L^2(\Omega)}^{1-\theta_r} \|v\|_{W^{1,r}(\Omega)}^{\theta_r} \quad \forall\, v \in W^{1,r}(\Omega)$$

Theorem 1.7.24. *Assume that either that the above unique continuation property holds or that strong stabilization holds. Assume moreover that g satisfies (H1) and H satisfies (H2). Let (u^0, u^1) be in $(H^4(\Omega) \cap H_0^2(\Omega)) \times H_0^2(\Omega)$. Let r be any real in (N, q_N) and θ_r be defined as above. Then the energy E satisfies the lower estimate*

$$\left(\frac{1}{\gamma_p C_{\gamma_p}} \left((H')^{-1} \left(\frac{1}{t - T_0} \right) \right) \right)^{1/(1 - \theta_r)} \le E(t), \quad \forall\, t \ge T_1 + T_0\,.$$

where $T_0 \ge 0$.

The proof is based as before on the dissipation relation, interpolation estimates (Gagliardo–Nirenberg) and the key comparison Lemma 1.5.6 we used before.

We proved lower energy estimates for the $1 - D$ either locally or boundary damped wave equation. These results can be extended to radial solutions in annulus type domains. What can be said on the wave equations in general domains in $\mathbb{R}^N$? Indeed, the situation is more difficult to handle. We can state a *formal* result based on an assumption of regularity and a priori estimate of the velocity in a suitable functional space. This result is the natural generalization of the lower estimates we proved in the $1 - D$ case. It can be formulated as follows. We consider the nonlinearly damped wave equation

$$\begin{cases} \partial_{tt}u - \Delta u + a(x)g(u_t) = 0 & \text{in} \quad \Omega \times \mathbb{R}, \\ u = 0 & \text{on} \quad \Sigma = \Gamma \times \mathbb{R}, \\ (u, \partial_t u)(0) = (u^0, u^1) & \text{on} \quad \Omega, \end{cases} \tag{1.201}$$

where Ω is a bounded domain of $\mathbb{R}^N$ with sufficiently smooth boundary Γ. We define the energy of solutions by

$$E(t) = \frac{1}{2} \int_\Omega (|u_t|^2 + |\nabla u|^2)\, dx.$$

We recall that this problem is well-posed in the energy space $H_0^1(\Omega) \times L^2(\Omega)$. In all the sequel, we define H by (1.118) extending the definition over $[0, \infty)$. We prove the following *formal* result.

Theorem 1.7.25 (Alabau-Boussouira [7]). *We assume that g satisfies (H1) and that H satisfies (H2). We moreover assume that Ω and g are such that there exist solutions u of (1.201) such that $u_t \in L^\infty([0, \infty); W^{1,\infty}(\Omega))$. Then the energy satisfies the lower estimate*

$$C_1\left((H')^{-1}(\frac{1}{t - T_0})\right)^{(N+2)/2} \le E(t), \quad \forall\, t \ge T_1, \tag{1.202}$$

where $C_1 > 0$ and T_1 depend on $\|\nabla u_t\|_{L^\infty([0,\infty)\times\Omega)}$.

The proof of this result requires the following interpolation result. We give a proof of this result in [8]. (see also Sect. 1.43 in [83] for interpolation between Hölder functions and functions in $L^1(U)$).

Theorem 1.7.26 (Alabau-Boussouira [8]). *Let U be a bounded open set in $\mathbb{R}^N$, with $N \ge 1$. Then there exists a constant $C_- > 0$ such that*

$$\|v\|_{L^\infty(U)}^{(N+2)} \le C_1 \|v\|_{L^2(U)}^2 \|\nabla v\|_{L^\infty(U)}^N, \quad \forall\, v \in W^{1,\infty}(U) \cap H_0^1(U). \tag{1.203}$$

Proof (of Theorem 1.7.25). Thanks to the dissipation relation and defining $\widetilde{H}$ as

$$\widetilde{H}(x) = \frac{H(x)}{x}, \quad x > 0, \widetilde{H}(0) = 0, \tag{1.204}$$

we have

$$- E'(t) \leq \beta_a \widetilde{H}(\|u_t\|^2_{L^\infty(\Omega)}) E(t), \quad t \geq 0, \tag{1.205}$$

where $\beta_a = 2\|a\|_{L^\infty(\Omega)}$.

On the other hand, thanks to the regularity results for u_t, we can apply Theorem 1.7.26 to $v = u_t$, $U = \Omega$ and N. Therefore, we have

$$\|u_t\|^2_{L^\infty(\Omega)} \leq \gamma E^{2/(N+2)}(t), \quad t \geq 0.$$

where

$$\gamma = \gamma\big(\|\nabla u_t\|_{L^\infty([0,\infty)\times\Omega)}\big)$$

is a constant which depends on the norm of $\|\nabla u_t\|_{L^\infty([0,\infty)\times\Omega)}$.

On the other hand, thanks to our hypotheses on g and a, we can apply Theorem 1.3.5 (due to Dafermos, see Proposition 2.3 in [46]), saying that $\lim_{t\to\infty} E(t) = 0$. Hence, there exists $T_0 \geq 0$ such that

$$E(t) \leq \Big(\frac{r_0^2}{\gamma}\Big)^{(N+2)/2}, \quad \forall\, t \geq T_0.$$

Thanks to our hypotheses, $\widetilde{H}$ is increasing on $[0, r_0^2]$, we obtain

$$- E'(s) \leq \beta_a E(s) \widetilde{H}\big(\gamma(E(s))^{2/(N+2)}\big), \quad s \geq T_0. \tag{1.206}$$

Thus, integrating with respect to s between T_0 and t, we deduce that

$$K_m(\gamma\, (E(t))^{2/(N+2)}) \leq \frac{2}{(N+2)} \beta_a (t - T_0). \tag{1.207}$$

where K_m is defined by

$$K_m(\tau) = \int_\tau^{\gamma E^{2/(N+2)}(T_0)} \frac{1}{H(y)}\, dy, \quad \tau \in (0, \gamma E^{2/(N+2)}(T_0)], \tag{1.208}$$

We set $z_0 = \gamma E^{2/(N+2)}(0)$, $\kappa = \frac{2}{(N+2)} \beta_a$ and we denote by z the solution of (1.98), where H replaces $\widetilde{H}$. We set $\tilde{z}(t) = z(t - T_0)$ for $t \geq T_0$. Then, we have

$$\tilde{z}(t) = K_m^{-1}\big(\kappa\, (t - T_0)\big) = K_m^{-1}\Big(\frac{2\beta_a}{(N+2)}\, (t - T_0)\Big), \quad \forall\, t \geq T_0.$$

Thus, thanks to (1.207), we have

$$\Big(\frac{1}{\gamma}\tilde{z}(t)\Big)^{(N+2)/2} \leq E(t), \quad \forall\, t \geq T_0.$$

We now use our comparison Lemma 1.5.6 to conclude. Applying this lemma to H for $R = 1$, we deduce that there exists $C_\gamma > 0$ depending on γ, (and in addition of μ if the second alternative of the present theorem holds), such that

$$\left(H'\right)^{-1}\left(\frac{1}{t - T_0}\right) \le C_\gamma \tilde{z}(t), \quad \forall\, t \ge T_1 + T_0 .$$

for a sufficiently large T_1. The desired estimate then easily follows. $\qquad\square$

Remark 1.7.27. Smoothness properties of the solutions of the multi-dimensional wave equations in general domains is an open question since the sixties Lions and Strauss [77]. In particular, the required above smoothness on the velocity is an open question for general multi-dimensional domains. In $1 - D$ the required regularity is assured by Haraux' smoothness property for sufficiently smooth initial data.

We would like to further extend the above methodology to systems of coupled equations such as Timoshenko beams for instance. Haraux' regularity proof cannot be extended to this case, but we have already seen that it still may be possible to derive lower estimates for less regular solutions, that is for strong solutions.

We consider the following Timoshenko system:

$$\begin{cases} \rho_1 \varphi_{tt} - k(\varphi_x + \psi)_x = 0 & t > 0,\ 0 < x < L, \\ \rho_2 \psi_{tt} - b\psi_{xx} + k(\varphi_x + \psi) + a(x)g(\psi_t) = 0 & t > 0,\ 0 < x < L, \end{cases}$$

where $a \in L^\infty(\Omega)$ and $a \ge 0$ a.e. on $\Omega = (0, L)$ with $a > 0$ on an open subset ω of Ω. In this latter case ad assuming that g and H satisfy respectively $(H1)$ and $(H2)$, we can prove a similar result that the one for the $1 - D$ wave equation for strong solutions, that is

$$C(E_1(0))\left(\left(H'\right)^{-1}\left(\frac{1}{t - T_0}\right)\right)^2 \le E(t)$$

where $E_1(0)$ is the energy of first order. We refer to [4] for the detailed result.

Weak velocity lower estimates. Haraux' lower estimates [57] relies on an indirect argument, which is a weak lower estimate on the velocity, for very regular solutions of the one dimensional locally damped wave equation, that is

$$\limsup_{t\to\infty}(t^{1/(p-1)}\|u_t\|_{L^\infty((0,\infty)\times\Omega)}) > 0 .$$

from which he deduces the above weak lower energy estimate.

We can generalize this weak lower velocity estimate to general dampings and other PDE's such as Petrowsky equation. Let us state the result.

Theorem 1.7.28 (Alabau-Boussouira [7]). *We assume that Ω and g are such that there exist solutions u of the above multi-dimensional wave equation, such that $u_t \in L^\infty([0,\infty); W^{1,\infty}(\Omega))$. We also assume that g and H satisfy respectively*

(H1) and (H2). Then the velocity u_t satisfies the following weak lower estimate.

$$\left\{ \limsup_{t \to \infty} \frac{1}{(H')^{-1}\left(\frac{D_0}{t}\right)} \|u_t\|^2_{L^\infty(\Omega)} \geq 1, \right. ,$$

The proof of this result relies on the following technical Lemma.

Lemma 1.7.29 (Alabau-Boussouira [7]). *We assume that g and H satisfies respectively* (H1) *and* (H2). *We set* $\mu = 1$ *if the first alternative of* (H2) *holds, otherwise* $\mu > 0$ *is the constant involved in the second alternative of* (H2). *Assume that* $E : [0, \infty) \mapsto (0, \infty)$ *is in* $W^{1,1}_{loc}([0, \infty)$ *and satisfies the two inequalities*

$$- E'(t) \leq \beta_a \widetilde{H}\left(\|u_t\|^2_{L^\infty(\Omega)}\right) E(t), \forall\, t \geq T_-, \tag{1.209}$$

where $\beta_a > 0$ *and* $T_- \geq 0$ *are given constants, and*

$$E(t) \leq \beta_{E(0)}\left(H'\right)^{-1}\left(\frac{D}{t}\right), \quad \text{for t sufficiently large}. \tag{1.210}$$

Define the solution z of the ordinary differential equation:

$$z'(t) + H(\mu\, z(t)) = 0, z(0) = z_0 > 0, \tag{1.211}$$

where z_0 *is given. Then we have*

$$\limsup_{t \to \infty}\left(\frac{\widetilde{H}\left(\|u_t(t)\|^2_{L^\infty(\Omega)}\right)}{\widetilde{H}(\mu\, z(t))}\right) \geq \frac{\mu}{\beta_a} > 0. \tag{1.212}$$

Remark 1.7.30. In the one-dimensional case and for polynomial feedback close to the origin, that is for $g(s) = |s|^{p-1}s$ for s close to 0, one recovers Haraux's weak lower velocity estimate

$$\limsup_{t \to \infty}(t^{1/(p-1)}\|u_t(t)\|_{L^\infty(\Omega)}) > 0.$$

On the other hand, using then the interpolation estimate given in Theorem 1.7.26 for $v = u_t$, $U = \Omega$ and $N = 1$, we have

$$\|u_t(t)\|_{L^\infty(\Omega)} \leq C E^{1/3}(t)$$

Inserting this last estimate in the previous one, we obtain

$$\limsup_{t \to \infty}(t^{3/(p-1)} E(t)) > 0,$$

so that we recover Haraux' result, that is a weak lower energy estimate through the above indirect argument, as a peculiar case. We also extend these weak velocity lower estimates to Petrowsky equations in [4].

The above lower estimates depend on the dimension. Let us now give an example of extension of the one-dimensional case to radial solutions in annulus domains in the three-dimensional space. We assume that the space dimension is $N = 3$ and we consider $\Omega = B(0, R_2) \setminus B(0, R_1)$ in $\mathbb{R}^N$. For the sake of simplicity, we assume that $a \equiv 1$. We consider (1.201). Well-posedness in the energy space $H_0^1(\Omega) \times L^2(\Omega)$ holds. For initial data that depend only on the radial component, we can apply our previous results. Let $u^0(.) = u^0(r)$ and $u^1(.) = u^1(r)$ in Ω where $(u^0, u^1) \in (H_0^1(R_1, R_2) \cap H^2(R_1, R_2)) \times H_0^1(R_1, R_2)$. Then the solution u of (1.201) depends only on r. Its energy with respect to spherical coordinates is given by

$$E_u(t) = \frac{1}{2} \left(\int_{R_1}^{R_2} \int_0^{2\pi} \int_0^\pi \left(u_t^2 + \left| \frac{1}{r} \partial_r (r\, u) \right|^2 \right) r^2 \sin(\theta)\, dr\, d\theta\, d\phi \right). \tag{1.213}$$

We make the change of unknown

$$v(t, r) = ru(t, r), \quad t > 0, r \in (R_1, R_2). \tag{1.214}$$

Then, v satisfies the equation

$$\begin{cases} v_{tt} - v_{rr} + rg(\dfrac{v_t}{r}) = 0, & 0 < t, r \in (R_1, R_2), \\ v(t, R_1) = v(t, R_2) = 0, & \text{for } 0 < t, \\ v(0, r) = ru^0(r), \, v_t(0, r) = ru^1(r), & r \in (R_1, R_2). \end{cases} \tag{1.215}$$

Moreover we have

$$E_u(t) = 4\pi E(t) =: 4\pi \int_{R_1}^{R_2} \frac{1}{2} \left(v_t^2 + v_r^2 \right) dr. \tag{1.216}$$

We prove the following result.

Theorem 1.7.31 (Alabau-Boussouira [7]). *Let $u^0(.) = v^0(r)$ and $u^1(.) = v^1(r)$ in Ω where $(v^0, v^1) \in W^{2,\infty}(R_1, R_2) \times W^{1,\infty}(R_1, R_2)$. Assume also that g and H satisfy respectively $(H1)$ and $(H2)$. Then the energy satisfies the lower estimate*

$$\frac{4\pi}{\gamma^2} \left((H')^{-1} (\frac{1}{t - T_0}) \right)^{3/2} \leq E_u(t), \quad \forall\, t \geq T_1 + T_0, \tag{1.217}$$

where γ depends on a smoother norm of the initial data.

Proof. Thanks to the regularity of the initial data, we can apply Haraux' Proposition 2.3 in [57] stating that is if $(u^0, u^1) \in W^{2,\infty}((R_1, R_2)) \times W^{1,\infty}((R_1, R_2))$, then the solution of (1.201) is such that u_t is in $L^\infty([0, \infty); W^{1,\infty}((R_1, R_2))$ and is bounded above in this space by a quantity involving only the initial data in a smoother space than the energy space. Thanks to these results, we can apply Theorem 1.7.25, which concludes the proof. □

Remark 1.7.32. This result is to be compared with the corresponding three dimensional one in Theorem 1.7.25 for which the resulting estimate is weaker, since the exponent $3/2$ on the left hand side of (1.217) has to be replaced by 2. The previous results hold true for a damping term of the form $a(|x|)g(u_t)$ in (1.201), where $|\cdot|$ stands for the euclidian norm in $\mathbb{R}^N$, and where $a \in L^\infty(R_1, R_2)$, $a \geq 0$ on (R_1, R_2) and a is nonvanishing on a subset of (R_1, R_2). We give in [7] results for less regular (indeed strong) solutions. For the boundary stabilization, one cannot assert the a priori regularity and a priori estimates of the velocity in $L^\infty([0, \infty); W^{1,\infty}(\Omega)$ even in the one-dimensional situation. Nevertheless it is still possible to derive lower energy estimates (see [7]) as the result presented in Theorem 1.7.23 for the one-dimensional boundary wave equation (1.194) and further extensions to radial cases in two and three dimensional annulus type domains in [7].

Let us now illustrate these results with concrete feedback examples to conclude this section.

Theorem 1.7.33 (Alabau-Boussouira [7]). *We consider* (1.201). *We make the hypotheses of Theorem 1.7.31, then the energy of solutions with initial data in* $W^{2,\infty}(\Omega) \times W^{1,\infty}(\Omega)$ *satisfies, for t sufficiently large, the estimates*

$$
\begin{cases}
C_1(E(0), E_1(0))t^{-3/(p-1)} \leq E(t) \leq C_2(E(0))t^{-2/(p-1)} \\
\text{for } g(x) = x^p, x \in [0, r_0], p > 1, \\[2ex]
C_1(E(0), E_1(0))(\ln(t))^{-3/2} \leq E(t) \leq C_2(E(0))(\ln(t))^{-1} \\
\text{for } g(x) = e^{-1/x^2}, x \in [0, r_0], \\[2ex]
C_1(E(0), E_1(0))t^{-3/(p-1)}(\ln(t))^{-3q/(p-1)} \leq E(t) \leq C_2(E(0))t^{-2/(p-1)}(\ln(t))^{-2q/(p-1)}, \\
\text{for } g(x) = x^p(\ln(\frac{1}{x}))^q, p > 1, q > 1, x \in [0, r_0], \\[2ex]
C_1(E(0), E_1(0))e^{-3(\ln(t))^{1/p}} \leq E(t) \leq C_2(E(0))e^{-2(\ln(t))^{1/p}}, \\
\text{for } g(x) = e^{-(\ln(\frac{1}{x}))^p}, 1 < p < 2, x \in [0, r_0],
\end{cases}
$$

Remark 1.7.34. Contrarily to the ODE case, we cannot deduce optimality from the above lower estimates. We refer the reader to Martinez and Vancostenoble [101] and to Vancostenoble [102] for an optimality result in case of a one-dimensional boundary damped wave equation and for initial data with zero velocity.

1.8 Memory Stabilization

Abstract The purpose of this section is to present energy decay results for memory-damped wave-like equations. These equations model viscoelastic materials which keep traces of their deformations. The damping operator is nonlocal and involves the convolution in time between a time-dependent kernel and the action of the infinitesimal generator of the undamped system on the unknown state. We shall first consider the case of exponentially and polynomially decaying kernels. We then extend the results to the case of general decaying kernels, using the optimal-weight convexity method and prove in particular that the energy decays as fast as the kernel.

1.8.1 Introduction and Motivations

Many applications in the stabilization of mechanical structures involve viscoelastic materials, that is materials that keep memory of their deformations. Various models (see [53, 75, 92]) exist for describing their behavior. These models involve damping phenomena which is no longer of local—i.e. frictional—nature, but is now nonlocal in time and involves a convolution with respect to time with a decaying kernel, also called relaxation function.

We will consider here a simple model example, which is the viscoelastic membrane equation

$$\begin{cases} u_{tt} - \Delta u + k * \Delta u = 0 \\ u = 0 \text{ in } [0, \infty) \times \Gamma \\ (u, u_t)(0, .) = (u^0, u^1) \text{ in } \Omega, \end{cases} \quad (1.218)$$

in a bounded open domain $\Omega \subset \mathbb{R}^N$ with smooth boundary Γ (see, e.g. [30, 47, 48, 53, 75, 88, 92]) and where $*$ is defined by

$$(k * v)(t) = \int_0^t f(t - s)v(s)\, ds.$$

and the kernel k is a positive, nonincreasing differentiable function which decays at infinity. We refer to [11] for models with singular kernels and to [13] for an example of combined boundary memory and frictional dampings. We refer to [14, 90] for well-posedness of (1.218) in the energy space $H_0^1(\Omega) \times L^2(\Omega)$ and regularity of its solutions for smoother initial data. In particular, for initial data in $(H^2(\Omega) \cap H_0^1(\Omega)) \times H_0^1(\Omega)$, solutions of (1.218) are strong solutions. For this system, the natural energy is defined by

$$E_u(t) = \frac{1}{2}\|u_t(t)\|_{L^2(\Omega)}^2 \, dx + \frac{1}{2}\left(1 - \int_0^t k(s)\, ds\right)\|\nabla u(t)\|_{L^2(\Omega)}^2 \quad (1.219)$$

$$+ \frac{1}{2}\int_0^t k(t - s)\|\nabla u(t) - \nabla u(s)\|_{L^2(\Omega)}^2 \, ds$$

The term $k * \Delta u$ is a damping term. Indeed, one can show that the energy of strong solutions satisfies the dissipation relation (see [14, 88] for a proof):

$$E_u'(t) = -\frac{1}{2}k(t)\|\nabla u(t)\|^2 \tag{1.220}$$

$$+\frac{1}{2}\int_0^t k'(t-s)\|\nabla u(s) - \nabla u(t)\|^2\, ds \leq 0$$

The main purpose is to obtain decay estimates for (1.218) and also for more general abstract equation of the same nature under minimal assumptions on the kernel k. Another important aspect is to "guarantee" in some way, that our decay estimates are optimal or quasi-optimal. This will be done through comparison with the behavior of the kernel at infinity. A lot of results in the literature do not allow to recover the "optimal" polynomial decay for polynomially decaying kernels. Thus, optimality and introduction of appropriate mathematical tools to obtain satisfactory results in this direction are of important concerns. Also, when aiming to optimality, identifying physical properties of the model and their use from a mathematical point of view become essential.

1.8.2 Exponential and Polynomial Decaying Kernels

We shall assume that $k : [0,\infty) \rightarrow [0,\infty)$ is a locally absolutely continuous function satisfying the assumptions

Assumptions (M1):   1. $k : [0,\infty) \rightarrow [0,\infty)$ is a locally absolutely continuous function such that

$$\int_0^\infty k(t)\, dt < 1\,, k(0) > 0 \qquad k'(t) \leq 0 \quad \text{for a.e. } t \geq 0\,.$$

2. There exist $p \in (2,\infty]$ and $k_0 > 0$ such that

$$k'(t) \leq -k_0 k^{1+\frac{1}{p}}(t) \qquad \text{for a.e. } t \geq 0\,.$$

where we set $\frac{1}{p} = 0$ whenever $p = \infty$. We prove the following result.

Theorem 1.8.1 (Alabau–Cannarsa–Sforza [14]). *Assume first that hypothesis* (M1) *holds with* $p = \infty$. *Then, for all initial data* $(u^0, u^1) \in H_0^1(\Omega) \times L^2(\Omega)$, *the energy* $E_u(t)$ *of the mild solution* u *of* (1.218) *decays as*

$$E_u(t) \leq E_u(0)\exp(1 - Ct) \qquad \forall t \geq 0\,.$$

Assume that hypothesis (M1) *holds with* $p \in (2,\infty)$. *Then for all initial data* $(u^0, u^1) \in H_0^1(\Omega) \times L^2(\Omega)$, *the energy* $E_u(t)$ *of the mild solution* u *of* (1.218) *decays as*

$$E_u(t) \leq E_u(0) \left(\frac{C(1+p)}{t+pC} \right)^p \quad \forall t \geq 0.$$

where C is a positive constant.

Remark 1.8.2. We prove the above result for general abstract hyperbolic evolution equations and in the presence of source terms in [14]. This result is known in the case of the wave equation, but with different techniques, the main point here is in the proof which is new as far as the proof of a weighted dominant memory energy and the use of weighted nonlinear polynomial Gronwall inequalities are concerned. This is important for the further generalization to other types of growth of the kernel with a *conjectured* optimal energy decay rate which is the one of the kernel which will be presented below.

Proof. We will give only an overview of the proof. It is based on the choice of appropriate multipliers and on the linear (resp. polynomial) Gronwall inequalities of Sect. 1.2 (resp. 3) for exponentially (resp. polynomially) decaying kernels k. Given a solution u of (1.218), we define the *memory energy* of u as

$$E_u^{km}(t) = \int\limits_0^t k(t-s)\|\nabla u(s) - \nabla u(t)\|^2 \, ds \qquad t \geq 0. \tag{1.221}$$

The first part, based on multipliers Mu of the form $Mu = E^r(s)(c_1(\beta * u)(s) + c_2(s)u)$ where c_1 is a suitable constant, whereas c_2 may be chosen dependent on k, allows us to prove that the memory energy is dominant. More precisely, we prove that for any $S_0 > 0$, there exists constants $C_{S_0} > 0$ and $\delta_4 = \delta_4(S_0) > 0$ such that for any $T \geq S \geq S_0$, the solutions u of (1.218) satisfy

$$\int\limits_S^t E_u^r(t)E_u(t)\, dt \leq C_{S_0} E_u^r(S)E_u(S) + \delta_4 \int\limits_S^T E_u^r(t)$$

$$\left(\int\limits_0^t k(t-s)\|\nabla u(t) - \nabla u(s)\|^2 \, ds \right) dt, \tag{1.222}$$

where δ_4, C_{S_0} are positive constants and $r > 0$ is arbitrary and will be chosen later on. We assume $p \in (2, \infty)$ Using then this estimate with successively $r = 2/p$ and then $r = 1/p$ combined with an argument of Cavalcanti and Oquendo [41], we prove that E_u satisfies the polynomial Gronwall inequality (1.66) of Theorem 9.1 [65], that is Theorem 1.5.1 in the present Notes, with $r = 2/p$ first and then with $r = 1/p$. $\qquad\square$

1.8.3 General Decaying Kernels and Optimality

Considering the former results and the existing literature on the subject, a natural question which raises is: what can be said for general decaying kernels? In particular, does it hold true that the energy decays at least as fast as the kernel, as time goes to infinity? It is still possible to prove that the memory energy dominates the behavior of the total energy with a general time-dependent weight function. The difficulty, a real one, is to prove that the energy satisfies a nonlinear Gronwall inequality for an appropriate choice of the time-dependent weight function. Cavalcanti and Oquendo's argument no longer applies for general decaying kernel. Indeed, we will be able to answer these questions using the optimal-weight convexity method [17] and in particular its simplification in [8]. Nevertheless, one has to introduce new tools to prove these results. We will consider a more general PDE. Let X be a real Hilbert space X be, with scalar product $\langle \cdot, \cdot \rangle$ and norm $\| \cdot \|$, We consider the second order integro-differential equation

$$u''(t) + Au(t) - \int_0^t k(t-s)Au(s)\,ds = 0, \qquad t \in (0,\infty), \qquad (1.223)$$

where $A : D(A) \subset X \to X$ is a densely defined self-adjoint linear operator satisfying

$$\langle Ax, x \rangle \geq \omega \|x\|^2 \qquad \forall x \in D(A) \qquad (1.224)$$

for some $\omega > 0$. We recall that, for any nonincreasing locally absolutely continuous function $k : [0,\infty) \to (0,\infty)$ satisfying

$$\int_0^\infty k(t)\,dt < 1, \qquad (1.225)$$

(1.223), complemented with the initial conditions

$$\begin{cases} u(0) = u^0 \in D(A^{1/2}) \\ u'(0) = u^1 \in X, \end{cases} \qquad (1.226)$$

has a unique mild solution $u \in \mathscr{C}^1([0,T]; X) \cap \mathscr{C}([0,T]; D(A^{1/2}))$, see [90] (see also [14]). We now consider kernels satisfying a general condition of the form

$$k'(t) \leq -\chi(k(t)) \quad \text{for a.e. } t \geq 0 \qquad (1.227)$$

which includes the polynomial case and many other interesting cases. We make the following assumptions on the function χ, which we do below.

Assumption (M2): χ is a nonnegative measurable function on $[0, k_0]$, for some $k_0 > 0$, strictly increasing and of class $\mathscr{C}^1$ on $[0, k_1]$, for some $k_1 \in (0, k_0]$, such that

$$\chi(0) = \chi'(0) = 0 \qquad (1.228)$$

$$\exists\, \chi_0 > 0 \text{ such that } \chi \geq \chi_0 \text{ on } [k_1, k_0] \qquad (1.229)$$

$$\int\limits_{0}^{k_0} \frac{dx}{\chi(x)} = \infty \tag{1.230}$$

$$\int\limits_{0}^{k_0} \frac{x}{\chi(x)}\, dx < 1 \tag{1.231}$$

Remark 1.8.3. Condition (1.231), together with (1.227) and the change of variable $x = k(t)$, yields (1.225) which ensures in turn that (1.223)–(1.226) is well-posed. This implies a natural restriction on the growth of the memory damping: k cannot decay too slowly at infinity or, equivalently, χ cannot be too flat at the origin. Note that this type of restriction does not hold for the frictionally damped wave equation. We note that our approach could be useful in situations that look even more general than (1.227), such as

$$k'(t) \leq -\xi(t)\, \chi\big(k(t)\big) \quad \text{for a.e.} \quad t \geq 0. \tag{1.232}$$

Indeed, if k is strictly decreasing—hence, invertible—one can define $\widetilde{\chi}(x) = (\xi \circ k^{-1})(x)\chi(x)$ and then check if $\widetilde{\chi}$ satisfies $(M2)$. Also χ does not need to be explicitly known. Our result applies to the wave equation, Petrovsky system, anisotropic elasticity and other models.

Let us define the energy of a given solution u of (1.223) on $[0, \infty)$ by

$$E_u(t) := \frac{1}{2}\|u'(t)\|^2 + \frac{1}{2}\Big(1 - \int\limits_{0}^{t} k(s)\, ds\Big)\|A^{1/2}u(t)\|^2 \tag{1.233}$$

$$+\frac{1}{2}\int\limits_{0}^{t} k(t-s)\|A^{1/2}u(s) - A^{1/2}u(t)\|^2\, ds \qquad t \geq 0.$$

In the above expression, we define the *memory-energy* of u by

$$E_u^m(t) = \int\limits_{0}^{t} k(t-s)\|A^{1/2}u(s) - A^{1/2}u(t)\|^2\, ds \qquad t \geq 0. \tag{1.234}$$

We prove the following result.

Theorem 1.8.4 (Alabau–Cannarsa [10]). *Assume that the convolution kernel k is a locally absolutely nonnegative continuous function such that (1.227) holds for a given function χ satisfying hypothesis $(M2)$. Assume, moreover, that χ is strictly convex on an interval of the form $(0, \delta]$ and*

$$\liminf_{x \to 0^+} \Lambda(x) > \frac{1}{2}, \tag{1.235}$$

where

$$\Lambda(x) = \frac{\chi(x)/x}{\chi'(x)}, \quad x \in (0, \delta]. \tag{1.236}$$

Let u be a solution of (1.223) with initial data in $D(A^{1/2}) \times X$. Then

$$E_u(t) \le \kappa(E_u(0)) \, \rho_1(t) \,, \quad \forall \, t \ge T_1 \,, \tag{1.237}$$

where $T_1 > 0$ is an explicit universal constant, $\kappa(E_u(0))$ is a constant which depends on the initial data, and ρ_1 is a function which decays to 0 as $t \to \infty$.

For kernel which are not close to the exponential decaying case, we prove a stronger result.

Theorem 1.8.5 (Alabau–Cannarsa [10]). *Assume the hypotheses of Theorem 1.8.4 and suppose moreover that*

$$\limsup_{x \to 0^+} \Lambda(x) < 1 \quad \text{and} \quad k'(t) = -\chi(k(t)) \quad \text{for a.e. } t \ge 0 \,.$$

Then

$$E_u(t) \le \kappa(E_u(0))k(t) \,, \quad \forall \, t \ge T_1 \tag{1.238}$$

where $T_1 > 0$ and $\kappa(E_u(0))$ are as above.

Proof. The proof is based on the optimal-weight convexity method [8, 17] and a pseudo-iterative process to build optimal-weights for suitable nonlinear Gronwall inequalities to handle the nonautonomous system.   □

Remark 1.8.6. The function ρ_1 can be precised but one cannot assert that the energy decays at least as fast as the kernel if $\limsup_{x \to 0^+} \Lambda(x) = 1$. This case is analogous to the frictional nonlinear damped wave equation for feedbacks close to a linear behavior.

Theorem 1.8.5 is, to our knowledge, the first result which provides for general kernels, a comparison criteria between the energy and the kernel asymptotically. This was known only for kernels decaying polynomially at infinity. The question of optimality is completely open as far as we know. This theorem contains the polynomial case as a peculiar case and other general examples of kernels with for instance polynomial-logarithmic decay at infinity (see [10] for examples). Its proof strongly relies on the optimal-weight convexity method but requires additional ideas to handle the nonautonomous character of the problem.

1.9 Indirect Stabilization for Coupled Systems

Abstract We give in this section an introduction to indirect stabilization of coupled systems. For such type of stabilization, only some equations or components of the vector state are directly damped, the other ones being undamped. One then hopes that the coupling effects will be sufficient so that the full system is damped. Many questions arise for such analysis. How the interactions between the involve diffusion

operators, the damping operators and the coupling operators may influence the
indirect stabilization phenomena? If stabilization holds, is it possible to determine
in which sense? We shall present in this section three model examples: one with
boundary damping and coercive coupling, one with distributed damping and coercive coupling, and one with locally distributed damping and noncoercive damping.
The results are based on an integral inequality (Alabau, Compt. Rendus Acad. Sci.
Paris I 328: 1015–1020, 1999) which is not of differential nature contrarily to the
approach of the previous sections, and on suitable crossed multipliers, which are not
geometric multipliers but are adapted to the coupling.

1.9.1 Introduction and Motivations

More and more phenomena in mechanics, physics and biology are modelized by
coupled systems. For engineering or biological constraints, an important issue is to
stabilize such coupled systems by a reduced number of feedbacks. This is called
indirect damping. This notion has been introduced by Russell [95] in 1993.

As an example, we consider the following coupled wave system:

$$\begin{cases} u_{tt} - \Delta u + u_t + \alpha v = 0 \\ v_{tt} - \Delta v + \alpha u = 0 \end{cases} \quad \text{in } (0, \infty) \times \Omega, \qquad u = 0 = v \quad \text{on } (0, \infty) \times \partial\Omega.$$

$$(1.239)$$

where α is a coupling parameter, which may be constant or not. Here, the first
equation is damped through a linear distributed feedback, while no feedback is
applied to the second equation. This second equation is conservative when $\alpha = 0$.
The question is to determine if the full coupled system is stabilized for nonzero
values of the coupling parameter α thanks to the stabilization of the first equation
only. Note that this question has already received attention in the finite dimensional
case. For stabilization or control of coupled ODE's, one has an algebraic criteria,
named Kalman rank condition. The situation is much more involved in the case of
coupled PDE's. One can first show that the above system fails to be exponentially
stable (see [20] for a proof). Hence if stabilization holds, it cannot be uniform since
the system is linear. Is it stable in a weaker sense? What can be said for other types
of PDE's, in case of boundary damping . . . ?

More generally, we study the stability of abstract system

$$\begin{cases} u'' + A_1 u + B u' + \alpha P v = 0 \\ v'' + A_2 v + \alpha P^\star u = 0, \end{cases}$$

$$(1.240)$$

where A_1, A_2 and B are self-adjoint positive linear operators in a Hilbert space H
with norm $|\cdot|$. Moreover, B is assumed to be a bounded or unbounded operator in
H. It can be coercive or not, modelling the case of globally or locally supported
feedbacks. The coupling operator P is supposed to be a bounded operator, not
necessarily coercive on H. So, this formulation includes systems with internal

damping supported in the whole domain Ω such as (1.239), or localized and boundary dampings. It also includes the case of localized couplings. The reader is referred to Alabau-Boussouira [21, 23] for related results concerning boundary stabilization problems and to Beyrath [31, 32] (see also [104]) for localized indirect dampings. Extensions to boundary and localized dampings together with noncoercive coupling operators are given in Alabau-Boussouira and Léautaud [2]. System (1.240) fails to be exponentially stable, at least when H is infinite dimensional and A_1 has a compact resolvent as in (1.239) even in the case of $P = Id$. A common feature of these systems is that they possess an energy which is dissipated, but only some of the components are directly dissipated. From the point of view of applications, one hopes that this *partial* dissipative action will be compensated by the coupling effects, so that the full system is stabilized. If so, one also wants to determine which type of stabilization occurs: uniform or not, exponential or of weaker type. Since the fact that some components of the vector state are not directly damped has to be counterbalanced by the coupling effects, the answer to the above questions will depend on the different assumptions on the given data of the system. It will depend of the involved operators A_1, A_2 and in particular if they allow a *good transmission* of the information from the damped equation to the undamped one, the strength of the damping operator B in the sense that it may be coercive or not, bounded or unbounded corresponding to the case of globally or not distributed dampings and to locally distributed or boundary dampings. It will depend also on the properties of the coupling operator P, and in particular if it is coercive or partially coercive, this corresponding respectively to the case of globally distributed positive (or negative) couplings and to the case of locally distributed couplings. An interesting aspect is the one which allows us to identify classes of such systems which can be analyzed with common tools to be found, even if the technicity of the proofs may vary under the different assumptions. Due to the property that the coupling action is compact in the energy space, these systems cannot be uniformly stable as shown in Cannarsa and Komornik [20]. We will see that for such classes of coupled systems, one can prove under different hypotheses that polynomial stabilization holds for sufficiently smooth initial data. One important ingredient for proving such stability properties is to show that the energy of the system satisfies an integral inequality, which is not of differential nature and relies on the underlying semigroup property. This nondifferential integral inequality [23] (see also [20, 21]) is interesting in itself and leads to polynomial decay. The technical aspects and properties of transmission of the information from the damped to the undamped equations have to be understood in particular by determining suitable coupling multipliers to prove that the energy of the full system satisfies such nondifferential integral inequalities. This is where the proofs vary under the different assumptions on A_1, A_2, B, P.

We shortly present below some of the results on concrete examples for indirect stabilization in case of boundary damping, then globally distributed damping, with coercive coupling and then the case of noncoercive couplings. We refer the reader to the indicated references for the general abstract results.

1.9.2   A Nondifferential Integral Inequality

Let A be the infinitesimal generator of a continuous semi-group $\exp(tA)$ on an Hilbert space $\mathscr{H}$, and $D(A)$ its domain. For U^0 in $\mathscr{H}$ we set in all the sequel $U(t) = \exp(tA)U^0$.

Theorem 1.9.1 (Alabau [23]). *Assume that there exists a functional E defined on $C([0, +\infty), \mathscr{H})$ such that for every U^0 in $\mathscr{H}$, $E(\exp(.A))$ is a non-increasing, locally absolutely continuous function from $[0, +\infty)$ on $[0, +\infty)$. Assume moreover that there exist an integer $k \in \mathbb{N}^*$ and nonnegative constants c_p for $p = 0, \ldots k$ such that*

$$\int_S^T E(U(t))\, dt \leq \sum_{p=0}^k c_p E(U^{(p)}(S)) \quad \forall\, 0 \leq S \leq T, \forall\, U^0 \in D(A^k). \quad (1.241)$$

Then the following inequalities hold for every U^0 in $D(A^{kn})$, where n is any positive integer:

$$\int_S^T E(U(\tau)) \frac{(\tau - S)^{n-1}}{(n-1)!}\, d\tau \leq c \sum_{p=0}^{kn} E(U^{(p)}(S)), \quad (1.242)$$

$$\forall\, 0 \leq S \leq T, \forall\, U^0 \in D(A^{kn}),$$

and

$$E(U(t)) \leq c \sum_{p=0}^{kn} E(U^{(p)}(0)) t^{-n} \quad \forall t > 0 \quad, \forall\, U^0 \in D(A^{kn}),$$

where c is a constant which depends on n.

Proof. We first prove (1.242) by induction on n. For $n = 1$, it reduces to the hypothesis (1.241). Assume now that (1.242) holds for n. and let U^0 be given in $D(A^{k(n+1)})$. Then we have

$$\int_S^T \int_t^T E(U(\tau)) \frac{(\tau - t)^{n-1}}{(n-1)!}\, d\tau\, dt \leq c \sum_{p=0}^{kn} \int_S^T E(U^{(p)}(t))\, dt$$

$$\forall\, 0 \leq S \leq T, \forall\, U^0 \in D(A^{kn}).$$

Since U^0 is in $D(A^{k(n+1)})$ we deduce that $U^{(p)}(0) = A^p U^0$ is in $D(A^k)$ for $p \in \{0, \ldots kn\}$. Hence we can apply the assumption (1.241) to the initial data $U^{(p)}(0)$. This together with Fubini's Theorem applied on the left hand side of the above inequality give (1.242) for $n + 1$. Using the property that $E(U(t))$ is non increasing in (1.242) we easily obtain the last desired inequality.   □

Remark 1.9.2. This generalized integral inequality is a key point for the obtention of polynomial decay rates for smooth solutions of coupled equations when only one of the equation is stabilized. One can remark that this inequality is not of differential nature as the nonlinear Gronwall inequalities proved in Sect. 1.3. Note that this inequality has been later used by several authors for other polynomial stabilization results. Theorem 1.9.1 has been first announced in [23] and its proof given in [21] (see also [20]).

1.9.3   The Case of Coercive Couplings

1.9.3.1   Coupled Boundary Damped Wave Equations

We consider the following system

$$\begin{cases} u_{tt} - \Delta u + \alpha v = 0 & \text{in} \quad (0, \infty) \times \Omega, \\ v_{tt} - \Delta v + \alpha u = 0 & \text{in} \quad (0, \infty) \times \Omega, \\ u = v = 0 & \text{on} \quad \Sigma_0 = (0, \infty) \times \Gamma_0, \\ \dfrac{\partial u}{\partial \nu} + au + \ell u_t = 0, v = 0 & \text{on} \quad \Sigma_1 = (0, \infty) \times \Gamma_1, \\ (u, u_t)(0) = (u^0, u^1), (v, v_t)(0) = (v^0, v^1) & \text{on} \quad \Omega, \end{cases} \tag{1.243}$$

where

$$a = (N - 1)m \cdot v / 2R^2, \; l = m \cdot v / R. \tag{1.244}$$

For the sake of clearness we will assume that $a \neq 0$ or meas $(\Gamma_0) > 0$, where the measure stands for the Lebesgue measure. We set $H = L^2(\Omega)$ and $V_1 = H^1_{\Gamma_0}(\Omega)$ equipped respectively with the L^2 scalar product and the scalar product $(u, z)_{V_1} = \int_\Omega \nabla u \cdot \nabla z + \int_{\Gamma_1} auz$ and the corresponding norms. Moreover we set $V_2 = H^1_0(\Omega)$ equipped with the scalar product $(u, z)_{V_2} = \int_\Omega \nabla u \cdot \nabla z$ and the associated norm. We define the duality mappings A_1 and A_2. Moreover we define a continuous linear operator B from V_1 to V_1' by

$$< Bu, z >_{V_1', V_1} = \int_{\Gamma_1} \ell u z \, d\gamma.$$

We also set $P = P^* = \text{Id}_H$. Then the system (1.243) can be rewritten under the form (1.240) with the above notation. The energy of a solution $U = (u, v, w, z) = (u, v, u_t, v_t)$ is then given by

$$E(U(t)) = \frac{1}{2}(||u||^2_{V_1} + ||v||^2_{V_2} + ||u_t||^2_H + ||v_t||^2_H) + \alpha(u, v)_H. \tag{1.245}$$

We now turn back to the weakly coupled system (1.243). We set $V = V_1 \times V_2$. This space is equipped with the usual scalar product $((u, v), (\tilde{u}, \tilde{v}))_V = (u, \tilde{u})_{V_1} + (v, \tilde{v})_{V_2}$ and the corresponding norm $||\ ||_V$, where $(u, v) \in V$ and $(\tilde{u}, \tilde{v}) \in V$. We have $V \subset H \times H \subset V'$ with continuous, dense and compact injections. We also define a linear continuous operator A_α from V on V' by

$$A_\alpha(u, v) = (A_1 u + \alpha P v, A_2 v + \alpha P^\star u), (u, v) \in V.$$

Moreover, we consider on V the continuous bilinear form

$$((u, v), (\tilde{u}, \tilde{v}))_\alpha = ((u, v), (\tilde{u}, \tilde{v}))_V + \alpha(P v, \tilde{u})_H + \alpha(P^\star u, \tilde{v})_H, (u, v), (\tilde{u}, \tilde{v}) \in V.$$

Proposition 1.9.3. *There exists $\alpha_0 > 0$, such that for all $0 \leq |\alpha| < \alpha_0$, there exist constants $c_1(\alpha) > 0$ and $c_2(\alpha) > 0$ such that,*

$$c_1(\alpha)||(u, v)||_V \leq ((u, v), (u, v))_\alpha^{1/2} \leq c_2(\alpha)||(u, v)||_V \quad \forall\ (u, v) \in V.$$

Hence, for all $0 \leq |\alpha| < \alpha_0$, the application

$$(u, v) \in V \mapsto ||(u, v)||_\alpha = ((u, v), (u, v))_\alpha^{1/2}$$

defines a norm on V which is equivalent to the norm $||\ ||_V$. Moreover, for all $0 \leq |\alpha| < \alpha_0$, A_α is the duality mapping from V on V' when V is equipped with the scalar product $(\ ,\ )_\alpha$.

We set $\mathscr{H} = V \times H^2$ equipped with the scalar product

$$(U, \tilde{U})_{\mathscr{H}} = ((u, v), (\tilde{u}, \tilde{v}))_\alpha + ((w, z), (\tilde{w}, \tilde{z}))_{H \times H},$$

and the corresponding norm

$$||U||_{\mathscr{H}} = (||(u, v)||_\alpha^2 + ||(w, z)||_{H \times H}^2)^{1/2},$$

where $U = ((u, v), (w, z)) \in \mathscr{H}$. We also define the unbounded linear operator $\mathscr{A}$ on $\mathscr{H}$ by

$$\mathscr{A}_\alpha U = (-w, -z, A_\alpha(u, v) + (Bw, 0)),$$
$$D(\mathscr{A}_\alpha) = \{U = ((u, v), (w, z)) \in V \times V,\ A_\alpha(u, v) + (Bw, 0) \in H \times H\}.$$

One can easily prove that

$$U = ((u, v), (w, z)) \in D(\mathscr{A}_\alpha) \iff (u, w) \in D(\mathscr{A}_1), (v, z) \in D(\mathscr{A}_2).$$

We can now reformulate the system (1.243) as the abstract first order equation

$$\begin{cases} U' + \mathscr{A}_\alpha U = 0, \\ U(0) = U^0 \in \mathscr{H}. \end{cases} \qquad (1.246)$$

Well-posedness thanks to semigroup theory can easily be proved for this abstract equation. Indeed, one can show that $\mathscr{A}_\alpha$ generates a continuous semigroup of contractions on $\mathscr{H}$.

Theorem 1.9.4 (Alabau [21, 23]). *There exists $\alpha_1 \in (0, \alpha_0]$, such that for all $0 < |\alpha| < \alpha_1$, the solution $U(t) = \exp(-\mathscr{A}_\alpha t) U^0$ of (1.243) satisfies*

$$E(U(t)) \leq \frac{c}{t^n} \sum_{p=0}^{2n} E(U^{(p)}(0)) \quad \forall\, t > 0,\ \forall\, U^0 \in D(\mathscr{A}_\alpha^{2n}).$$

Moreover strong stability holds in the energy space $\mathscr{H} = V_1 \times V_2 \times H^2$.

To prove this Theorem, we first prove the following result [21], which proof is technical and obtained through multiplier techniques and using estimates due to the coercivity properties of the coupling operator.

Theorem 1.9.5. *There exists $\alpha_1 \in (0, \alpha_0]$, such that for all $0 < |\alpha| < \alpha_1$, the solution $U(t) = \exp(-\mathscr{A}_\alpha t) U^0$ of (1.243) satisfies*

$$\int_S^T E(U(t))\, dt \leq \sum_{p=0}^{2} c_p E(U^{(p)}(S)) \quad \forall\, 0 \leq S \leq T,\ \forall\, U^0 \in D(\mathscr{A}_\alpha^2). \quad (1.247)$$

We then apply the above nondifferential integral inequality [23], that is Theorem 1.9.1 [23] in the present Notes.

Remark 1.9.6. The above results have been stated in a more general abstract form in [21] together with applications on wave–wave, wave-Petrowsky equations and various concrete examples.

1.9.3.2 Coupled Internally Damped Wave Equations

Let us study the problem

$$\begin{cases} u_{tt} - \Delta u + \beta u_t + \kappa u - \kappa v = 0 \\ v_{tt} - \Delta v + \kappa v - \kappa u = 0 \end{cases} \quad \text{in} \quad \Omega \times (0, +\infty) \qquad (1.248)$$

with boundary conditions

$$u(\cdot, t) = 0 = v(\cdot, t) \quad \text{on} \quad \Gamma \quad \forall t > 0 \qquad (1.249)$$

and initial conditions

$$\begin{cases} u(x,0) = u^0(x), & u'(x,0) = u^1(x) \\ v(x,0) = v^0(x), & v'(x,0) = v^1(x) \end{cases} \qquad x \in \Omega. \qquad (1.250)$$

The above model can be used to describe the evolution of a system consisting of two elastic membranes subject to an elastic force that attracts one membrane to the other with coefficient $\kappa > 0$. The term βu_t, $\beta > 0$, is a stabilizer of the first membrane. We set

$$H = L^2(\Omega), \quad B = \beta I, \quad \alpha = -\kappa$$

and let A be given by

$$D(A) = H^2(\Omega) \cap H_0^1(\Omega), \qquad Au = -\Delta u + \kappa u \quad \forall u \in D(A).$$

One can prove that (see [20])

$$D(\mathscr{A}^n) = D(A^{\frac{n+1}{2}}) \times D(A^{\frac{n}{2}}) \times D(A^{\frac{n+1}{2}}) \times D(A^{\frac{n}{2}})$$

where

$$D(A^{\frac{m}{2}}) = \left\{ u \in H^m(\Omega) \; : \; u = \Delta u = \cdots = \Delta^{[\frac{m-1}{2}]} u = 0 \text{ on } \Gamma \right\}$$

for $m = 1, 2, \dots$. For this system, we prove (see [20] for the proof)

Theorem 1.9.7 (Alabau–Cannarsa–Komornik [20]). *Let us denote by (u, v) a solution of system (1.248). If*

$$u^0, v^0 \in H^2(\Omega) \cap H_0^1(\Omega), \qquad u^1, v^1 \in H_0^1(\Omega),$$

then

$$\int_\Omega \left(|u_t|^2 + |\nabla u|^2 + |v_t|^2 + |\nabla v|^2 \right) dx$$

$$\leq \frac{c}{t} \left(\|u^0\|_{2,\Omega}^2 + \|u^1\|_{1,\Omega}^2 + \|v^0\|_{2,\Omega}^2 + \|v^1\|_{1,\Omega}^2 \right) \qquad \forall t > 0.$$

Moreover if

$$u^0, v^0 \in H^{n+1}(\Omega) \qquad and \qquad u^1, v^1 \in H^n(\Omega)$$

are such that

$$u^0 = \cdots = \Delta^{[\frac{n}{2}]} u^0 = 0 = v^0 = \cdots = \Delta^{[\frac{n}{2}]} v^0 \quad on \quad \Gamma,$$

$$u^1 = \cdots = \Delta^{[\frac{n-1}{2}]} u^1 = v^1 = \cdots = \Delta^{[\frac{n-1}{2}]} v^1 = 0 \quad on \quad \Gamma,$$

then

$$\int_{\Omega} \left(|u_t|^2 + |\nabla u|^2 + |v_t|^2 + |\nabla v|^2 \right) dx$$

$$\leq \frac{c_n}{t^n} \left(\|u^0\|_{n+1,\Omega}^2 + \|u^1\|_{n,\Omega}^2 + \|v^0\|_{n+1,\Omega}^2 + \|v^1\|_{n,\Omega}^2 \right) \qquad \forall t > 0.$$

This result can be generalized to (1.240) assuming an additional compatibility condition between the operators A_1 and A_2 as follows. We define the total energy of the system by

$$E(U) = \frac{1}{2} \left(|A_1^{1/2} u|^2 + |u'|^2 \right) + \frac{1}{2} \left(|A_2^{1/2} v|^2 + |v'|^2 \right) + \alpha \langle u, v \rangle.$$

We will denote the energy either by $E(U(t))$ or by $E(t)$ when there is no ambiguity. Strong solutions of (1.240) satisfy the dissipation relation

$$E'(t) = -|B^{1/2} u'(t)|^2$$

We make the following assumptions:

- For $i = 1, 2$, $A_i : D(A_i) \subset H \to H$ is a densely defined closed linear operator such that

$$A_i = A_i^*, \qquad \langle A_i x, x \rangle \geq \omega_i |x|^2 \quad \forall x \in D(A_i) \qquad (1.251)$$

  for some $\omega_i > 0$.
- B is a bounded linear operator on H such that

$$B = B^*, \qquad \langle Bx, x \rangle \geq \beta |x|^2 \quad \forall x \in H \qquad (1.252)$$

  for some $\beta > 0$.
- α is a real number such that

$$0 < |\alpha| < \sqrt{\omega_1 \omega_2}. \qquad (1.253)$$

We assume further that there is a compatibility condition between the two operators A_1 and A_2, i.e. we assume there exists an integer $j \geq 2$ such that

$$|A_1 u| \leq c |A_2^{j/2} u| \qquad \forall u \in D(A_2^{j/2}). \qquad (1.254)$$

Then we prove the following result.

Theorem 1.9.8 (Alabau–Cannarsa–Komornik [20]). *We assume the above hypotheses. Let U_0 be in $D(\mathscr{A}^{nj})$ for some integer $n \geq 1$, then there exists a constant $c_n > 0$ such that the solution U satisfies*

$$E(U(t)) \leq \frac{c_n}{t^n} \sum_{k=0}^{nj} E(U^{(k)}(0)) \qquad \forall t > 0.$$

For every $U_0 \in \mathcal{H}$ we have

$$E(U(t)) \to 0 \quad as \quad t \to \infty.$$

We give just a very brief sketch of the proof and refer the reader to [20] for the complete proof.

Proof. The proof is based on a finite iteration scheme and suitable coupling multipliers for proving that the full energy of the system satisfies an estimate of the form

$$\int_0^T E(U(t))dt \leq c \sum_{k=0}^j E(U^{(k)}(0)) \qquad \forall \, T \geq 0 \,,$$

where j is as above. This estimate also relies on the coercivity of the coupling operator and on the compatibility condition (1.254) between the operators. Once this estimate is proved. Then we apply Theorem 1.9.1 [23] and we conclude. □

1.9.3.3 The Case of Noncoercive Couplings

We now consider the example of two coupled wave equations with a localized coupling, that is

$$\begin{cases} u_1'' - \Delta u_1 + pu_2 + bu_1' = 0 & \text{in } (0,\infty) \times \Omega, \\ u_2'' - \Delta u_2 + pu_1 = 0 & \text{in } (0,\infty) \times \Omega, \\ u_1 = u_2 = 0 & \text{on } (0,\infty) \times \Gamma, \\ u_j(0,\cdot) = u_j^0(\cdot), \; u_j'(0,\cdot) = u_j^1(\cdot), \; j = 1,2 \text{ in } \Omega, \end{cases}$$

The damping function b and the coupling function p are two bounded real valued functions on Ω, satisfying

$$\begin{cases} 0 \leq b \leq b^+ \text{ and } 0 \leq p \leq p^+ \text{ on } \Omega, \\ b \geq b^- > 0 \text{ on } \omega_b, \\ p \geq p^- > 0 \text{ on } \omega_p, \end{cases} \tag{1.255}$$

for ω_b and ω_p two non-empty open subsets of Ω. So here, the damping as well as the coupling are localized. The previous results use the coercitivity of the coupling in a crucial way. What can be said about stability in this situation and under which geometric conditions when the coupling operator is no longer coercive, but only partially coercive? We suppose that ω_b and ω_p satisfy the piecewise multiplier condition and that $b, p \in W^{q,\infty}(\Omega)$.

Theorem 1.9.9 (Alabau-Boussouira–Léautaud [2]). *Assume the above hypotheses. For p sufficiently small, we have*

$$E(U(t)) \leq \frac{c}{t^n} \sum_{i=0}^{n} E(U^{(i)}(0)) \quad \forall t > 0,$$

$$U^0 = (u_1^0, u_2^0, u_1^1, u_2^1) \in (H^{n+1} \cap H_0^1)^2 \times (H^n \cap H_0^1)^2,$$

where $\mathcal{H}_n = (\mathcal{D}(A^{\frac{n+1}{2}}))^2 \times (\mathcal{D}(A^{\frac{n}{2}}))^2 \subset \mathcal{H}$

Remark 1.9.10. In particular, in the one-dimensional case, the above result holds true for subdomains ω_b and ω_p that are disjoint arbitrary non empty open sets in Ω. In higher dimensions, the fact that the two sets should satisfy the piecewise multiplier geometric condition imply that they necessarily intersect. The best geometric condition is obtained by selecting opposite observer positions for ω_b and ω_p. We generalize recently this result in [1, 5] to subsets ω_b and ω_p satisfying the Geometric Control Condition of Bardos et al. [27], so that the above polynomial decay is still valid for coupling and damping regions which do not intersect in multi-dimensional domains. This result also holds true for general abstract systems of the form (1.240) with $A_1 = A_2$ and for which B is bounded or unbounded and for partially coercive coupling operators P. The controllability of coupled systems by a reduced number of controls have been addressed in [1, 5, 19, 22].

The proof of this result and its abstract version is based on proving that the full energy satisfies a nondifferential integral inequality and to apply Theorem 1.9.1 [23]. The difficulty is to determine suitable coupling multipliers to overcome the lack of coercitivity of the coupling operator. This requires geometric hypotheses on the localization region of the coupling.

Remark 1.9.11. All the above results have been stated in a more general abstract form in [20, 21] together with applications on wave–wave, wave-Petrowsky equations and various concrete examples. Extensions to the case of hybrid boundary conditions, corresponding to operators A_1 and A_2 that do not satisfy the compatibility condition (1.254), have been obtained by Alabau, Cannarsa and Guglielmi in [3]. This allows to treat cases of coupled wave equations for which one is subjected to homogeneous Dirichlet conditions whereas the second one is subjected to homogeneous Robin type boundary condition. Also using interpolation theory, we prove polynomial decay estimates for initial data which are only in the domain of the underlying generator [3] (see also [28]).

1.10   Bibliographical Comments

The above results have also been studied later on from functional analysis point of view by Batkai, Engel, Prüss and Schnaubelt [28] using resolvent and spectral criteria for polynomial stability of abstract semigroups. More precise results linking resolvent estimates and stability properties in the case of nonuniform stability have been later on obtained by Batty and Duyckaerts [29] and Borichev and Tomilov [33].

Resolvent estimates and their link to stability properties in the framework of control theory have been earlier developed, in a different context—that is for geometric purposes and not for indirect stabilization of coupled systems—in the works of Lebeau [74] and Burq [37]. The first results on polynomial stability for indirect stabilization of coupled systems on a general abstract point of view have been obtained by Alabau et al. [20, 21, 23]. The use of resolvent estimates in the context of indirect stabilization came later on with the work of Batkai et al. [28].

Another approach have been introduced by Bader and Ammar Khodja [26] for one-dimensional wave systems, using a factorization method to first order operators, and deriving spectral properties on the essential spectrum.

Liu [79] and later on by Liu and Rao [81], Loreti and Rao [82] have used spectral conditions for peculiar abstract systems and in general for coupled equations only of the *same* nature (wave–wave for instance), so that a dispersion relation for the eigenvalues of the coupled system can be derived. Also these last results are given for internal stabilization only. From the above limitations, Z. Liu-Rao and Loreti-Rao's results are less powerful in generality than the ones given Alabau [21, 23] and by Alabau, Cannarsa and Komornik [20]. Moreover results through energy type estimates and integral inequalities can be generalized to include nonlinear indirect dampings as shown in [15]. On the other side spectral methods are very precise for the obtention of optimal decay rates provided that one can determine in an optimal way at which speed the eigenvalues approach the imaginary axis for high frequencies. A similar remark holds when one uses the resolvent estimates approach and in particular [33]. Optimality for the PDE evolution system can be deduced if and only if optimality of the required resolvent estimates is proved.

1.11   Open Problems

Abstract We indicate below some open problems on the subject of these Notes, that is on nonlinear stabilization and its numerical aspects, on memory-stabilization and on indirect stabilization for coupled systems.

Concerning nonlinear stabilization, the question of optimality in infinite dimensions of the sharp upper energy estimates is largely open. How much is optimality linked to the regularity of solutions? We prove a lower energy estimate for solutions of the multi-dimensional wave equation for general domains provided that the velocity is sufficiently smooth and a priori bounded in spaces such as $L^\infty([0, \infty); W^{1,r}(\Omega))$ with some $r > N$ where N stands for the spatial dimension for arbitrary domains. Is it possible to establish the velocity regularity and a priori estimates in such spaces with some $r > N$ and for arbitrary domains? Is the decay rate depending on initial data and in this case is it possible to identify which properties of the initial data are determinant for the optimal energy decay rates? Many other questions exist in this direction. In particular, is it possible to extend our lower energy estimates in Sect. 1.5?

The semi-discretization of nonlinearly damped wave-like equations lead to nonlinearly damped finite dimensional systems. Is it possible to build numerical schemes so that the optimal discretized energy rate does not depend on the discretization parameter. Such an approach has been performed in the linear stabilization case by Tebou and Zuazua [96]. Can their results be extended to the nonlinear damping case?

Many problems are still open for memory stabilization such as a deep understanding for localized memory-damped equations. The non autonomous character of the damping makes Lasalle principle fails to hold at least by direct arguments. Is it possible to somehow find a way to compare the loss of energy produced by frictional and memory dissipations? As far as we know the question of the optimality of our estimates presented in Sect. 1.6 is completely open.

Concerning indirect stabilization of coupled systems, still many very interesting questions have to be answered. We just indicate some of them. Is it possible to extend our stabilization results to large values of the parameter, to other types of couplings such as first order couplings (an example is given in [15]) but also to couplings that may change sign in the domain? What happens for the general case of coupled hyperbolic equations with different speeds of propagation? Is it possible to prove optimality of the decay rates? Does there exist examples of indirectly coupled systems for which the decay rates are weaker than polynomial? We can still exhibit examples of indirectly coupled systems which are not solved by the above results. So there are still a large number of open and challenging questions on coupled complex systems. We hope these Notes will give some insight on this subject and those of the other sections.

References

1. F. Alabau-Boussouira, M. Léautaud, *Indirect controllability of locally coupled wave-type systems and applications* (submitted)
2. F. Alabau-Boussouira, M. Léautaud, Indirect stabilization of locally coupled wave-type systems. ESAIM Contr. Optim. Calc. Var. Prépublication HAL, hal-00476250 (in press) DOI: 10.1051/cocv/2011106
3. F. Alabau-Boussouira, P. Cannarsa, R. Guglielmi, *Indirect stabilization of weakly coupled systems with hybrid boundary conditions.* Mathematical Control and Related Fields **1**, 413–436 (2011)
4. F. Alabau-Boussouira, Strong lower energy estimates for nonlinearly damped Timoshenko beams and Petrowsky equations. Nonlinear Differ. Equat. Appl. **18**, 571–597 (2011)
5. F. Alabau-Boussouira, M. Léautaud, Indirect controllability of locally coupled systems under geometric conditions. Compt. Rendus Math. Acad. Sci. Paris I **349**, 395–400 (2011)
6. F. Alabau-Boussouira, K. Ammari, Sharp energy estimates for nonlinearly locally damped PDE's via observability for the associated undamped system. J. Funct. Anal. **260**, 2424–2450 (2011)
7. F. Alabau-Boussouira, New trends towards lower energy estimates and optimality for nonlinearly damped vibrating systems. J. Differ. Equat. **249**, 1145–1178 (2010)
8. F. Alabau-Boussouira, A unified approach via convexity for optimal energy decay rates of finite and infinite dimensional vibrating damped systems with applications to semi-discretized vibrating damped systems. J. Differ. Equat. **248**, 1473–1517 (2010)

9. F. Alabau-Boussouira, K. Ammari, Nonlinear stabilization of abstract systems via a linear observability inequality and application to vibrating PDE's. Compt. Rendus Math. Acad. Sci. Paris **348**, 165–170 (2010)

10. F. Alabau-Boussouira, P. Cannarsa, A general method for proving sharp energy decay rates for memory-dissipative evolution equations. Compt. Rendus Math. Acad. Sci. Paris **347**, 867–872 (2009)

11. F. Alabau-Boussouira, J. Prüss and R. Zacher, *Exponential and polynomial stability of a wave equation for boundary memory damping with singular kernels*. Comptes Rendus Mathématique, Compt. Rendus Acad. Sci. Paris I **347**, 277–282 (2009)

12. F. Alabau-Boussouira, P. Cannarsa, in *Control of Partial Differential Equations*. Springer Encyclopedia of Complexity and Systems Science (Springer, New-York, 2009), pp. 1485–1509

13. F. Alabau-Boussouira, Asymptotic stability of wave equations with memory and frictional boundary dampings. Appl. Math. **35**, 247–258 (2008)

14. F. Alabau-Boussouira, P. Cannarsa, D. Sforza, Decay estimates for second order evolution equations with memory. J. Funct. Anal. **254**, 1342–1372 (2008)

15. F. Alabau-Boussouira, Asymptotic behavior for Timoshenko beams subject to a single nonlinear feedback control. Nonlinear Differ. Equat. Appl. **14**(5–6), 643–669 (2007)

16. F. Alabau-Boussouira, Piecewise multiplier method and nonlinear integral inequalities for Petrowsky equations with nonlinear dissipation. J. Evol. Equ. **6**(1), 95–112 (2006)

17. F. Alabau-Boussouira, Convexity and weighted integral inequalities for energy decay rates of nonlinear dissipative hyperbolic systems. Appl. Math. Optim. **51**(1), 61–105 (2005)

18. F. Alabau-Boussouira, Une formule générale pour le taux de décroissance des systèmes dissipatifs non linéaires. Compt. Rendus Acad. Sci. Paris I Math. **338**, 35–40 (2004)

19. F. Alabau-Boussouira, A two-level energy method for indirect boundary observability and controllability of weakly coupled hyperbolic systems. SIAM J. Contr. Optim. **42**, 871–906 (2003)

20. F. Alabau, P. Cannarsa, V. Komornik, Indirect internal damping of coupled systems. J. Evol. Equat. **2**, 127–150 (2002)

21. F. Alabau-Boussouira, Indirect boundary stabilization of weakly coupled systems. SIAM J. Contr. Optim. **41**(2), 511–541 (2002)

22. F. Alabau, Observabilité et contrôlabilité frontière indirecte de deux équations des ondes couplées. Compt. Rendus Acad. Sci. Paris I Math. **333**, 645–650 (2001)

23. F. Alabau, Stabilisation frontière indirecte de systèmes faiblement couplés. Compt. Rendus Acad. Sci. Paris I **328**, 1015–1020 (1999)

24. K. Ammari, M. Tucsnak, Stabilization of second order evolution equations by a class of unbounded feedbacks. ESAIM Contr. Optim. Calc. Var. **6**, 361–386 (2001)

25. K. Ammari, M. Tucsnak, Stabilization of Bernoulli-Euler beams by means of a pointwise feedback force. SIAM. J. Contr. Optim. **39**, 1160–1181 (2000)

26. A. Bader, F. Ammar Khodja, Stabilizability of systems of one-dimensional wave equations by one internal or boundary control force. SIAM J. Contr. Optim. **39**, 1833–1851 (2001)

27. C. Bardos, G. Lebeau, J. Rauch, *Sharp sufficient conditions for the observation, control, and stabilization of waves from the boundary*. SIAM J. Contr. Optim. **30**, 1024–1065 (1992)

28. A. Bátkai, K.J. Engel, J. Prüss, R. Schnaubelt, Polynomial stability of operator semigroups. Math. Nachr. **279**(13, 14), 1425–1440 (2006)

29. C. Batty, T. Duyckaerts, Non-uniform stability for bounded semi-groups on Banach spaces. J. Evol. Equat. **8**, 765–780 (2008)

30. S. Berrimi, S. A. Messaoudi, Existence and decay of solutions of a viscoelastic equation with a nonlinear source. Nonlinear Anal. **64**, 2314–2331 (2006)

31. A. Beyrath, Stabilisation indirecte localement distribué de systèmes faiblement couplés. Compt. Rendus Acad. Sci. Paris I Math. **333**, 451–456 (2001)

32. A. Beyrath, Indirect linear locally distributed damping of coupled systems. Bol. Soc. Parana. Mat. **22**, 17–34 (2004)

33. A. Borichev, Y. Tomilov, Optimal polynomial decay of functions and operator semigroups. Math. Ann. 347, 455–478 (2010)
34. H. Brézis, in *Analyse Fonctionnelle. Théorie et Applications* (Masson, Paris, 1983)
35. N. Burq, Contrôlabilité exacte des ondes dans des ouverts peu réguliers. Asymptot. Anal. **14**, 157–191 (1997)
36. N. Burq, P. Gérard, Condition nécessaire et suffisante pour la contrôlabilité exacte des ondes. Compt. Rendus Acad. Sci. Paris I Math. **325**, 749–752 (1997)
37. N. Burq, Decay of the local energy of the wave equation for the exterior problem and absence of resonance near the real axis. Acta Math. **180**, 1–29 (1998)
38. N. Burq, G. Lebeau, Mesures de défaut de compacité, application au système de Lamé. Ann. Sci. Ecole Norm. Sup. (4) **34**(6), 817–870 (2001)
39. N. Burq, M. Hitrik, Energy decay for damped wave equations on partially rectangular domains. Math. Res. Lett. **14**(1), 35–47 (2007)
40. A. Carpio, Sharp estimates of the energy for the solutions of some dissipative second order evolution equations. Potential Anal. **1**, 265–289 (1992)
41. M.M. Cavalcanti, H.P. Oquendo, Frictional versus viscoelastic damping in a semilinear wave equation. SIAM J. Contr. Optim. **42**, 1310–1324 (2003)
42. G. Chen, A note on boundary stabilization of the wave equation. SIAM J. Contr. Optim. **19**, 106–113 (1981)
43. I. Chueshov, I. Lasiecka, D. Toundykov, Long-term dynamics of semilinear wave equation with nonlinear localized interior damping and a source term of critical exponent. Discrete Contin. Dyn. Syst. **20**, 459–509 (2008)
44. F. Conrad, B. Rao, Decay of solutions of the wave equation in a star-shaped domain with nonlinear boundary feedback. Asymptot. Anal. **7**, 159–177 (1993)
45. J.-M. Coron, in *Control and Nonlinearity*. Mathematical Surveys and Monographs, 136 (American mathematical society, Providence, 2007)
46. C. Dafermos, in *Asymptotic Behavior of Solutions of Evolution Equations*. Nonlinear Evolution Equations (Proc. Sympos. Univ. Wisconsin, Madison, Wis.), Publ. Math. Res. Center University of Wisconsin, 40, (Academic, New York, 1978), pp. 103–123
47. C. Dafermos, Asymptotic stability in viscoelasticity. Arch. Ration. Mech. Anal. **37**, 297–308 (1970)
48. C. Dafermos, An abstract Volterra equation with applications to linear viscoelasticity. J. Differ. Equat. **7**, 554–569 (1970)
49. R. Dáger, E. Zuazua, in *Wave Propagation, Observation and Control in* $1 - d$ *Flexible Multistructures*. Mathematics & Applications, 50 (Springer, Berlin, 2006)
50. M. Daoulatli, I. Lasiecka, D. Toundykov, Uniform energy decay for a wave equation with partially supported nonlinear boundary dissipation without growth conditions. Discrete Contin. Dyn. Syst. S **2**, 67–94 (2009)
51. M. Eller, J. Lagnese, S. Nicaise, Decay rates for solutions of a Maxwell system with nonlinear boundary damping. Special issue in memory of Jacques-Louis Lions. Comput. Appl. Math. **21**, 135–165 (2002)
52. S. Ervedoza, E. Zuazua, Uniformly exponentially stable approximation for a class of damped systems. J. Math. Pure. Appl. **91**, 20–48 (2008)
53. M. Fabrizio, A. Morro, Viscoelastic relaxation functions compatible with thermodynamics. J. Elasticity **19**, 63–75 (1988)
54. X. Fu, J. Yong, X. Zhang, Exact controllability for multidimensional semilinear hyperbolic equations. SIAM J. Contr. Optim. **46**, 1578–1614 (2007)
55. R. Glowinski, C.H. Li, J.-L. Lions, A numerical approach to the exact boundary controllability of the wave equation. I. Dirichlet controls: description of the numerical methods. Jpn. J. Appl. Math. **7**(1), 1–76 (1990)
56. A. Haraux *Semi-groupes linéaires et équations d'évolutions linéaires périodiques*. Publications du Laboratoire d'Analyse Numérique 78011, Université Pierre et Marie Curie, Paris 1978

57. A. Haraux, L^p estimates of solutions to some nonlinear wave equation in one space dimension. Publications du laboratoire d'analyse numérique, Université Paris VI, CNRS, Paris 1995
58. A. Haraux, in *Nonlinear Evolution Equations—Global Behavior of Solutions*. Lecture Notes in Mathematics 841 (Springer, Berlin, 1981)
59. A. Haraux, Stabilization of trajectories for some weakly damped hyperbolic equations. J. Differ. Equat. **59**, 145–154 (1985)
60. A. Haraux, in *Systèmes Dynamiques Dissipatifs et Applications*. Collection Recherches en Mathématiques Appliquées (Masson, Paris, Milan, Barcelone, 1991)
61. A. Haraux, Une remarque sur la stabilisation de certains systèmes du deuxième ordre en temps. Portugal. Math. **46**, 245–258 (1989)
62. L.F. Ho, Observabilité frontière de l'équation des ondes. Compt. Rendus Acad. Sci. Paris I Math. **302**, 443–446 (1986)
63. J.A. Infante, E. Zuazua, Boundary observability for the space semi-discretizations of the 1-D wave equation. M2AN Math. Model. Numer. Anal. **33**(2), 407–438 (1999)
64. V. Komornik, E. Zuazua, A direct method for the boundary stabilization of the wave equation. J. Math. Pure. Appl. **69**(1), 33–54 (1990)
65. V. Komornik, in *Exact Controllability and Stabilization; The Multiplier Method*. Collection RMA, vol 36 (Wiley, Masson, Paris, 1994)
66. J. Lagnese, Decay of solutions to wave equations in a bounded region. J. Differ. Equat. **50**, 163–182 (1983)
67. J.-P. Lasalle, The extent of asymptotic stability. Proc. Symp. Appl. Math. **13**, Amunozsalva, 299–307 (1962)
68. J.-P. Lasalle, Asymptotic stability criteria. Proc. Nat. Acad. Sci. USA **46**, 363–365 (1960)
69. I. Lasiecka, Stabilization of waves and plate like equations with nonlinear dissipation on the boundary. J. Differ. Equat. **79**, 340–381 (1989)
70. I. Lasiecka, D. Tataru, Uniform boundary stabilization of semilinear wave equation with nonlinear boundary damping. Differ. Integr. Equat. **8**, 507–533 (1993)
71. I. Lasiecka I, J.-L. Lions, R. Triggiani, Nonhomogeneous boundary value problems for second order hyperbolic operators. J. Math. Pure. Appl. **65**(2), 149–192 (1986)
72. I. Lasiecka I, R. Triggiani, in *Control Theory for Partial Differential Equations: Continuous and Approxiamtion Theories*. I. Encyclopedia of Mathematics and its Applications, vol 74 (Cambridge University Press, Cambridge, 2000)
73. I. Lasiecka I, R. Triggiani, *Control Theory for Partial Differential Equations: Continuous and Approximation Theories*. II. Encyclopedia of Mathematics and its Applications, vol 75 (Cambridge University Press, Cambridge, 2000)
74. G. Lebeau, in *Equations des ondes amorties*, ed. by A. Boutet de Monvel et al. Algebraic and Geometric Methods in Mathematical Physics. Math. Phys. Stud. 19, 73–109 (Kluwer Academic Publishers, Dordrecht, 1996)
75. G. Lebon, C. Perez-Garcia, J. Casas-Vazquez, On the thermodynamic foundations of viscoelasticity. J. Chem. Phys. **88**, 5068–5075 (1988)
76. J.-L. Lions, in *Contrôlabilité exacte et stabilisation de systèmes distribués*, vol 1 (Masson, Paris, 1988)
77. J.-L. Lions, W. Strauss, Some non linear evolution equations. Bull. Soc. Math. Fr. **93**, 43–96 (1965)
78. K. Liu, Locally distributed control and damping for the conservative systems. SIAM J. Contr. Optim. **35**, 1574–1590 (1997)
79. Z. Liu, S. Zheng, in *Semigroups Associated with Dissipative Systems*. Chapman Hall CRC Research Notes in Mathematics, 398 (Chapman Hall/CRC, Boca Raton, 1999)
80. W.-J. Liu, E. Zuazua, Decay rates for dissipative wave equations. Ricerche. Matemat. **48**, 61–75 (1999)
81. Z. Liu, B. Rao, Frequency domain approach for the polynomial stability of a system of partially damped wave equations. J. Math. Anal. Appl. **335**, 860–881 (2007)

82. P. Loreti, B. Rao B, Optimal energy decay rate for partially damped systems by spectral compensation. SIAM J. Contr. Optim. **45**, 1612–1632 (2006)
83. A. Lunardi, in *Interpolation Theory*. Edizioni della Normale, Pisa (Birkhaüser, Basel, 2009)
84. P. Martinez, A new method to obtain decay rate estimates for dissipative systems with localized damping. Rev. Mat. Complut. **12**, 251–283 (1999)
85. P. Martinez, A new method to obtain decay rate estimates for dissipative systems. ESAIM Contr. Optim. Calc. Var. **4**, 419–444 (1999)
86. L. Miller, Escape function conditions for the observation, control, and stabilization of the wave equation. SIAM J. Contr. Optim. **41**(5), 1554–1566 (2002)
87. A. Munch, A. Pazoto, Uniform stabilization of a viscous numerical approximation for a locally damped wave equation. ESAIM Contr. Optim. Calc. Var. **13**, 265–293 (2007)
88. J.E. Muñoz Rivera, A. Peres Salvatierra, Asymptotic behaviour of the energy in partially viscoelastic materials Quart. Appl. Math. **59**, 557–578 (2001)
89. M. Nakao, Decay of solutions of the wave equation with a local nonlinear dissipation. Math. Ann. **305**, 403–417 (1996)
90. J. Prüss, in *Evolutionary Integral Equations and Applications*. Monographs in Mathematics, 87 (Birkhäuser, Basel, 1993)
91. J. Rauch, M. Taylor, Exponential Decay of Solutions to Hyperbolic Equations in Bounded Domains. Indiana Univ. Math. J. **24**, 79–86 (1974)
92. M. Renardy, W.J. Hrusa, J.A. Nohel, in *Mathematical Problems in Viscoelasticity*. Pitman Monographs in Pure and Applied Mathematics, vol 35 (Longman Scientific and Technical, Harlow, 1988)
93. R.T. Rockafellar, in *Convex Analysis* (Princeton University Press, Princeton, 1970)
94. A. Ruiz, Unique continuation for weak solutions of the wave equation plus a potential. J. Math. Pures Appl. **71**, 455–467 (1992)
95. D.L. Russell, A general framework for the study of indirect damping mechanisms in elastic systems. J. Math. Anal. Appl. **173**, 339–358 (1993)
96. L.R. Tebou, E. Zuazua, Uniform exponential long time decay for the space semi-discretization of a locally damped wave equation via an artificial numerical viscosity. Numer. Math. **95**, 563–598 (2003)
97. L.R. Tebou, A Carleman estimates based approach for the stabilization of some locally damped semilinear hyperbolic equations. ESAIM Contr. Optim. Calc. Var. **14**, 561–574 (2008)
98. G. Todorova, B. Yordanov, The energy decay problem for wave equation with nonlinear dissipative terms in $\mathbb{R}^n$. Indiana Univ. Math. J. **56**, 389–416 (2007)
99. M. Tucsnak, G. Weiss, in *Observation and Control for Operator Semigroups*. Birkhäuser Advanced Texts (Birkhäuser, Basel, 2009)
100. J. Valein, E. Zuazua, Stabilization of the wave equation on 1-D networks. SIAM J. Contr. Optim. **48**, 2771–2797 (2009)
101. J. Vancostenoble, P. Martinez, Optimality of energy estimates for the wave equation with nonlinear boundary velocity feedbacks. SIAM J. Contr. Optim. **39**, 776–797 (2000)
102. J. Vancostenoble, Optimalité d'estimation d'énergie pour une équation des ondes amortie. Compt. Rendus Acad. Sci. Paris I **328**, 777–782 (1999)
103. W.H. Young, On classes of summable functions and their Fourier series. Proc. Royal Soc. (A) **87**, 225–229 (1912)
104. A. Wehbe, W. Youssef, Stabilization of the uniform Timoshenko beam by one locally distributed feedback. Appli. Anal. **88**, 1067–1078 (2009)
105. E. Zuazua, Exponential decay for the semilinear wave equation with locally distributed damping. Comm. Part. Differ. Equat. **15**, 205–235 (1990)
106. E. Zuazua, Uniform stabilization of the wave equation by nonlinear feedbacks. SIAM J. Contr. Optim. **28**, 265–268 (1989)
107. E. Zuazua, Propagation, observation and control of wave approximation by finite difference methods. SIAM Rev. **47**, 197–243 (2005)

Chapter 2
Notes on the Control of the Liouville Equation

Roger Brockett

Abstract In these notes we motive the study of Liouville equations having control terms using examples from problem areas as diverse as atomic physics (NMR), biological motion control and minimum attention control. On one hand, the Liouville model is interpreted as applying to multiple trials involving a single system and on the other, as applying to the control of many identical copies of a single system; e.g., control of a flock. We illustrate the important role the Liouville formulation has in distinguishing between open loop and feedback control. Mathematical results involving controllability and optimization are discussed along with a theorem establishing the controllability of multiple moments associated with linear models. The methods used succeed by relating the behavior of the solutions of the Liouville equation to the behavior of the underlying ordinary differential equation, the related stochastic differential equation, and the consideration of the related moment equations.

2.1 Introduction

In these notes we describe a number of problems in automatic control related to the Liouville equation and various approximations of it. Some of these problems can be cast either in terms of designing a single feedback controller which effectively controls a particular system over repeated trials corresponding to different initial conditions or, alternatively, using a broadcast signal to simultaneously control many copies of a particular system. Sometimes these different points of view lead to problems that are identical from the mathematical point of view. In many cases a certain continuum limit can be formulated, either by considering an infinity of trials

R. Brockett (✉)
Division of Engineering and Applied Sciences, Harvard University, 345 Maxwell Dworkin, 33 Oxford Street, Cambridge, MA 02138, USA
e-mail: brockett@seas.harvard.edu

F. Alabau-Boussouira et al., *Control of Partial Differential Equations*,
Lecture Notes in Mathematics 2048, DOI 10.1007/978-3-642-27893-8_2,
© Springer-Verlag Berlin Heidelberg 2012

or an infinity of copies. In this situation we are often led to problems involving the control of an associated Liouville equation.

The use of feedback as part of a regulatory mechanism is a standard idea in engineering, biology, and even economics. This stands in contrast to the many other uses of feedback in communication, adaptive sensing, learning algorithms and, more typically in engineering, tracking problems where it is used to improve the speed and accuracy of the response of servomechanisms. Its main virtue is that it is a single mechanism capable of dealing with a great variety of disturbances.

Before introducing the controlled Liouville equation and some mathematical problems that go along with it, we will discuss some additional motivation.

2.2   Some Limitations on Optimal Control Theory

An optimal control problem, as usually formulated, assumes that one has exact knowledge of the equations of evolution. The problem is posed as that of finding a control that transfers the state of the system from a given initial condition to a final one, or possibly a manifold of final states, while minimizing some performance measure. This formulation fits well a number of real-world problems, such as finding the minimum fuel trajectory for getting a payload from the earth to Mars. On the other hand it is less useful as a tool for designing feedback compensators for tracking servomechanisms, a typical problem in robotics, and other path following problems. In these situations there is no fixed initial state and no fixed final state. We do not know what the initial condition will be at a particular time; it is as if the system needs to be ready for a wide variety of challenges.

The development of the various least squares methods for linear systems has led to tools that address more directly the issues raised by such tracking problems. By exploiting the linear structure and by assuming that the desired end state is the point 0, least squares theory produces a feedback control rule that is simultaneously optimal for all initial conditions. Of course the fact that the control can be expressed in feedback form is the key to the invariance with respect to initial conditions. However, the assumptions include the fact that there is a fixed desired steady state and this is a strong limitation.

Moreover, and here we are beginning to discuss a second major point, there are a great many applications in which the payoff for implementing a linear relationship between sensed signals and control variables does not justify the cost of the equipment needed to achieve it. For example, in high volume consumer goods, such as dish washers and clothes dryers, it is inexpensive to sense the temperature of the water or air but the benefits associated with implementing a linear relationship between the temperature of the mixed water and the flow from the hot and cold water lines do not justify the cost. Acceptable performance is obtainable using on-off control which can be implemented much more cheaply. Even in the case of audio equipment, where there is a significant payoff for building systems that are very close to linear, the benefits of linearity are confined to finite range of amplitudes and a subset of frequencies. Standard optimal control theory provides no mechanisms to incorporate implementation costs. This is a major reason why we can not consider the usual optimal control formulation, even when robustness is taken into account, to be completely satisfactory.

Finally, we might ask why optimal control theory has not been more useful in understanding the control mechanisms found in biology. The questions there range from understanding control of the operation of an individual cell to the motor control of the complete organism. In particular, given that evolution has had as long as it has to optimize the neuroanatomy and the muscle/skeletal structures, why is that we don't find optimal control theory to be more effective in explaining these structures?

2.3 Measuring Implementation Cost

The expense required to implement a control policy in an industrial setting where each control signal is generated by a box requiring both a capital investment and continuing maintenance cost, can be accounted in a straightforward way. Unfortunately, such costs are strongly dependent of the technology being used. We wish to focus instead on measures which are intrinsic in the sense that they might apply, at least to some degree, to a range of situations including both those found in engineering and those found in biology. Some considerations that are relevant here have been discussed in our paper [1] which we now paraphrase.

Our point of view is that the easiest control law to implement is a constant input. Anything else requires some attention. The more frequently the control changes, the more effort it takes to implement it. Because the control law will depend on the state x and the time t, it can be argued that the cost of implementation is linked to the rate at which the control changes with changing values of x and t. This rate of change may also affect the effort required to compute the desired control or some suitable approximation to it. In any case, solutions that require less frequent adjustments as x and t change are to be preferred over those that require more frequent adjustments. From the point of view of an animal controlling its body, or a systems engineer allocating the cpu cycles of a computer controlling a machine tool, control laws with small values of $\|\partial u/\partial t\|$ and $\|\partial u/\partial x\|$ require less frequent updating and will be more robust with respect to small changes in the data. These considerations suggest that a suitable quantification of what is meant by "attention" might include a measure of the size of the partial derivatives, $\partial u/\partial x$ and $\partial u/\partial x$. For example, the numerical measure of the attention of a given control law might be might be a weighted Sobolev norm of $u(t, x)$.

This reasoning suggests a class of optimization problems associated with selecting the architecture of a control system. The general structure of the optimization problem will involve minimizing functionals of the form

$$\eta_a = \int_\Omega \phi\left(x, t, \frac{\partial u}{\partial x}, \frac{\partial u}{\partial t}\right) dx dt$$

subject to constraints on u such as will insure that the performance is adequate for the task. We can think of η as an *attention functional* and use it as a guide to suggest which control laws might be more or less expensive to implement. To this we may add the observation that although textbooks on control often discuss the difference between open-loop and closed-loop control, the distinction is either vague or applicable only in highly restrictive situations. In many cases, e.g., fixed end-point linear-quadratic optimal control on finite time intervals, it is unclear what might be meant by a closed-loop solution. This makes it difficult for researchers in other fields to discuss the distinction in a precise way. At an intuitive level, it seems that biological motor control involves not only "pure" open-loop control but also a gradation of modalities spanning a range between open-loop and closed-loop operation. Intuitively, one thinks that large values of $\|\partial u/\partial x\|$ indicate closed-loop control and that large values of $\|\partial u/\partial t\|$ indicate open-loop control. By modifying the attention functional we can change the ratio of the penalty put on the closed-loop $\|\partial u/\partial x\|$ terms relative to the penalty put on the open-loop $\|\partial u/\partial t\|$ terms. In this way we create a continuum and arrive at a characterization which makes possible a quantitative study of the trade-offs between open-loop and closed-loop control.

Example 2.3.1. To give some indication about where these ideas can lead it may be helpful to have an example. Consider the scalar control problem $\dot{x} = u$ with the distribution of initial conditions given by a density $\rho_0(x)$. Our goal is to minimize

$$\eta = \int_0^\infty \int_{\mathbb{R}} \rho(t,x)ax^2\,dx\,dt + \int_{\mathbb{R}} \left(\frac{\partial u}{\partial x}\right)^2 dx$$

where $\rho(t,x)$ denotes the density at time t. To avoid complication, we constrain u to be a function of x alone. The calculations leading to a characterization of u now follow.

We rewrite the equation of motion as $dx/u = dt$. Using this we see that if $\int_0^\infty x^2 dt$ is finite then

$$\int_0^\infty x^2\,dt = \int_0^{x(0)} \frac{-x^2}{u(x)}\,dx \quad \text{and} \quad \int_0^\infty u^2\,dt = \int_0^{x(0)} -u(x)\,dx$$

Thus

$$\int_0^\infty \int_{\mathbb{R}} \rho(t,x)ax^2\,dx\,dt = \int_{\mathbb{R}} \rho_0(x_0)\left(\int_0^{x(0)} a\frac{x^2\,dx}{u(x)} + bu(x)\right)dx_0$$

and the functional to be minimized can be written as

$$\eta = \int_{\mathbb{R}} \rho_0(x)\left(\int_0^x a\frac{w^2}{u(w)}\,dw\right)dx + \int_{\mathbb{R}} \left(\frac{\partial u}{\partial x}\right)^2 dx$$

If ρ_0 is a delta function this expression can be further simplified. Let $\rho(0,x)$ be a delta function centered at $x_0 > 0$. In this case the term involving ρ_0 can be simplified giving

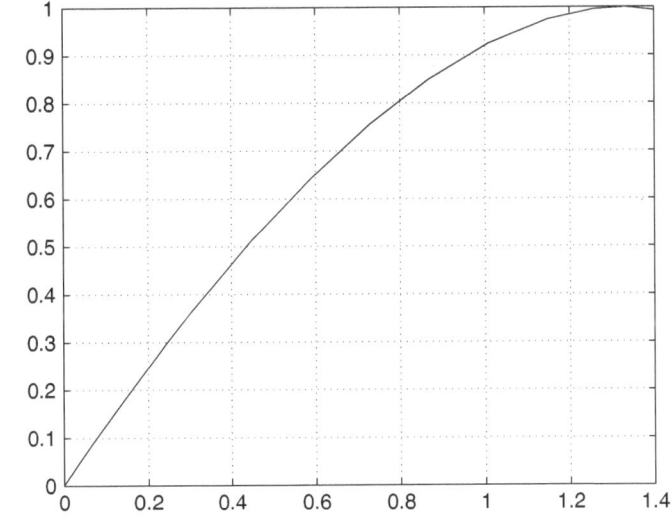

Fig. 2.1 The graph of the optimal gain function for $x > 0$

$$\eta = \int_0^{x_0} \left(a \frac{x^2}{u(w)} \right) dx + \int_{\mathbb{R}} \left(\frac{\partial u}{\partial x} \right)^2 dx$$

The indicated partial derivative is actually a total derivative and so an application of the Euler–Lagrange operator gives

$$\frac{d^2 u}{dx^2} + a \frac{x^2}{u^2} = 0$$

Because the solution should remain at zero when reaching zero, it is necessary that $u(0) = 0$. Because the support of the density will be confined to the interval $[0, x_0]$, and because we are minimizing the square of du/dx, the optimal u will have $du/dx = 0$ at $x = x_0$. A numerical solution of these equations corresponding to $a = 1$ and $x_0 = 1$ is shown in Fig. 2.1.

This control law is nearly linear near 0 and approaches saturation as x approaches 1, reflecting the fact that we are putting a penalty on the derivative.

2.4   Ensemble Control

There are several areas of work that have been called "ensemble control" but generally this term applies to problems involving a large number of more or less identical subsystems which are being manipulated by a single source of command signals. (See [2–4].)

A finite number of copies of a system, controlled by an t-dependent function of time can be investigated both as an approximation to the Liouville equation and as something of interest in its own right. Such collections are of interest as models for flocks, swarms, and ensembles of various kinds. Their study gives rise a number of interesting questions, centering around the topics of controllability and stabilizability, but also the control of various averages.

Special aspects of the replicated systems include the degeneracy that will occur when two or more elemental systems are in the same state. In addition, a direct application of Lie algebraic controllability conditions, while in principle quite routine, can be tedious because of the large number of subsystems.

Example 2.4.1. Consider k copies of the scalar system $\dot{x} = -x^3 + u$ and note that the lie bracket of two power law vector fields is given by

$$[x^m \frac{\partial}{\partial x}, x^n \frac{\partial}{\partial x}] = (n - m)x^{m+n-1} \frac{\partial}{\partial x}$$

Thus the Lie algebra generated by the drift vector field and the control vector field is infinite dimensional and contains all polynomial vector fields. To investigate the controllability of k copies of the scalar system we need to look at the distribution generated by bracketing

$$[\sum x_i^3 \frac{\partial}{\partial x_i}, \sum \frac{\partial}{\partial x_i}] = -3 \sum x_i^2 \frac{\partial}{\partial x_i}$$

In this case the Lie algebra contains all the vector fields of the form

$$L_i^j = \sum x^j \frac{\partial}{\partial x_i}; \quad j = 0, 1, 2, \ldots, k$$

The distribution generated by these vector fields at the point $x_1 = p_1, x_2 = p_2, \ldots,$ $x_k = p_k$ is the range space of the Vandermonde matrix

$$V = \begin{bmatrix} 1 & p_1 & p_1^2 & \cdots & p_1^{k-1} \\ 1 & p_2 & p_2^2 & \cdots & p_2^{k-1} \\ 1 & p_3 & p_3^2 & \cdots & p_3^{k-1} \\ \cdots & \cdots & \cdots & \cdots & \cdots \\ 1 & p_k & p_k^2 & \cdots & p_k^{k-1} \end{bmatrix}$$

Because the Vandermonde matrix is nonsingular if and only if the p_i are distinct, we see that this distribution spans $\mathbb{R}^k$ at all points corresponding to unrepeated values of the p_i.

Any given ordering of the x_i, for example $x_1 < x_2 < \cdots < x_k$, defines a connected, open subset of $\mathbb{R}^k$ in which the Vandermonde matrix does not vanish. Each of the $n - 1$ co-dimension one planar subsets of the boundary is an integral manifold for both the control vector field and the drift and so these can not be crossed. Natural questions then arise about the reachable set in such a cone. In particular, can any point in the cone be reached from any other point in it, independent of the value of k?

Example 2.4.2. Consider a system consisting of a pair of identical second order systems with nonlinear restoring forces.

$$\ddot{x} + x + x^2 = u; \quad \ddot{y} + y + y^2 = u$$

Clearly the set $\{(x, y, \dot{x}, \dot{y}) \mid x = y ; \ \dot{x} = \dot{y}\}$ is an invariant set. We may ask if there are other invariant sets and, if not, can we drive an arbitrary initial state to this invariant set. Rewriting the system as

$$\frac{d}{dt} \begin{bmatrix} x \\ \dot{x} \\ y \\ \dot{y} \end{bmatrix} = \begin{bmatrix} \dot{x} \\ -x - x^2 \\ \dot{y} \\ -y - y^2 \end{bmatrix} + \begin{bmatrix} 0 \\ 1 \\ 0 \\ 1 \end{bmatrix} u$$

we see that in addition to the vector fields defined by f and g the Lie algebra contains

$$[f, g] = \begin{bmatrix} 1 \\ 0 \\ 1 \\ 0 \end{bmatrix} ; \quad [f, [f, g]] = \begin{bmatrix} 0 \\ -1 - 2x \\ 0 \\ -1 - 2y \end{bmatrix} ; \quad [f, [f, [f, g]]] = \begin{bmatrix} -1 - 2x \\ 0 \\ -1 - 2y \\ 0 \end{bmatrix}$$

Thus the distribution associated with the vector fields in the Lie algebra generated by drift and the control vector field spans a four-dimensional space at all points except those on the co-dimension one hyperplane defined by $x = y$.

Our language will be to call the overall system "the system" and to refer to the individual subsystems as being "the elemental systems". Examples of what we have in mind can be found in the literature the on the following topics:

A. *Classical thermodynamics* deals with the control of ensembles, usually modeled as collections of identical particles. Viewed as a control problem, the conversion of heat into work concerns the control of various averages such as temperature and pressure (the intensive variables) using heat flow and adjustable volume. Here the elemental systems consist, for example, of gas molecules; the over all system would be described by a combination of intensive and extensive variables. One might take the controls to be heat flow and volume. Formulated as a control problem, a possible goal is to extract as much mechanical work as possible given constraints on the path. In elementary thermodynamics the system is described in terms of the thermodynamic "state". It is typical to assume that the controls are applied in such a way as to keep the system in thermodynamic equilibrium; which is to say, all paths are adiabatic.

B. *Quantum control of ensembles of identical, weakly interacting, particles.* This arises in the model used in many discussions of nuclear magnetic resonance (NMR) problems. The control is an electromagnetic field consisting of short bursts, or pulses, consisting of different frequencies of controlled duration. The goal is usually to manipulate the orientation of a collection of quantum mechanical spins, say those of the hydrogen ions coming from water molecules, such that the majority of the elemental systems align in a particular nonequilibrium configuration.

C. *Quantum control of a parameterized family* of nearly identical systems using a common control. Here again, a well studied model comes from NMR. Because of slight variations in the magnetic field the resonant frequencies of the individual hydrogen ions differ over the ensemble. Because of this, the control has a different effect on the various elemental systems. Consequently, even if the elemental systems were to start from the same state it requires great care to steer the largest possible fraction of them to a desired end state.

D. *Control of flocks*: It is of interest to understand the extent to which a leader can shape and stabilize the motion of the elemental systems comprising a flock using a broadcast signal. A natural constraint would be to ask that any feedback signal be based on a symmetric function of the states of the elemental systems, for example on the average velocity of the elemental systems.

These applications have in common the goal of controlling a (large) number of weakly interacting individual systems with a single, or perhaps small number, of control inputs. In some quantum mechanical applications the Liouville–von Neumann density equation is appropriate to describe the situation; in other situations the Fokker–Planck equation, or even many copies of a finite state model may serve better.

2.5 The Liouville Equation

Given an ordinary differential equation, $\dot{x}(t) = f(x(t))$ defined on a manifold X, and having the property that there exists a unique solution through each point, there is an associated partial differential equation which describes the evolution of an initial density of points. Let $\rho(0, \cdot)$ be the initial density, thought of as a probability density for $x(0)$. As such it is nonnegative and normalized

$$\int_X \rho(0, x)\, dx = 1$$

Let ψ be a smooth function $\psi : X \rightarrow \mathbb{R}^+$ having compact support. The expected value of ψ at some future time is

$$\mathscr{E}\psi(x(t)) = \int_X \psi(x)\rho(t, x)\, dx$$

The time derivative of this expression can be expressed in two ways. It is expressible in terms of $\partial \rho / \partial t$ but also in terms of $\langle \partial \psi / \partial x, f(x) \rangle \rho(t, x)$. These possibilities are the basis for the expression

$$\int_X \psi(x) \frac{\partial \rho}{\partial t} \, dx = \int_X \langle \partial \psi / \partial x, f(x) \rangle \rho(t, x) \, dx$$

Integrating the right-hand side by parts we have

$$\int_X \psi(x) \frac{\partial \rho}{\partial t} \, dx = - \int_X \psi(x) \langle \partial / \partial x, f(x) \rho(t, x) \rangle \, dx$$

Because ψ is arbitrary this implies, subject to mild additional assumptions, that

$$\frac{\partial \rho(t, x)}{\partial t} = - \left\langle \frac{\partial}{\partial x}, f(x) \rho(t, x) \right\rangle$$

We can think of this as a Cauchy problem to be solved, subject to an initial condition $\rho(0, x) = \rho_0(x)$. It describes how the density evolves in time under the flow defined by the given deterministic equation. It is easy to verify that if $\rho(t, x)$ is nonnegative and satisfies this equation then

$$\frac{d}{dt} \int_{\mathbb{R}^n} \rho(t, x) dx = 0$$

The solution of the Liouville equation can be expressed in terms of the general solution of $\dot{x} = f(x, t)$. If the solution of $\dot{x} = f(x, t)$ is such that the initial value x_0 goes to $\phi(t, x_0)$ at time t then the solution of the Liouville equation is

$$\rho(t, x) = \rho_0(\phi^{-1}(t, x)) / \det J_\phi(x)$$

where ϕ^{-1} denotes the result of solving $x = \phi(t, x_0)$ for x_0 and J is the Jacobian of this map; its determinant is necessarily positive. Thus the properties of the Liouville equation reflect quite closely the properties of the underlying ordinary differential equation. The example that follows uses a special case of the fact that $\det \partial f / \partial x$ is the exponential of an integral of the trace of a Jacobian.

Example 2.5.1. Using the fact that the solution of $\dot{x} = Ax + f(t)$ can be written as

$$x(t) = e^{At} x(0) + \int_0^t e^{A(t-\sigma)} f(\sigma) \, d\sigma$$

For the corresponding Liouville equation and initial density ρ_0 the solution is

$$\rho(t, x) = \frac{1}{e^{\text{tr} At}} \rho_0 \left(e^{-At} \left(x - \int_0^t e^{A(t-\sigma)} f(\sigma) \, d\sigma \right) \right)$$

If we have a control present, as in $\dot{x} = f(x, u)$, the Liouville equation keeps this same form but because the control is now operated on by the partial derivative operator, feedback controls and open loop controls lead to different solutions.

Example 2.5.2. Consider the scalar equation $\dot{x} = u$; $x(0) = 1$. If we let $u(t, x) = -x$ then of course $x = e^{-t}$ and $u(t) = -e^{-t}$. We get the same solution if we set $u(t, x) = -e^{-t}$. On the other hand, if we have an initial density ρ_0 the solution of the Liouville equation corresponding $u(t) = -e^{-t}$ is

$$\rho(t, x) = \rho_0(x - e^t)$$

whereas the solution corresponding to $u(x) = -x$ is

$$\rho(t, x) = e^t \rho_0(e^{-t} x)$$

2.6 Comparison with the Fokker Planck Equation

We have suggested that one interpretation of the Liouville equation is that it provides a description of the evolution of a probability density under the deterministic flow defined by $\dot{x} = f(x, u)$. Of course there is also an evolution equation for the density associated with stochastic equations containing Wiener processes, such as those of the Itô form

$$dx = f(x, u)dt + \sum g_i(x)dw_i$$

The effect of the $g_i dw_i$ terms is to introduce a diffusion, something completely absent in the model provided by the Liouville equation. For the scalar equation $dx = axdt + cdw$ the Fokker–Planck equation is

$$\frac{\partial \rho(t, x)}{\partial t} = -\frac{\partial ax\rho}{\partial x} + \frac{1}{2}c^2 \frac{\partial^2 \rho(t, x)}{\partial x^2}$$

If the initial density is Gaussian, $\rho_0(x) = \left(1/\sqrt{2\pi s(0)}\right) e^{(x-\bar{x}(0))^2/2s(0)}$ then the solution of this equation remains Gaussian for all time

$$\rho(t, x) = \frac{1}{\sqrt{2\pi s(t)}} e^{(x-\bar{x})^2/2s(t)}$$

where $\bar{x}$ is $e^{at}\bar{x}(0)$ and $s(t)$ is the solution of the variance equation given $s(0)$

$$\dot{s} = 2as + c^2$$

This can be compared with the solution of the Liouville equation corresponding to $c = 0$ which, for the same initial condition, is

$$\rho(t, x) = \frac{1}{\sqrt{2\pi t e^{2at}}} e^{(x-\bar{x})^2 / 2e^{at}}$$

2.7 Sample Problems Involving the Liouville Equation

In the case where

$$\dot{x}(t) = f(x) + \sum u_i(t) g_i(x(t))$$

the Liouville equation is

$$\frac{\partial \rho(t, x)}{\partial t} = -\left\langle \frac{\partial}{\partial x}, f(x)\rho(t, x) \right\rangle - \sum \left\langle \frac{\partial}{\partial x}, u_i(t, x) g_i(x)\rho(t, x) \right\rangle$$

When solving a standard control problem modeled as $\dot{x} = f(x, u)$ one seeks a control u defined on the interval of interest. Often u will be found through the use of variational principles and may be found as a function of t or as a function of the optimal trajectory x. Whether u is expressed as a function of t alone or as some combination of t and x is regarded as being of secondary importance. However, the situation is quite different for the Liouville equation because now $f(x, u)$ is acted on by the partial derivatives with respect to x. The value of $\rho(t, x)$ depends on whether u is expressed as an open loop function ($u = u(t)$ or as a closed loop function ($u = u(t, x)$).

We now briefly describe a number of problems which can be phrased in terms of the Liouville equation even though they fall outside the usual theory of optimal control.

Problem 2.7.1. *The regulator in a box:* Just as one of the basic examples in quantum mechanics is the charged particle in a square well potential, we can consider control problems where the domain of interest is limited for technological reasons to a sharply defined interval. Suppose that there exist limitations such that values of x and u that lie outside a certain range are of no interest. We seek a control that has good performance and is easy to implement. Building on our earlier example, we consider the scalar control problem $\dot{x} = u$ with the distribution of initial conditions given by a density $\rho_0(x)$ which is uniform on $[-1, 1]$ and zero outside that interval. Find u as a function of x such as to minimize

$$\eta = \int_0^\infty \int_{-1}^1 \rho(t, x) a x^2 \, dx \, dt + \int_{\mathbb{R}} \left(\frac{\partial u}{\partial x} \right)^2 dx$$

Problem 2.7.2. *Maximizing the Domain of Attraction:* Consider the system $\dot{x} = f(x, u)$ with $f(0, 0) = 0$ having the property that the solution $(x, u) = (0, 0)$ is unstable. It is often of interest to determine u so as to make the null solution of $\dot{x} = f(x, u(x))$ asymptotically stable and to make its domain of attraction as large as possible. We can formulate this in terms of a controlled Liouville equation in the following way. For the equation

$$\frac{\partial \rho(t, x)}{\partial t} = -\left\langle \frac{\partial}{\partial x}, f(x, u)\rho(t, x) \right\rangle$$

Find u as a function of x so as to minimize

$$\eta = \int\limits_0^\infty \int\limits_{\mathbb{R}^n} \tanh(k\|x\|)\rho(t, x)\, dx dt$$

Notice that for large positive values of k this assigns zero cost to trajectories that go to zero as t goes to infinity.

Problem 2.7.3. *Trajectory Confinement:* In most models concerned with discretized control signals, asymptotic stability is not possible. In [5] we discussed the possibility of confinement to a region about the target value. This can be restated as requiring that the support of the density should be limited to some neighborhood of the target. If the target is $x = 0$ we might also reformulate the problem in terms of minimizing a measure such as

$$\eta = \int\limits_X x^2 \rho(T, x)\, dx$$

Problem 2.7.4. *Enhancing Controllability:* As we have seen, identical linear systems are not ensemble controllable in any reasonable sense. Yet with nonlinear feedback they can become so. We can ask about the nonlinearities that make the linear system ensemble controllable. Of course we need the Lie algebra generated by $Ax + bg(x), b$ to have enough independent components so as to achieve controllability. Moreover, we would like $[Ax, b(g \cdot b)]$ to be "strongly independent" in some sense, probably involving an average over the domain of interest. As noted in the example above, replicated systems are not controllable along the walls of the cone defined by the planes characterizing equality of components, but in the interior they can be.

Problem 2.7.5. *Restricted Range Feedback:* In our paper [5] we discussed the possibility of controlling a linear system with outputs that are generated by a finite state machine. The idea was to model the feedback controller as a Markov process and to adjust the transition rates of the Markov process in such a way as to achieve control. This can be contrasted with the older idea of pulse-width modulated control, commonly used in less sophisticated control systems, which operates in an on-off mode, with the switching times synchronized with a clock.

2.8 Controllability

Suppose that $\dot{x} = f(x, u)$ and that there is a density of initial conditions for x with support of ρ_0 being the set X_0. Suppose, further, that we would like to find $u(t, x)$ so as to steer ρ from its initial value to $\rho_1(x)$ whose support is X_1. For example, if, in fact, we have a regulator problem then X_1 could be a small set containing 0. If we have a cost function involving u we could arrive at a problem of the form

$$\frac{\partial \rho(t, x)}{\partial t} = -\left\langle \frac{\partial}{\partial x}, \ f(x, u)\rho(t, x) \right\rangle; \quad \rho(0, x) = \rho_0; \ \text{Supp } \rho(T, x) \in X_1$$

$$\eta = \int\limits_0^T \int\limits_X L(u(x))\rho(t, x)dxdt$$

with the goal of minimizing η.

In other situations the final density might be completely specified or it might be that certain linear functionals of it are to satisfy some inequalities. It might happen that L depends on x as well as u, etc. Some concrete examples appear elsewhere in these notes.

Let X be an oriented differentiable manifold with a fixed, nondegenerate, volume form dv. Let $\phi : X \to X$ be a diffeomorphism. If ρdv is a nonnegative measure on X and if ϕ is orientation preserving, then ϕ acts on densities according to

$$\rho(\cdot) \mapsto \rho(\phi^{-1}(\cdot))/\det J_\phi$$

where J_ϕ is the Jacobian of ϕ. In this sense $\text{Diff}_O(X)$, the set of orientation preserving diffeomorphisms, generates an orbit through a given ρ.

If the manifold is compact and we restricted discussion to strictly positive densities then this action is transitive, see Moser [6] and Dacorogna and Moser [7]. If the densities are only assumed to be nonnegative the situation is much more complicated.

A natural question to ask is then, given two nonnegative densities, ρ_0 and ρ_1, each of which integrates to one, does there exist a control vector $u(t, x)$ defined on $[0, T]$ that steers ρ_0 to ρ_1? From the point of view that the Liouville equation defines an evolution equation on $L_1(\mathbb{R}^n)$, It might be expected that in considering this question the Lie algebra generated by the first order linear operators

$$\mathcal{L} = \left\{ -\left\langle \frac{\partial}{\partial x}, f(x)\rho(t, x) \right\rangle, \ \left\langle \frac{\partial}{\partial x}, u_i(t, x)g_i(x)\rho(t, x) \right\rangle \right\}_{LA}$$

should play a role. However, because the bracket

$$\left[\left\langle \frac{\partial}{\partial x}, u_i(t, x)g_i'(x)\rho(t, x) \right\rangle, \left\langle \frac{\partial}{\partial x}, u_j(t, x)g_j(x)\rho(t, x) \right\rangle \right]$$

involves the partial derivatives of $u(x)$, and deeper brackets involve successively higher partial derivatives, this line of attack leads to complications.

Of course the set of operators of the form

$$L = \left\langle \frac{\partial}{\partial x}, u_i(t, x) g_i(x) \right\rangle$$

as u varies over the set of $\mathscr{C}^\infty$ functions of x is an infinite dimensional set. We could reformulate the problem in this way. Let $\psi_i(x) \in \mathscr{C}^\infty(\mathbb{R}^n)$ be a basis for some subset of $\mathscr{C}^\infty$ and consider vector fields of the form

$$L_j = \left\langle \frac{\partial}{\partial x}, \sum \psi_j(x) g_i(x) \right\rangle$$

This is to be compared with the controllability of the system

$$\dot{x} = f(x) + \sum u_i g_i(x)$$

for which the relevant Lie algebra is

$$\mathscr{L} = \{f, g_1, g_2, \dots, g_m\}_{LA}$$

In our paper [8] we studied the problem of controllability of the density equation associated with linear systems. More recently Agrachev and Caponigro [9] published a study phrased in terms of controlling diffeomorphisms, not restricted to linear systems.

2.9 Optimization with Implementation Costs

Not surprisingly, the addition of an implementation term usually complicates the mathematics required to solve a trajectory optimization problem.

Example 2.9.1. Consider the problem of minimizing the quantity

$$\eta = \int\limits_0^\infty x^2 + u^2 \, dt$$

while steering the solution of $\dot{x} = -x + u$ from $x(0) = 10$ to $x(1) = 0$. Of course a variational argument implies immediately that

$$\ddot{x} - 2x = 0$$

and together with the boundary conditions on x this determines the optimal trajectory. But another way to solve this problem is to find a solution of the Riccati equation

$$\dot{k} = 2k - 1 + k^2$$

on the interval $[0, 1]$ and to make the substitution $u = v - kx$. It then follows that the original trajectory optimization problem is equivalent to a modified one for which the evolution equation is

$$\dot{x} = (-1 - k)x + v \; ; \quad \eta = \int_0^1 v^2 \, dt$$

and the performance measure is

$$\eta = \int_0^1 v^2 \, dt$$

The optimal v is then expressible in terms of the controllability Gramian W associated with the new system. Matters being so, optimizing v leads to an expression for u. In more detail,

$$v(t) = e^{\int_0^t (-1 - k(\tau)) d\tau} p \implies u = -k(t)x(t) + v(t)$$

This solution has both open loop, and closed loop terms. Their relative size depends on which solution of the Riccati equation is chosen. The above construction works for any solution of the Riccati equation and includes the possibility that we choose an equilibrium solution. This choice could be made with the goal of minimizing some functional of the form

$$\eta = \int_0^1 L(\dot{k}, \dot{v}) \, dt$$

such as

$$\eta = \int_0^1 (\partial u / \partial t)^2 \, dt$$

Example 2.9.2. As an example of a problem in this setting that is solvable in special cases, consider

$$\dot{x} = f(x) + g(x)u$$

For this system

$$\frac{\partial \rho(t, x)}{\partial t} = -\left\langle \frac{\partial}{\partial x}, (f(x) + g(x)u)\rho(t, x) \right\rangle$$

with a given initial probability density $\rho_0(x)$. Suppose we consider a trajectory term

$$\eta_p = \int_0^\infty \int_{\mathbb{R}^n} (x^T L x + u^T u) \rho(t, x) \, dx \, dt$$

and an implementation penalty that favors a linear control.

$$\eta_i = \int_0^\infty \int_{\mathbb{R}^n} \left\| \frac{\partial u}{\partial x} x - u(x) \right\|^2 \rho(t, x) \, dx \, dt$$

It is obvious that in the special case where $f(x) = Ax$ and $g(x) = b$ the optimal solution is

$$u = -B^T K x$$

where K satisfies $A^T K + KA - L + KBB^T K = 0$. More interesting is the suggestion that if $\| f(x) - Ax \|$ and $\| g(x) - b \|$ are not too large in the region of interest then we can use the known solution as the initial guess in a successive approximation scheme.

In the context of this example there are three distinct aspects of the linear case are worth noting. i) The pure trajectory optimization is solvable in feedback form, ii) the implementation term adds no additional cost at precisely at the optimal feedback control, and iii) the form of the initial distribution is irrelevant. Generalizing the problem in such a way as to take away any of these will yield more interesting solutions.

2.10   Controlling the Variance

We now turn our attention to questions involving the simultaneous use of open loop and closed loop terms to shape the first and second moments of the density. This can be thought of as part of the larger problem of controlling the Liouville equation. For linear stochastic systems this amounts to controlling the mean and the variance and represents a compromise between controlling one individual trajectory associated with $\dot{x} = f(x, u)$ and controlling the entire density. It is, perhaps, the simplest set of problems illustrating how the parametrization of the control as a sum of an open loop part plus a closed loop part can provide additional controllability beyond what is available using open loop control alone. For simplicity, we suppose that the uncontrolled system is linear and time invariant; the extension to the time varying case presents little additional difficulty.

Consider the stochastic system

$$dx = Ax \, dt + Bu \, dt + G \, dw$$

Let $\bar{x}$ and Σ denote the corresponding mean and variance so that with the control law $u(t) = K(t)x(t) + u_0(t)$ we have

$$\frac{d}{dt}\bar{x} = (A + BK)\bar{x} + Bu_0$$

$$\dot{\Sigma} = (A + BK(t)\Sigma + \Sigma(A + BK(t))^T + Q$$

with $Q = GG^T$. We now investigate the set of reachable values for $\bar{x}$ and Σ, considering K and u_0 to be controls.

In thinking about controlling the variance, it is helpful to keep in mind that the set of positive semidefinite matrices is both a cone and an additive semigroup and that any vector field of the form $F(\Sigma) = A\Sigma + \Sigma A^T$ maps this cone into itself. Moreover, the general linear group acts transitively on the set of positive definite matrices in accordance with the group action

$$(T, Q) \mapsto TQT^T$$

Of course there is a large literature devoted to the steady state solution of the variance equation, going back to Wiener's work on filtering and continuing with the celebrated linear-quadratic-Gaussian theory developed in the context of modern control theory . Much of this work is devoted to questions about how to minimize the variance through the choice of constant K. Here we are interested in treating $K(t)$ as a control and focusing on the transient behavior.

Remark 2.10.1. As motivation consider the following type of problem. Suppose that an athlete has an objective such as placing the ball with a tennis serve or gaining a certain height as a pole vaulter. The penalty for missing the objective may be highly nonlinear and the number of tries limited. Thus the best policy typically involves a tradeoff between controlling the mean and controlling the variance. If the only uncertainty enters through the initial state, the problem can be phrased in the terms described above.

The feedback gains K enter the variance equation multiplicatively and hence this is an example of what has come to be called bilinear control. The presence of the bias term Q and the constraint imposed by the fact that the variance is automatically positive semidefinite sets this problem apart from much of the literature. In the appendix we give some results on the general bilinear problem but here we focus on the variance equation itself. We will make use of the idea that when studying controllability for systems with a drift term, if the drift vector field generates a periodic motion then the effect of moving backwards along the drift vector field can be achieved by letting the system flow along the drift vector field for something less than a full period. This idea was used by Jurdjevic and Sussmann [10] in the context of control on Lie groups and later, without the Lie group hypothesis, in [11].

Lemma 2.10.2. *Let A be a real n-by-n matrix and let B be n-by-m. If A, B is a controllable pair in the sense that the rank of $[B, AB, \ldots, A^{n-1}B]$ is n then, considering K as a time varying control, the system*

$$\dot{\Sigma} = (A + BK(t))\Sigma + \Sigma(A + BK(t))^T + Q; \quad \Sigma(0) \geq 0$$

has the property that the reachable set from any $\Sigma(0) > 0$ has nonempty interior in the space of symmetric positive definite matrices.

Proof. Step 1: Clearly the Lie algebra generated by the matrices A and BK for all possible constant K, contains A and every matrix whose range space is contained in the range space of B. It also contains all matrices of the form $ABK - BKA$. However, the range space of BKA is contained in the range space of B and so we see that the Lie algebra in question contains all matrices of the form ABK. It also contains all matrices whose range space is AB as well as those whose range space is contained in the range space of B. Continuing with $[A, ABK] = A^2 BK - ABKA$, etc. we see the Lie algebra contains all matrices whose range space is contained in the sum of the ranges of $B, AB, \ldots, A^n B$ which is the entire Lie algebra of n-by-n matrices.

Step 2: In the case where $Q = 0$ and K is piecewise constant on $[0, t]$ we have

$$\Sigma(t) = M\Sigma(0)M^T$$

where

$$M = e^{(A+BKr)t_r} e^{(A+K_{r-1})t_{r-1}} \cdots e^{A+BK_1)t_1}$$

Thus with $Q = 0$ the given equation is controllable on the space of symmetric matrices with rank and a signature matching that of $\Sigma(0)$, provided that the matrix equation $\dot{X} = (A+BK)X(t)$ is controllable on the space of nonsingular matrices. In particular, it is controllable on the space of symmetric, positive definite matrices.

Step 3: The effect of Q is simply to offset the solution in accordance with the variation of constants formula

$$\Sigma(t) = \Phi(t, 0)\Sigma(0)\Phi^T(t, 0) + \int_0^t \Phi(t, \tau)Q\Phi^T(t, \tau)d\tau$$

and thus even with $Q \neq 0$ the reachable set retains the property of containing an open set. $\qquad\square$

Remark 2.10.3. Theorem 1 of [12] provides a complete characterization of the Lie algebra generated by A and bc^T, under the assumption that (A, b, c) is controllable and observable. In particular, it is established there that if the trace of $A + \alpha bc^T$ is nonzero for some α and if $c^T(I(s+\alpha)-A)^{-1}B$ is not equal to $c^T(-I(s+\alpha)-A)^{-1}b$ for any α then the Lie algebra generated by A and bc^T is the set of all of n-by-n matrices. Observe that in the present situation we can choose K such that the trace of BK is nonzero and by virtue of the controllability assumption we can select a rank one matrix K such that $BK = bc^T$ meets these requirements. We give a general result later (Theorem 2.10.6) but perhaps a concrete example will be helpful at this point.

Example 2.10.4. Consider the two-by-two variance equation associated with

$$\begin{bmatrix} dx_1 \\ dx_2 \end{bmatrix} = u \begin{bmatrix} 0 & 1 \\ 0 & 0 \end{bmatrix} \begin{bmatrix} x_1 \\ x_2 \end{bmatrix} dt + \begin{bmatrix} 0 \\ 1 \end{bmatrix} u + \begin{bmatrix} 0 \\ dw \end{bmatrix}$$

If we let $u = k_1 x_1 + k_2 x_2$ the corresponding variance equation is

$$\frac{d}{dt} \begin{bmatrix} \sigma_{11} & \sigma_{12} \\ \sigma_{21} & \sigma_{22} \end{bmatrix} = \begin{bmatrix} 0 & 1 \\ k_1 & k_2 \end{bmatrix} \begin{bmatrix} \sigma_{11} & \sigma_{12} \\ \sigma_{21} & \sigma_{22} \end{bmatrix} + \begin{bmatrix} \sigma_{11} & \sigma_{12} \\ \sigma_{21} & \sigma_{22} \end{bmatrix} \begin{bmatrix} 0 & k_1 \\ 1 & k_2 \end{bmatrix} + \begin{bmatrix} 0 & 0 \\ 0 & 1 \end{bmatrix}$$

We want to show that this equation is controllable on $\Sigma > 0$.

Write the equations in component form

$$\dot{\sigma}_{11} = 2\sigma_{12}$$

$$\dot{\sigma}_{12} = k_1\sigma_{11} + k_2\sigma_{12} + \sigma_{22}$$

$$\dot{\sigma}_{22} = 2k_1\sigma_{12} + 2k_2\sigma_{22} + 1$$

Positive definiteness can be characterized by $\sigma_{11} > 0$ and $\sigma_{11}\sigma_{22} > \sigma_{12}^2$. Observe that given (u_1, u_2), the simultaneous equations

$$\begin{bmatrix} u \\ v \end{bmatrix} = u \begin{bmatrix} \sigma_{11} & \sigma_{12} \\ 2\sigma_{12} & 2\sigma_{22} \end{bmatrix} \begin{bmatrix} k_1 \\ k_2 \end{bmatrix} dt + \begin{bmatrix} \sigma_{22} \\ 0 \end{bmatrix}$$

can be solved for (k_1, k_2) in the set $\Sigma > 0$ and if we make the corresponding replacements we have

$$\dot{\sigma}_{11} = 2\sigma_{12}$$

$$\dot{\sigma}_{12} = u$$

$$\dot{\sigma}_{22} = v$$

Now the first two of these equations depend on u alone; $\ddot{\sigma}_{11} = 2u$ and $\sigma_{12} = \dot{\sigma}_{11}/2$. It is clear, for example from the classical treatment of the time-optimal control of $\ddot{x} = u$, the point $(\sigma_{11}(0), \sigma_{12}(0))$ can be steered to any point in the half-plane $\sigma_{11} > 0$

Fig. 2.2 Showing possible
trajectories respecting
$\sigma_{11} > 0$ in the $(\sigma_{11}, \sigma_{12})$-plane

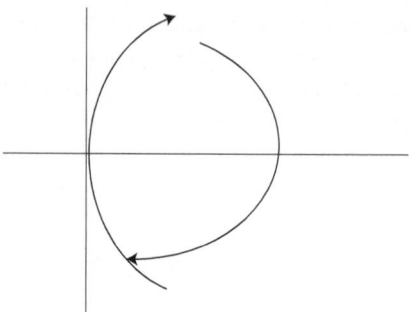

without leaving that half-plane. Suppose we select some u that accomplishes this
transfer in T units of time (Fig. 2.2). Define h by

$$\frac{\sigma_{12}^2(T)}{\sigma_{11}(T)} - \frac{\sigma_{12}^2(0)}{\sigma_{11}(0)} = h/T$$

Finally choose v to be the time derivative of

$$\sigma_{22}(t) = \sigma_{22}(0) + \frac{\sigma_{12}^2(0)}{\sigma_{11}(0))} + \frac{t}{T}\left(\frac{\sigma_{12}^2(t)}{\sigma_{11}(t)} - \frac{\sigma_{12}^2(0)}{\sigma_{11}(0))}\right)$$

More generally, consider the variance equation associated with an nth order
system with scalar control. Let e_i denote the standard basis vectors in $\mathbb{R}^n$ $dx =
(A + e_n k^T)x dt + e_n dw$. If we partition the variance in terms of blocks compatible
with e_n and its complement we have

$$\frac{d}{dt}\begin{bmatrix} \Sigma_{11} & \Sigma_{12} \\ \Sigma_{12}^T & \sigma_{22} \end{bmatrix} = \begin{bmatrix} S & 0 \\ k^T & k_n \end{bmatrix}\begin{bmatrix} \Sigma_{11} & \Sigma_{12} \\ \Sigma_{12}^T & \sigma_{22} \end{bmatrix} + \begin{bmatrix} \Sigma_{11} & \Sigma_{12} \\ \Sigma_{12}^T & \sigma_{22} \end{bmatrix}\begin{bmatrix} S^T & k \\ 0 & k_n \end{bmatrix} + \begin{bmatrix} 0 & 0 \\ 0 & 1 \end{bmatrix}$$

Using a linear transformation and a suitable offset for k we can arrange matters
so that A and $e_n k^T$ take the form

$$A = \begin{bmatrix} 0 & 1 & 0 & \cdots & 0 \\ 0 & 0 & 1 & \cdots & 0 \\ 0 & 0 & 0 & \cdots & 0 \\ \cdots & \cdots & \cdots & \cdots & \cdots \\ 0 & 0 & 0 & \cdots & 1 \\ 0 & 0 & 0 & \cdots & 0 \end{bmatrix} ; \quad e_n k^T = \begin{bmatrix} 0 & 0 & 0 & \cdots & 0 \\ 0 & 0 & 0 & \cdots & 0 \\ 0 & 0 & 0 & \cdots & 0 \\ \cdots & \cdots & \cdots & \cdots & \cdots \\ 0 & 0 & 0 & \cdots & 0 \\ k_1 & k_2 & k_3 & \cdots & k_n \end{bmatrix}$$

Observe that if Σ is positive definite then the equation

$$u = \Sigma k + b$$

can be solved for k and that in terms of u the variance equation can be written as

$$\dot{\Sigma}_{11} = S\Sigma_{12} + \Sigma_{11}S^T$$

$$\dot{\Sigma}_{12} = S\Sigma_{12} + u$$

$$\dot{\sigma}_{22} = u$$

This is a more general formulation of the example. In this notation the problem is that of showing that for

$$\dot{\Sigma} = A\Sigma + \Sigma A^T + e_n u^T + u e_n^T$$

it is possible to steer Σ from $\Sigma_0 > 0$ to $\Sigma_1 > 0$.

The proof of the following theorem shows that this is possible for a general controllable linear system.

Theorem 2.10.5. *Let (A, B) be a controllable pair and let Σ satisfy*

$$\dot{\Sigma} = (A + BK(t))\,\Sigma + \Sigma\,(A + BK(t))^T + Q; \quad \Sigma(0) > 0; \quad Q \geq 0$$

Considering K to be a control, any $\Sigma_1 > 0$ can be reached from any $\Sigma(0) > 0$.

Proof. Clearly the variance equation is linear and the operator mapping real symmetric matrices into real symmetric matrices defined by

$$L(\Sigma) = (A + BK)\Sigma + \Sigma(A + BK)^T$$

has eigenvalues which are all possible pairs of the form $\lambda_i + \lambda_j$ where λ_i and λ_j are eigenvalues of $A + BK$. Thus if there exists a K such that the eigenvalues of $(A + BK)$ are integer multiples of $\mu\sqrt{-1}$ then $\exp L$ is periodic and Theorem A1 of the appendix applies, provided that $e^{(A+BK_0C)t}$ is periodic for some choice of K_0 □

Theorem 2.10.6. *Assume that (A, B) is a controllable pair. The system of equations*

$$\dot{x}(t) = (A + BK(t))x + Bu(t)$$

$$\dot{\Sigma}(t) = (A + BK(t))\Sigma(t) + \Sigma(t)(A + BK(t))^T$$

is controllable in the sense that given any two pairs (x_0, Σ_0) and (x_1, Σ_1) with $\Sigma_0 = \Sigma_0^T > 0$ and $\Sigma_1 = \Sigma_1^T > 0$ and given any time $T > 0$ there exists a control (u, K) defined on $[0, T]$ steering the system from (x_0, Σ_0) to (x_1, Σ_1).

Proof. Select K in accordance with Theorem 1 so as to steer Σ to the desired state. Having selected K, select u by standard controllability arguments to steer x.

Going beyond controllability, there are a variety of optimization questions that arise this context. The most basic are the extensions of the problem considered in the previous section involving the minimization of

$$\eta = \int_0^T L(u_0, k, \dot{k}) \, dt$$

while using the control law $u = u_0 + kx$ to force the solution of $\dot{x} = f(x, u)$ to move from $x(0) = x_0$ to $x(T) = x_1$. Extending this idea to a stochastic setting, we can, for example, consider controlling the mean and variance equation as in Theorem 2.10.6, while minimizing

$$\eta = \int_0^T L(u_0, K, \dot{K}) \, dt \qquad \qquad \square$$

2.11   Ensembles, Symmetric Functions and Thermodynamics

This section is adapted from our paper [2]. It can be seen as taking the idea of simultaneous control of the mean and variance in a new direction.

Let u be a m-dimensional vector, let x_i for $i = 1, 2, \ldots, k$ be a n-dimensional vector, and let y be a p-dimensional vector. Consider a system consisting of k copies of a first order model, each with the same input

$$\dot{x}_i = f(x_i, u); \quad i = 1, 2, \ldots, k$$

We limit our attention to outputs of the form

$$y = c(x_1, x_2, \ldots, x_k)$$

with c being a symmetric function in the sense that for any permutation of the index set $\{1, 2, \ldots, k\} \rightarrow \{\pi(1), \pi(2), \ldots, \pi(k)\}$ we have $c(x_1, x_2, \ldots, x_k) = c(x_{\pi(1)}, x_{\pi(2)}, \ldots, x_{\pi(k)})$. If the system is stochastic we replace this model with a family of Itô equations of the form

$$dx_i = f(x_i, u)dt + g(x_i, u)dw_i; \quad i = 1, 2, \ldots k$$

with the Wiener processes $w_1, w_2, \ldots, w_k$ being independent.

Of course there are significant limitations that arise in the control of such systems because u acts on each system in the same way and y is constrained to be a symmetric function. In particular, linear systems of this type are never controllable or observable if $k > 1$.

If the elemental systems are linear then the overall system obtained by applying feedback $u = \sum C x_i$ is described by

$$\begin{bmatrix} \dot{x}_1 \\ \dot{x}_2 \\ \cdots \\ \dot{x}_k \end{bmatrix} = \begin{bmatrix} A+BC & BC & \cdots & BC \\ BC & A+BC & \cdots & BC \\ \cdots & \cdots & \cdots & \cdots \\ BC & BC & \cdots & A+BC \end{bmatrix} \begin{bmatrix} x_1 \\ x_2 \\ \cdots \\ x_k \end{bmatrix}$$

The matrix on the left is similar to a block triangular matrix with $n-1$ diagonal blocks of the form $A+BC$ and hence can not be controllable. A similar limitations apply to stochastic models of the form

$$dx_i = Ax_i\, dt + Bu\, dt + Gdw_i; \quad dy = C(x_1 + x_2 + \cdots + x_k)dt + dv$$

We discuss these in Theorem 2.11.1 after giving a few additional definitions. In a probabilistic setting it is meaningful to discuss statistical properties such as the mean and variance. In the case of many copies of a given system we can consider various averages taken over the ensemble. Of course the sample statistics, as opposed to the statistics themselves, are random variables. In the present situation, with $x_1, x_2, \ldots, x_k$ being described by identical probability laws, we have an interest in a particular type of sampling leading to what can be termed *ensemble sample-statistics*. This refers to averages over the variables $x_1, x_2, \ldots x_k$. For example, we refer to the random variable

$$a(t) = \frac{x_1(t) + x_2(t) + \cdots + x_k(t)}{k}$$

as the ensemble sample-mean.

We say that a homogeneous function $\phi(x_1, x_2, \ldots x_k)$ is *centered* if the sum of its partial derivatives vanishes, i.e.,

$$\sum_{i=1}^{k} \frac{\partial \phi}{\partial x_i} = \sum_{i=1}^{k} \begin{bmatrix} \partial\phi/\partial x_{i1} \\ \partial\phi/\partial x_{i2} \\ \cdots \\ \partial\phi/\partial x_{in} \end{bmatrix} = 0$$

Theorem 2.11.1. *Consider the linear stochastic ensemble*

$$dx_i = Ax_i\, dt + Bu\, dt + Gdw_i; \quad i = 1, 2, \ldots, k$$

The application of feedback in the form $u = \alpha(x)$ does not change the evolution equation of any centered homogeneous function of $x_1, x_2, \ldots x_k$.

Proof. Let ϕ be homogeneous and centered. Applying the Itô rule to ϕ we see that

$$d\phi(x) = \sum \langle \nabla\phi, Ax_j\, dt + Bu\, dt + Gdw_j \rangle + \frac{1}{2} \sum \left\langle \frac{\partial^2 \phi}{\partial x_{ij}}, GG^T \right\rangle$$

Clearly under the given hypothesis the effect of u disappears. □

Corollary. The ensemble sample variance

$$\Sigma_{esv} = \frac{1}{n} \sum \left(x_i - \frac{1}{n}(x_1 + x_2 + \cdots + x_k) \right) \left(x_i - \frac{1}{n}(x_1 + x_2 + \cdots + x_k) \right)^T$$

associated with the system

$$dx_i = Ax_i dt + Bu dt + G dw_i; \quad i = 1, 2, \ldots k$$

is not altered by feedback of the form $u = \phi(x)$.

Proof. It is easy to see that each term in the sum defining Σ_{esv} is homogenous and centered and therefore the sum is as well.

These results show that it is necessary to go beyond linear theory if we are to find any benefit from the use of control in the ensemble setting. The following theorem applies to multiplicative control. □

Theorem 2.11.2. *Consider the ensemble*

$$dx_i = Ax_i + uBx_i dt + G dw_i; \quad i = 1, 2, \ldots, k$$
$$y = \frac{1}{n}(x_1^T Lx_1 + x_2^T Lx_2 + \cdots + x_k^T Lx_k)$$

with $L = L^T > 0$. If there exists a symmetric matrix Q such that $QB + B^T Q$ is negative definite and the eigenvalues of Q all have the same sign, then there exists $\beta > 0$ such that for any real c between 0 and β there is a feedback control law $u = \phi(y)$ which stabilizes the trace of the variance of the sample variance at c.

Proof. The variance of the sample variance, i.e.

$$\Sigma_{esv} = \sum \left(x_i - \frac{1}{n}(x_1 + x_2 + \cdots + x_k) \right) \left(x_i - \frac{1}{n}(x_1 + x_2 + \cdots + x_k) \right)^T$$

satisfies the equation

$$\dot{\Sigma}_{esv} = (A + uB)\Sigma_{esv} + \Sigma_{esv}(A + uB)^T + GG^T$$

Let $QB + B^T Q = R < 0$. Then $Q(A+uB)+(A+uB)^T Q$ is negative definite for suitable choice of u. Thus we see that for a semi-infinite range of u the eigenvalues of $A + uB$ have negative real parts. In fact, $Q(A + uB) + (A + uB)^T Q$ can be made more negative definite than $-\alpha I$ for any α and so the eigenvalues of $A + uB$ can be placed to the left of any vertical line in the complex plane. This means that a steady state variance exists and satisfies

$$0 = (A + uB)\Sigma_{esv} + \Sigma_{esv}(A + uB)^T + GG^T$$

Over the range of u for which the system is stable let $ly_{A+uB}^{-1}(GG)$ denote the solution of this equation. Clearly $\mathrm{tr}(ly_{A+uB}^{-1}(GG))$ goes to zero as the eigenvalues of $A + uB$ go to minus infinity and so, as u varies $\mathrm{tr}\,\Sigma_{esv}$ sweeps out a range of values of the form $(0, c)$ as required. We can let the feedback control law be a constant, independent of y. □

The control of heat engines and provides a good example of ensemble control. The mathematical description consists of a family of identical scalar linear systems with multiplicative control driven by independent Brownian motion terms.

$$dx_i = (u_1 - u_2)x_i(t)dt + u_3 dw_i; \quad i = 1, 2, \ldots, k$$

$$y = \sum_{i=1}^{k} x_i^2$$

In this case the ensemble equations are supplemented by two auxiliary equations which complete the description and serve to distinguish u_1 from u_2. These are

$$\dot{x}_{k+1} = u_1$$

$$\dot{x}_{k+2} = u_1 y$$

The physical interpretation is as follows. The x's represent (one dimensional) velocities of individual particles in the ensemble. The controls represent the time rate of change of the volume occupied by the gas (u_1), the type of contact the gas has with the available heat sources (u_2), and the selection of a heat source with a particular temperature (u_3). Further details will emerge from the discussion. If we had the services of a Maxwell demon we could observe each of the x_i individually but in reality only certain ensemble averages are observable. Likewise, if we had access to a demon we could generate individual controls for each of the state variables but in reality we can only apply controls which influence all elements of the ensemble in the same way. In the context of the elementary thermodynamics of gases, we are able to change the volume of the gas by moving a piston and to alter the internal energy of the gas by adding or removing heat. Such actions translate into choices of u_1, u_2, u_3 in the above model. The objective of the control action might be, for example, to cause the development of a given quantity of work over a period $[0, P]$. Here work corresponds to the integral

$$w = \int_0^P u_1(t)y(t)dt$$

The relevant summary of the behavior of the population is, in this case, provided by the ensemble sample variance. From the equations for $x_1, x_2, \ldots, x_k$ we see that the ensemble sample variance satisfies the stochastic differential equation

$$d\sigma_{esv} = 2(u_1 - u_2)\sigma_{esv}dt + u_3 \frac{1}{n}\sum dw_i + u_3^2 dt$$

Because we have assumed that the w_i are independent the sum appearing here may be replaced by $dw/\sqrt{k}$ without changing the statistical properties of the solution. That is, we may as well adopt the model

$$d\sigma_{esv} = 2(u_1 - u_2)\sigma_{esv}dt + u_3\frac{1}{\sqrt{k}}dw + u_3^2dt$$

This stochastic differential equation represents a sample statistic obtained from k samples. If we are dealing with a mole of gas then $k \approx 6 \times 10^{23}$! If we assume that the Brownian motion term is insignificant we are led to the set of deterministic equations

$$\frac{d}{dt}\sigma_{esv}(t) = 2(u_1 - u_2)\sigma_{esv}(t) + u_3^2$$

$$\dot{x}_{k+1} = u_1$$

$$\dot{x}_{k+2} = u_1 y$$

as a reduced model for the stochastic ensemble.

For the sake of simplicity we rename the variables $(\sigma_{esv}, x_{k+1}, x_{k+2})$ as (x_1, x_2, x_3). The control terms enter these equations in such a way as to define three vector fields as brought out by the notation

$$\begin{bmatrix} \dot{x}_1 \\ \dot{x}_2 \\ \dot{x}_3 \end{bmatrix} = u_1 \begin{bmatrix} 2x_1 \\ 1 \\ 2x_1 \end{bmatrix} - u_2 \begin{bmatrix} 2x_1 \\ 0 \\ 0 \end{bmatrix} + u_3^2 \begin{bmatrix} 1 \\ 0 \\ 0 \end{bmatrix}$$

subject to the constraint that u_2 should be nonnegative. The three vector fields appearing here are

$$A = 2x_1\frac{\partial}{\partial x_1} + \frac{\partial}{\partial x_2} + 2x_1\frac{\partial}{\partial x_3}; \quad B = -2x_1\frac{\partial}{\partial x_1}; \quad C = \frac{\partial}{\partial x_1}$$

Together with the pair

$$D = x_1\frac{\partial}{\partial x_3}; \quad E = \frac{\partial}{\partial x_3}$$

they obey the commutation relations

$$[A, B] = 4D; \quad [A, C] = -2C - 2E; \quad [B, C] = -2C$$

$$[A, D] = -[B, D] = 2D; \quad [A, E] = 0$$

and these five vector fields define a basis for a Lie algebra. This algebra is a solvable subalgebra of the algebra corresponding to the three dimensional affine group. One might say that it is *the* Lie algebra of the Carnot cycle. Constraining $u_2(t)$ to be nonnegative, this system generates admissible flows.

Appendix

We collect here a few results on the bilinear controllability putting in a larger context the result of Sect. 2.10 on the control of the mean and variance. There is a large literature on this subject and we only touch a few points. References [10–15] are relevant.

As is well known, if the off-diagonal elements of $A(t)$ are nonnegative for all t then the solutions of the system $\dot{x}(t) = A(t)x(t)$ leave the positive orthant invariant. Thus if b is a vector with nonnegative entries and $\dot{x}(t) = (A(t) + U(t))x(t) + b(t)$ with U diagonal but otherwise unconstrained, and $A(t)$ is nonnegative off the diagonal then the positive orthant is an invariant set. This can be seen as a being a consequence of the direction of the vector field along the boundary of the positive orthant. In a similar way, if a symmetric matrix X satisfies $\dot{X} = A(t)X(t) + X(t)A^T(t) + B(t)$ with $B(t) = B^T(t)$ nonnegative definite then the cone of nonnegative definite matrices is an invariant set.

Thus, in the case of the scalar system $\dot{x} = (a + k)x + b$ with $b > 0$ the set $\{x | x > 0\}$ is positively invariant and x cannot leave the positive half-line. It is controllable there in the sense that any point in $\{x | x > 0\}$ can be steered in positive time to any other point in the set. In higher dimensions the situation is more complicated. For example, if b_1 and b_2 are positive then solutions of the system

$$\frac{d}{dt}\begin{bmatrix} x_1 \\ x_2 \end{bmatrix} = \begin{bmatrix} a + k_1 & 1 \\ 1 & b + k_2 \end{bmatrix}\begin{bmatrix} x_1 \\ x_2 \end{bmatrix} + \begin{bmatrix} b_1 \\ b_2 \end{bmatrix}$$

can never leave the first orthant regardless of the choice of (k_1, k_2) but if $b \neq 0$ the system

$$\frac{d}{dt}\begin{bmatrix} x_1 \\ x_2 \end{bmatrix} = \begin{bmatrix} 0 & a + k \\ -a - k & 0 \end{bmatrix}\begin{bmatrix} x_1 \\ x_2 \end{bmatrix} + \begin{bmatrix} b_1 \\ b_2 \end{bmatrix}$$

can be steered from any initial state in $\mathbb{R}^2$ to any final state.

In studying the controllability of an n-dimensional system

$$\dot{x} = (A + BK(t)C)x(t)$$

it is natural to appeal to Lie algebraic methods. In [12] it is shown that if (A, b, c) is a minimal triple in the sense that $[b, Ab, \dots, A^{n-1}b)]$ and $[c; cA; \cdots cA^{n-1}]$ are both of rank n then the Lie algebra generated by A and bc is either $gl(n)$, $sl(n)$, $sp(n/2)$ or $sp(n/2) \oplus I$, and is $gl(n)$ unless $g(s) = c(Is - A)^{-1}b$ has a reflection symmetry in the form $g(s + \sigma) = g(-s - \sigma)$ for some real number σ and $\text{tr}A = cb = 0$. Adapting that result to the present situation, we see that the Lie algebra generated by A and matrices of the form $BE_{ij}C$ is $gl(n)$ unless $CB = 0$, tr$A = 0$, and $C(Is - A - \sigma I)^{-1}B = C(-Is - A + \sigma I)^{-1}B$ for some real number σ. However, because the system has an irreversible drift term $\dot{x} = Ax$ the Lie algebra does not tell the whole story.

Theorem A1. *Let A, B, C be constant matrices with $A, B)$ being controllable and (A, C) being observable. Consider the system evolving in the space of n-by-n nonsingular matrices with positive determinant.*

$$\dot{X} = (A + BKC)X$$

with B, K, C being n-bym, m-by-p and p-by-n, respectively. Assume that the solution of $\dot{X} = (A + BKC)X$; $X(0) = I$ is periodic for some $K = K_0(\cdot)$. Then given any pair of nonsingular matrices with positive determinant there exists a K that steers one to the other provided that CB and tr$A$ are not both zero and $C(Is - A)^{-1}B$ does not have the reflection symmetry described above.

Proof. First of all the system is controllable in the sense that it is possible to reach an open set of nonsingular matrices because A and the possible values of BKC generate the entire Lie algebra $gl(n)$. Second, as is well known, form early work on controllability on Lie groups, if $\dot{X} = (A + BKC)X$ has a periodic solution with X nonsingular periodic and we have local controllability then we have global controllability.   □

Remark A1. Let A be n-by-n and b be 1-by-n. Observe that

$$M(t) = \exp \begin{bmatrix} At & bt \\ 0 & 0 \end{bmatrix} = \begin{bmatrix} e^{At} & e^{\int_0^t e^{A(t-\sigma)}d\sigma}b \\ 0 & 1 \end{bmatrix}$$

If e^{At} is periodic with period T then its eigenvalues lie on the imaginary axis. If none are zero then A is invertible and

$$e^{\int_0^T e^{A(T-\sigma)}d\sigma}b = A^{-1}(I - e^{-AT})b = 0$$

Thus M is periodic. If 0 is in the spectrum of A then the explicit form of the integration is not available. However, if b lies in the range space of A then we can write b as Av so that

$$e^{\int_0^t e^{A(t-\sigma)}d\sigma}b = e^{\int_0^t e^{A(t-\sigma)}d\sigma}Av = (I - e^{-AT})b = 0$$

and M is periodic. When restricted to evolution equations on $\mathbb{R}^n$ the conditions for controllability simplify because $sp(n/2)$ acts transitively on $\mathbb{R}^n$.

Theorem A2. *Let A, B, C be constant matrices with $A, B)$ being controllable and (A, C) being observable. Consider the system evolving in $\mathbb{R}^n$*

$$\dot{x} = (A + BKC)x + b$$

with B, K, C being n-bym, m-by-p and p-by-n, respectively. Assume that the solution of $\dot{X} = (A + BKC)X$; $X(0) = I$ is periodic for some $K = K_0(\cdot)$. Then the system is controllable in the sense that any pair $x_0 \neq 0$ and $x_1 \neq 0$ can be joined by a solution of the given equation.

References

1. R.W. Brockett, Minimum Attention Control, in *Proceedings of the 36th IEEE Conference on Decision and Control* (1997), pp. 2628–2632
2. R. Brockett, *Control of Stochastic Ensembles*, ed. by B. Wittenmark, A. Rantzer. Åström Symposium on Control (Studentlitteratur, Lund, 1999), pp. 199–216
3. R. Brockett, N. Khaneja, in *System Theory Modeling, Analysis, and Control* ed. by T. Djaferis, I. Schick. On the Stochastic Control of Quantum Ensembles (Kluwer Academic, Norwell, 1999), pp. 75–96
4. R. Brockett, On the control of a Flock by a leader. Proc. Skeklov Inst. Math. **268**(1), 49–57 (2010)
5. R.W. Brockett, Reduced Complexity Control Systems, in *Proceedings of the 17th World Congress*, The International Federation of Automatic Control, Seoul, 2008
6. J. Moser, On the volume elements of a manifold. Trans. Am. Math. Soc. **120**, 286–294 (1965)
7. B. Dacorogna, J. Moser, On a partial differential equation involving the Jacobian determinant. Ann. Inst. Henri Poincare C **7**(1), 1–26 (1990)
8. R. Brockett, in *Proceedings of the International Conference on Complex Geometry and Related Fields*, ed. by Z. Chen et al. Optimal Control of the Liouville Equation (American Mathematical Society, Providence, 2007), pp. 23–35
9. A.A. Agrachev, M. Caponigro, Controllability on the group of diffeomorphisms. Ann. Inst. Henri Poincare C Non Lin. Anal. **26**(6), 2503–2509 (2009)
10. V. Jurdjevic, H.J. Sussmann, Control systems on Lie groups, J. Differ. Equat. **12**(2), 313–329 (1972)
11. R. Brockett, Nonlinear Systems and Differential Geometry, in *Proceedings of the IEEE*, vol 64 (1976), pp. 61–72
12. R.W. Brockett, Linear feedback systems and the groups of Lie and Galois. Lin. Algebra Appl. **50**, 45–60 (1983)
13. Yu.L. Sachkov, *in Positive orthant scalar controllability of bilinear systems*. Springer Mathematical Notes, Mat. Zametki, **58**(3), pp. 666–669 (1995)
14. Y.L. Sachkov, On positive orthant controllability of bilinear systems in small codimensions. SIAM J. Contr. Optim. **35**(1), 29–35, (1997)
15. W.M. Boothby, Some comments on positive orthant controllability of bilinear systems. SIAM J. Contr. Optim. **20**(5), 634–644 (1982)

Chapter 3
Some Questions of Control in Fluid Mechanics

Olivier Glass

Abstract The goal of these lecture notes is to present some techniques of non-linear control of PDEs, in the context of fluid mechanics. We will consider the problem of controllability of two different models, namely the Euler equation for perfect incompressible fluids, and the one-dimensional isentropic Euler equation for compressible fluids. The standard techniques used to deal with the Cauchy problem for these two models are of rather different nature, despite the fact that the models are close. As we will see, this difference will also appear when constructing solutions of the controllability problem; however a common technique (or point of view) will be used in both cases. This technique, introduced by J.-M. Coron as the *return method*, is a way to exploit the nonlinearity of the equation for control purposes. Hence we will see its application in two rather different types of PDEs.

The plan of these notes is the following. In a first part, we recall in a very basic way some types of questions that can be raised in PDE control (in a non-exhaustive way). In a second part, we expose results concerning the controllability of the incompressible Euler equation. In a third part, we show how the techniques used to prove the controllability of the incompressible Euler equation can be used to prove some other controllability properties for this equation, namely the so-called Lagrangian controllability. In a fourth and last part, we consider the controllability of the isentropic Euler equation.

O. Glass (✉)
Ceremade, CNRS UMR 7534, Université Paris-Dauphine, Place du Maréchal de Lattre de Tassigny, 75775 Paris Cedex 16, France
e-mail: glass@ceremade.dauphine.fr

F. Alabau-Boussouira et al., *Control of Partial Differential Equations*,
Lecture Notes in Mathematics 2048, DOI 10.1007/978-3-642-27893-8_3,
© Springer-Verlag Berlin Heidelberg 2012

3.1 Introduction

In this first section, we give a short and elementary presentation of some questions in control theory as a general introduction before getting to some specific control problems in fluid mechanics.

3.1.1 Control Systems

We start with the definition of the basic object studied in control theory.

Definition 3.1.1. A *control system* is an evolution equation (an ODE or a PDE) depending on a parameter u, that we will write in a formal way as follows:

$$\dot{y} = f(t, y, u), \qquad\qquad (\text{CS})$$

where $t \in [0, T]$ is the time and:

- $y : [0, T] \to \mathscr{Y}$ is the unknown, called the *state* of the system.
- $u : [0, T] \to \mathscr{U}$ is the parameter called the *control*, that one can choose as a function of the time.

Of course, above, $\dot{y}$ stand for the time derivative of y.

The two standard examples that we have in mind with this definition are the following:

- The state $y(t)$ belongs to $\mathbb{R}^n$ (or to some finite-dimensional manifold), the control $u(t)$ to $\mathbb{R}^m$ (or again to some other finite-dimensional manifold), and (CS) is an ODE.
- Both the state $y(t)$ and the control $u(t)$ belong to some functional spaces, and (CS) is a PDE (so f is typically a differential operator acting on y).

The general question accompanying this definition is the following: how can one use the control to make the system fulfill some purpose that has been prescribed in advance? Before giving precise mathematical definitions corresponding to this general problem, let us give some examples of control systems.

3.1.2 Examples

To fix the ideas, we give examples of control systems both of finite and infinite dimensional type. In these lecture notes, we will be more interested in infinite-dimensional systems governed by PDEs.

1. *Finite dimensional linear autonomous control systems.* Here (CS) is as follows:

$$\dot{y} = Ay + Bu,$$

where the state $y \in \mathscr{Y} = \mathbb{R}^n$, the control $u \in \mathscr{U} = \mathbb{R}^m$, and $A \in \mathbb{R}^{n \times n}$ and $B \in \mathbb{R}^{n \times m}$ are fixed matrices.

2. *Driftless control-affine systems.* Here:

$$\dot{y} = \sum_{i=1}^{m} u_i f_i(y),$$

where the state $y \in \mathbb{R}^n$, the control $u \in \mathbb{R}^m$, and $f_1, \ldots, f_m$ are smooth vector fields on $\mathbb{R}^n$.

Let us now give some examples of infinite-dimensional control systems, connected to fluid mechanics. There are different classical ways to consider the action of a control on a distributed system governed by a PDE.

3. *Internal control of a PDE: the Navier–Stokes case.* Consider Ω a smooth bounded domain in $\mathbb{R}^n$, and a nonempty open set $\omega \subset \Omega$, see Fig. 3.1.

Here we consider an evolution PDE on Ω, e.g. the incompressible Navier–Stokes equations, the control acting as a source term located in ω:

$$\begin{cases} \partial_t v + (v \cdot \nabla)v - \Delta v + \nabla p = \mathbf{1}_\omega u \text{ in } [0, T] \times \Omega, \\ \operatorname{div} v = 0 \text{ in } [0, T] \times \Omega, \\ v = 0 \text{ on } [0, T] \times \partial\Omega. \end{cases}$$

Above, $v : \Omega \to \mathbb{R}^n$ is the velocity field, $p : \Omega \to \mathbb{R}$ is the pressure field. As well-known this equation describes the evolution of the velocity field of an incompressible, viscous fluid. Note that p is not a real unknown of the equation; as a matter of fact, the whole system could be reformulated without it. Here:

• The state is the velocity field v for instance taken in $L^2(\Omega; \mathbb{R}^n)$ (or a subspace in $L^?(\Omega; \mathbb{R}^n)$ in order to take div $v = 0$ and the boundary conditions into account).
• The control is the localized force $u = u(t, x)$, which we compel to be supported in ω, belonging for instance in $L^2(\omega; \mathbb{R}^n)$.

4. *Boundary control of the Navier–Stokes equation.* Consider $\Omega \subset \mathbb{R}^n$ a smooth bounded domain, and a non empty open part of the boundary $\Sigma \subset \partial\Omega$, see Fig. 3.2.

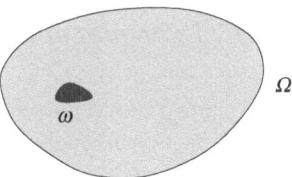

Fig. 3.1 Internal control

Fig. 3.2 Boundary control

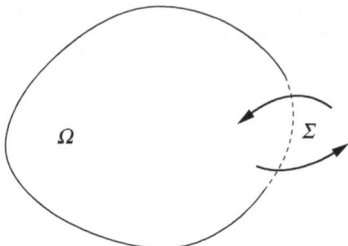

Consider the Navier–Stokes equations, the control acting as a boundary condition located in Σ:

$$\begin{cases} \partial_t v + (v \cdot \nabla)v - \nu \Delta v + \nabla p = 0 \text{ in } [0, T] \times \Omega, \\ \text{div } v = 0 \text{ in } [0, T] \times \Omega, \\ v = \mathbf{1}_\Sigma(x)u(t, x) \text{ on } [0, T] \times \partial \Omega, . \end{cases}$$

Here:

- The state is the velocity field v, for instance in $L^2(\Omega; \mathbb{R}^n)$.
- The control is the localized boundary term $u = u(t, x)$.

5. *Control by lower modes.* One may also consider finite-dimensional control in the context of PDEs, for instance:

$$\begin{cases} \partial_t v + (v \cdot \nabla)v - \Delta v + \nabla p = \sum_{i=1}^{m} u_i(t)e_i(x) \text{ in } [0, T] \times \Omega, \\ \text{div } v = 0 \text{ in } [0, T] \times \Omega, \\ v = 0 \text{ on } [0, T] \times \partial \Omega, \end{cases}$$

where:

- The state is again the velocity field v.
- The control is $(u_1, \ldots, u_m) \in \mathbb{R}^m$.

6. *Many other possibilities.* Let us underline that there are many other natural possibilities: u appearing in the coefficients, through an internal/a boundary operator, …

3.1.3 Examples of Control Problems

As explained above, the goal of control theory is to understand how one can use the control function in order to influence the dynamics of the system in a prescribed way. This general problem can take different forms and yields different mathematical problems. We list several of these questions below.

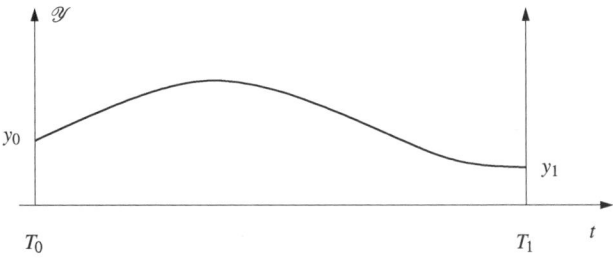

Fig. 3.3 Exact controllability

1. *Optimal control.* One looks for a control that minimizes some cost function, e.g.

$$J(u) = \|y(T;u) - \overline{y}\|^2 + \|u\|^2,$$

where $\overline{y}$ is some target and $y(T;u)$ is the state reached by the system at time T, starting from y_0 and with control u. The problem is to determine if such a control exists, is unique, and to try to characterize it.

2. *Exact controllability.* The question is the following: given two times $T_0 < T_1$, and y_0, y_1 two possible states of the system, does there exist $u : [T_0, T_1] \to \mathcal{U}$ such that

$$y_{|t=T_0} = y_0, \ \dot{y} = f(y,u) \implies y(T_1) = y_1?$$

In other words, is it possible to find, for each y_0 and y_1, a control which drives the system from the initial state y_0 to the target y_1? (Fig. 3.3).

Remark 3.1.2. For autonomous systems, this notion depends on $T_1 - T_0$ rather than on both T_0 and T_1.

3. *Approximate controllability.* The problem of approximate controllability is a relaxed version of the exact controllability. Instead of requiring that the state of the system reaches the target exactly, one may wonder if, at least, one can get arbitrarily close to the target. Mathematically speaking, this can be written as follows.

Given $T_0 < T_1$, y_0 and y_1 two possible states of the system *and* $\varepsilon > 0$, does there exist $u : [0, T] \to \mathcal{U}$ such that

$$y_{|t=0} = y_{T_0}, \ \dot{y} = f(y,u) \implies \|y(T_1) - y_1\| \leq \varepsilon?$$

Needless to say, the problem highly depends on the choice of the norm.

4. *Null controllability.* We suppose that $\mathcal{Y}$ is a vector space. Given $T_0 < T_1$, y_0 an initial state of the system, does there exist $u : [0, T] \to \mathcal{U}$ such that

$$y_{|t=T_0} = y_0, \ \dot{y} = f(y,u) \implies y(T_1) = 0?$$

Typically: can one use the control to put the fluid to rest?

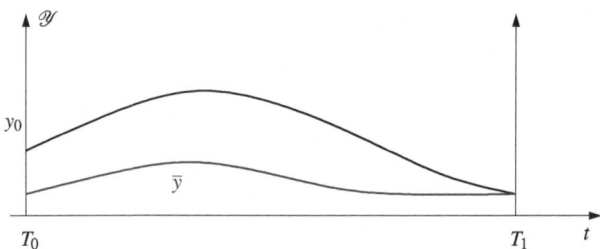

Fig. 3.4 Controllability to trajectories

5. *Controllability to trajectories.* Given $T_0 < T_1$, $y_0 \in \mathscr{Y}$ and $\overline{y} : [T_0, T_1] \to \mathscr{Y}$ a given trajectory of the system (corresponding to a control $\overline{u} : [T_0, T_1] \to \mathscr{U}$), does there exist $u : [T_0, T_1] \to \mathscr{U}$ such that (See Fig. 3.4)

$$y_{|t=T_0} = y_0, \quad \dot{y} = f(y, u) \implies y(T_1) = \overline{y}(T_1)?$$

Remark 3.1.3. The notions of zero-controllability and controllability to trajectories are particularly important for (irreversible) systems having a regularizing effect, since in that case, one cannot hope the exact controllability to hold. For instance, consider the internal control of the heat equation (or the Navier–Stokes equation with suitable assumptions), it can be proved that, whatever the choice of the control, the final state of the system is smooth when restricted to a part of Ω at a positive distance from ω.

Let us underline that many other types of controllability can be considered...

Now let us discuss another problem that one consider for a control system. To simplify the discussion, let us consider autonomous control systems:

$$\dot{y} = f(y, u).$$

We would like to ensure some *robustness* of the control. Indeed, the control which are considered in the above controllability problems are "open-loop", that is, depend on t, y_0 and y_1. But if the system deviates from its planned trajectory, the control may no longer be adapted to the situation. A way to find a control which is more robust to perturbations (which can come from noise, imprecisions of the model, etc.) is to look for a control in "closed-loop" form, that is, depending on the state $y(t)$ at time t, rather than on the memory of y_0. An important control problem connected to this discussion is the following.

6. *Asymptotic stabilization.* Given an equilibrium state (y_e, u_e) of the system (that is, a point such that $f(y_e, u_e) = 0$), can one find a *state feedback function* $u = u(y)$, such that $u_e = u(y_e)$ and that the so-called *closed-loop system*:

$$\dot{y} = f(y, u(y)), \tag{CLS}$$

is (globally) asymptotically stable at the point y_e, i.e.:

- For all $\varepsilon > 0$, there exists $\eta > 0$ such that all solution starting from $y_0 \in B(y_e, \eta)$ are global and satisfy that for all $t \geq 0$, $y(t) \in B(y_e, \varepsilon)$,
- Any maximal solution is global in time and satisfies $y(t) \to y_e$ as $t \to +\infty$?

Remark 3.1.4. The properties as described above, in both controllability and stabilization contexts, are called *global*. One can consider their *local* versions as follows. The local exact controllability near y_* allows to drive any y_0 to any y_1 in some neighborhood of y_*. The local zero controllability allows to drive any small y_0 to 0. The local controllability to trajectories allows to drive any initial y_0 sufficiently close to $\overline{y}(T_0)$ to $\overline{y}(T_1)$. The local asymptotic stabilization makes the closed-loop system merely locally asymptotically stable at y_e.

3.2 Controllability of the Euler Equation

In this section, we consider the problem of exact boundary controllability of the Euler equation for incompressible inviscid fluids. We begin by describing more precisely the control system under view.

3.2.1 The Control Problem

We consider a smooth bounded domain $\Omega \subset \mathbb{R}^n$, $n = 2$ or 3. For a positive time $T > 0$, we consider the Euler equation for perfect incompressible fluids in $[0, T] \times \Omega$:

$$\begin{cases} \partial_t v + (v.\nabla)v + \nabla p = 0 \text{ in } [0, T] \times \Omega, \\ \text{div } v = 0 \text{ in } [0, T] \times \Omega. \end{cases} \tag{3.1}$$

Here, $v : [0, T] \times \Omega \to \mathbb{R}^2$ (or $\mathbb{R}^3$) is the velocity field, $p : [0, T] \times \Omega \to \mathbb{R}$ is the pressure field. This equation describes the evolution of a homogeneous, incompressible and inviscid fluid. As is classical, the first equation stands for the conservation of momentum, and the second equation is the incompressibility constraint. Of course, the system needs boundary conditions to be determined. In general, to close the system, one adds the usual *impermeability condition* on the boundary:

$$v.n = 0 \text{ on } [0, T] \times \partial\Omega, \tag{3.2}$$

with n the unit outward normal on $\partial\Omega$. In other words, the fluid cannot cross the boundary (but it can slip on it).

As noticed before in the context of the Navier–Stokes equation, in incompressible fluid mechanics, the pressure is not a real unknown of the system, which can be reformulated in terms of v only. A way to look at the pressure is to consider it as a Lagrange multiplier associated to the incompressibility constraint.

This closed system (3.1)–(3.2) has been studied for a very long time. And it is known that the system in 2-D (respectively 3D), is well-posed globally (resp. locally) in time: see for instance the classical references: Lichtenstein [65], Wolibner [85], Yudovich [86], Kato [56], Ebin–Marsden [39], etc. The main condition is that the state space where $v(t, \cdot)$ is taken should be a Hölder or a Sobolev space which is continuously embedded in the Lipschitz space.

Now, we would like to understand the properties of this equation under the influence of a boundary control, see Fig. 3.2. To make a precise statement, we consider a nonempty open part Σ of the boundary $\partial\Omega$. Instead of imposing the impermeability condition (3.2) everywhere on $\partial\Omega$, we consider the possibility of choosing non-homogeneous boundary conditions on the "control zone" Σ as follows:

- On $\partial\Omega \setminus \Sigma$, the fluid does not cross the boundary, so

$$v.n = 0 \text{ on } [0, T] \times (\partial\Omega \setminus \Sigma), \tag{3.3}$$

- On Σ, we suppose that one can choose the boundary conditions, that is, use them as a control.

The non-homogeneous boundary value problem for the Euler system is not completely standard; for instance it is not sufficient to prescribe the normal velocity on Σ to determine the system. There are several possibilities to make the system determined. The most usual notion of non-homogeneous boundary conditions for the 2-D Euler equation is due to Yudovich [87] and consists in prescribing:

- The normal velocity on Σ, that is,

$$v(t, x).n(x) \text{ on } [0, T] \times \Sigma, \tag{3.4}$$

- The *entering vorticity*, that is, the vorticity (i.e. the curl of the velocity field) at points of Σ where the velocity field points *inside* Ω. In other words, one prescribes

$$\text{curl } v(t, x) \text{ on } \Sigma_T^- := \{(t, x) \in [0, T] \times \Sigma \; / \; v(t, x).n(x) < 0\}. \tag{3.5}$$

(Recall that n is the *outward* unit normal on $\partial\Omega$.)

Yudovich proves that under suitable assumptions on Σ and the boundary data, there exists a unique solution to the initial-boundary value problem. Let us underline that this result concerns quite regular solutions, not the solution of the celebrated reference [86] concerning the homogeneous case. In particular, this regularity is useful to prove uniqueness.

Concerning the 3D equation, Kazhikov [58] proved that one can prescribe as a natural boundary condition for the Euler equation:

- The normal velocity on Σ.
- The *tangential part* of entering vorticity

$$\operatorname{curl} v(t, x) \wedge n \text{ on } \Sigma_T^- := \{(t, x) \in [0, T] \times \Sigma \mathbin{/} v(t, x).n(x) < 0\}.$$

This difference with regards to the bidimensional case can be explained as follows. In 3-D, the vorticity is a divergence-free vector field, while in 2-D it is merely a scalar field. Due to the divergence-free condition of the vorticity, it is enough to prescribe the tangential part of curl v in order to recover this vector field completely.

In both cases the form of the boundary data seems rather involved, but as we will see later, we will express the controllability problem in a way that circumvents this difficulty.

Now, for what concerns the state of the system, the natural space will consist in smooth enough vector fields, satisfying the incompressibility condition div $(v) = 0$ and the constraint (3.3) on the wall. Since the regularity is not a real issue here (as long as the state v belongs to a Hölder or a Sobolev space contained in the Lipschitz space), we will consider velocities in $C^\infty(\overline{\Omega}; \mathbb{R}^n)$. The arguments could be adapted to less regular spaces.

The controllability problem becomes the following one: given a time $T > 0$, and two states v_0, v_1 in $C^\infty(\overline{\Omega}; \mathbb{R}^n)$ satisfying the compatibility conditions

$$\operatorname{div}(v_0) = \operatorname{div}(v_1) = 0 \text{ in } \Omega, \tag{3.6}$$

$$v_0.n = v_1.n = 0 \text{ on } \partial\Omega, \tag{3.7}$$

can one find a boundary control such that the corresponding solution v starting from v_0 satisfies

$$v_{|t=T} = v_1? \tag{3.8}$$

But as we saw, the form of the boundary control is a difficulty in itself. To overcome this difficulty, we reformulate the controllability problem as follows: given a time $T > 0$, and two states v_0, v_1 in $C^\infty(\overline{\Omega}; \mathbb{R}^n)$ satisfying the compatibility conditions (3.6)–(3.7), can one find a *solution* $v \in C^\infty([0, T] \times \overline{\Omega}; \mathbb{R}^n)$ of (3.1) starting from v_0, satisfying the constraint (3.3) and such that (3.8) holds?

This formulation can be found in many other contexts of PDE controllability problems. Let us underline that there is no real difference between the two formulations. Should one be able to construct a solution v satisfying the constraint (3.3), it suffices to take the appropriate trace of v on the boundary to get the control.

Note that the same reformulation works for the approximate controllability problem as well. The standard way to formulate the approximate controllability for the norm $\| \cdot \|$ is the following question: given $T > 0$, v_0, v_1 in $C^\infty(\overline{\Omega}; \mathbb{R}^n)$ satisfying (3.6)–(3.7) and $\varepsilon > 0$, does there exist a boundary control such that the corresponding solution v starting from v_0 satisfies

$$\|v_{|t=T} - v_1\| < \varepsilon? \tag{3.9}$$

And the way we will look to it is to ask if there exists a solution v starting from v_0, satisfying (3.3) and (3.9).

3.2.2 Controllability Results

The first result concerning the controllability of the Euler equation is the following, see [25].

Theorem 3.2.1 (Coron). *The 2-dimensional Euler equation is exactly controllable in arbitrary time if and only if Σ meets all the connected components of the boundary. In other words, under this condition, for all $T > 0$, for all $v_0, v_1 \in C^\infty(\overline{\Omega}; \mathbb{R}^2)$ satisfying (3.6)–(3.7), there exists $v \in C^\infty([0, T] \times \overline{\Omega}; \mathbb{R}^2)$, solution of (3.1), (3.3) and satisfying*

$$v_{|t=0} = v_0 \ \text{and} \ v_{|t=T} = v_1 \ \text{in} \ \Omega.$$

Note that this result is global, and that when the controllability holds, it holds for all time T. The fact that the controllability holds for all time is far from being true for all PDE controllability problems. For instance, it is well-known that the controllability of the wave equation can hold only for a sufficiently large time, due to the finite speed of propagation.

The controllability of the Euler equation was afterwards established in the 3-D case, see [44].

Theorem 3.2.2 (G.). *The previous result also holds in 3-D.*

An interesting fact concerning the 3-D case is that it is not known whether the regular (uncontrolled) solutions of the 3-D Euler equation are global in time or not. As a matter of fact, a possible blow-up is suspected. But here, the result states that, if Σ meets all the connected components of $\partial\Omega$, then one can use the control to make the solution "live" during any time interval $[0, T]$, and even, should one choose $v_1 = 0$, make the solution global in time. Hence the boundary control is strong enough to *prevent a possible blow-up*.

That the condition on Σ is necessary to get the exact controllability is not difficult to prove. Indeed, two different conservations prove that if Σ does not meet all the connected components of the boundary, then the controllability does not hold. Let us first discuss them in the 2-D case:

- First, Kelvin's law states that the circulation of velocity around a Jordan curve is constant as the curve follows the flow. Now, suppose that a certain connected component of the boundary, let us say γ, does not meet Σ. It follows that, whatever the solution v, this connected component is left (globally) invariant by the flow of v. Hence the circulation of velocity along this component is a conserved quantity, no matter the choice of the control. Hence it suffices to choose v_0 and v_1 having different circulations along γ to prove that the exact controllability does not hold.
- In 2D, the vorticity $\omega := \text{curl } v$ is constant along the flow of v. And again an uncontrolled component of $\partial\Omega$ is preserved by the flow. So it suffices to choose v_0 and v_1 having vorticity distributions on γ that cannot be driven one to another

by a smooth deformation to prove that the exact controllability does not hold. Note that this invariant is different from the previous one. One can easily construct velocity fields having the same first invariant and not the same second one.

These obstructions are in fact also valid in 3-D. For the first invariant, one considers a curve γ on an uncontrolled connected component of $\partial\Omega$, a velocity field v_0 with non trivial circulation along γ, and $v_1 = 0$. For what concerns the second obstruction, the vorticity does no longer follow the flow of the velocity in 3-D, but however the support of the vorticity does. Hence it suffices to choose v_0 such that Supp(curl v_0) meets an uncontrolled connected component of $\partial\Omega$ and $v_1 = 0$ to see that the exact controllability does not hold.

Now, one could wonder what happens when Σ does not meet all the connected components of the boundary. The exact controllability does not hold, but can one at least hope to get some approximate controllability? Here is a positive answer in dimension 2, see [25].

Theorem 3.2.3 (Coron). *If Σ is non empty (but does not meet all the connected components of the boundary), the system is approximately controllable for the norm $L^p(\Omega)$, $p < \infty$. Not for $p = \infty$.*

The same conserved quantity as previously (the velocity circulation along uncontrolled connected components of $\partial\Omega$) shows that the result is false in general if $p = +\infty$. Hence this result cannot be improved. But one could wonder whether this is the only obstruction. The following answer is given in [45].

Theorem 3.2.4 (G.). *If v_0 and v_1 have the same velocity circulation on the uncontrolled components of the boundary, then the approximate controllability occurs in $W^{1,p}(\Omega)$, $p < \infty$. Not for $p = \infty$.*

Here, the second conserved quantity (the distribution of vorticity along uncontrolled connected components of $\partial\Omega$, up to regular deformations) shows that the result is false in general if $p = +\infty$. It is natural to ask again whether this is the only obstruction to a better approximate controllability. This is also proven in [45].

Theorem 3.2.5 (G.). *If v_0 and v_1 have the same velocity circulation on the uncontrolled components of the boundary, and moreover there exist smooth deformations on these components sending the vorticity distributions of v_0 on the ones of v_1, then the approximate controllability occurs in $W^{2,p}(\Omega)$, $p < \infty$. Not for $p = \infty$.*

One can show that the case $p = \infty$ is not true by considerations on the derivatives of the vorticity. However, we are not able to describe the invariant properly. So the following open problem remains.

Open problem 1 *What happens next (e.g. for $W^{3,p}(\Omega)$) is open.*

For what concerns the 3-D case, we are not able to extend the above results. The main problem is a possibility of blow-up. It is not clear how to get rid of possible "germs of explosion" near the uncontrolled components of the boundary. This lefts us with the following.

Open problem 2 *The question of approximate controllability of the 3-D Euler equation in $L^p(\Omega)$ (when Σ does not meet all the connected components of the boundary), is still open.*

3.2.3   Proof of the Exact Controllability

In this paragraph, we try to explain the main ideas of the proof of Theorem 3.2.1. We will first consider the simpler case when Ω is simply connected, and then we will describe what is needed to extend the result to general bidimensional domains. We will also give a few ideas about Theorem 3.2.2.

3.2.3.1   Introduction

We are considering the controllability problem for a nonlinear PDE. Let us explain how this is often dealt with.

Standard approach to nonlinear PDE controllability problems

The most standard method to establish the controllability of a nonlinear PDE is the following:

1. Linearize the equation.
2. Prove a controllability result on the linearized equation.
3. Deduce a controllability result on the nonlinear system by a fixed point or an inverse mapping theorem.

Now to prove the controllability of the linearized equation, there is a standard approach by *duality* (D. Russell [76], J.-L. Lions [66]). This consists in proving an *observability inequality on the (homogeneous) dual system*. Roughly speaking, one has to prove the surjectivity of the control $\mapsto$ final state map, and the argument is somewhat close to the standard

$$A \text{ surjective } \iff \exists c > 0, \ \forall u, \ \|A^* u\| \geq c \|u\|.$$

But this is not (by far) the end of the story. Indeed, in general these observability inequalities (which measure the solution everywhere in terms of this solution measured in the control zone only) are very difficult to prove. Also, this is not the only way to establish controllability, even in the linear case. But this gives a good start: we have to prove some inequality, so the problem seems more standard than to find a control driving the solution from one place to another.

However, this method can have at least two drawbacks:

- Frequently, this merely leads to *local* results, unless the nonlinearity is nice (see e.g. Zuazua [88]).

- In many physical cases, and in particular for what concerns the Euler equation, the linearized equation is *not* (unconditionally) controllable.

Let us indeed consider the *linearized Euler equation* around 0:

$$\begin{cases} \partial_t v(t, x) + \nabla p(t, x) = 0 \text{ in } [0, T] \times \Omega, \\ \text{div } v(t, x) = 0 \text{ in } [0, T] \times \Omega. \end{cases} \qquad (3.10)$$

As noticed by J.-L. Lions [67], this equation is not controllable, because solutions of (3.10) satisfy

$$v_{|t=T} - v_{|t=0} \text{ is the gradient of a harmonic function.}$$

The return method

A method designed by J.-M. Coron to tackle this kind of situation is the *return method*. This method was introduced in the context of finite-dimensional control systems, see [23]. The idea is the following:

find a particular solution $\overline{y}$ of the (nonlinear) system (with control), such that $\overline{y}(0) = \overline{y}(T) = 0$ and such that the linearized system around $\overline{y}$ is controllable.

One may then hope to find a solution of the nonlinear controllability problem close to $\overline{y}$ (Fig. 3.5).

In general, it is not easy to construct such a solution of the nonlinear system. But it turns out that in many different physical situations, this method have proved very useful. It can be seen as a way to exploit the nonlinearity of the system. We refer to [29] for examples and references on that subject.

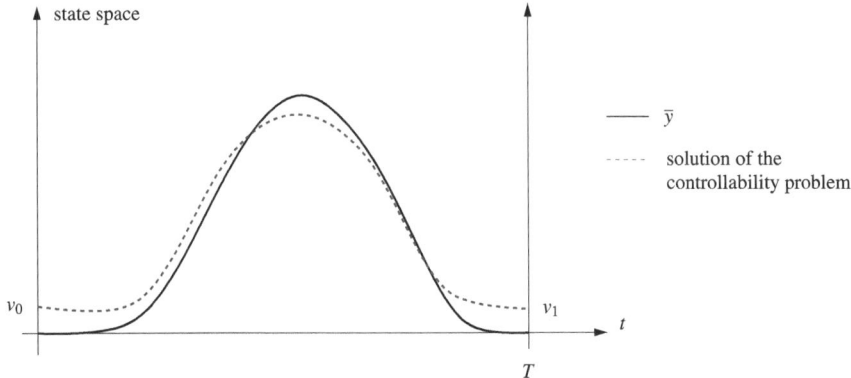

Fig. 3.5 The return method

3.2.3.2   The Solution $\overline{y}$

In this part, we explain the construction of the function $\overline{y}$ used to prove Theorem 3.2.1. The vast majority of the arguments here come from [25]; a slight part of the construction that we show is a little bit different from the one of [25] and uses arguments from [45]. First, in an informal manner, we explain how we are led to look for particular properties for $\overline{y}$.

The choice of $\overline{y}$: what should it do?

Let $\omega := \mathrm{curl}\,(v)$ the vorticity (either a scalar in 2D or a vector in 3D). It satisfies

$$\partial_t \omega + (v \cdot \nabla)\omega = 0 \quad (2D \text{ case}), \tag{3.11}$$

or

$$\partial_t \omega + (v \cdot \nabla)\omega = (\omega.\nabla)v \quad (3D \text{ case}). \tag{3.12}$$

We call Φ the flow of v, that is, the solution of the ODE associated to v:

$$\partial_t \Phi(t, s, x) = v(t, \Phi(t, s, x)) \text{ and } \Phi(t, t, x) = x. \tag{3.13}$$

Hence, in the 2D case, the vorticity follows the flow of v, that is to say, it is constant along flow lines:

$$\omega(t, \Phi(t, 0, x)) = \omega(0, x).$$

This is no longer true for what concerns the 3D case, but, at least, the support of the vorticity follows the flow in that case. A consequence of this is the following. If one wants to steer a state v_0 such that $\mathrm{curl}\,(v_0)$ does not vanish anywhere on Ω, to $v_1 = 0$, then even if $\|v_0\| \ll 1$, one should expect the following property to hold:

$$\textit{the flow of } \overline{y} \textit{ makes every point of } \overline{\Omega} \textit{ leave the domain.} \tag{3.14}$$

Indeed, we will look for a solution v close to $\overline{y}$; but v must satisfy such a property, because if it does not, there remains inside Ω at time T, points where the vorticity is inherited directly from $\mathrm{curl}\,(v_0)$, which contradicts $v(T) = 0$.

The choice of $\overline{y}$: what can it do?

But on another side, the question is: how can we construct a solution of the nonlinear system (with control)? In general, we have not so many particular solutions of a nonlinear system at our disposal. But for what concerns the Euler equations, a very classical form of particular solutions is known for a very long time. These are the *potential flows*:

$$\overline{y}(t, x) = \nabla_x \theta,$$

where

$$\Delta_x \theta(t, x) = 0 \text{ for all } (t, x) \in [0, T] \times \Omega.$$

These are solutions of the Euler equation: taking

$$p = -\theta_t - \frac{|\nabla \theta|^2}{2},$$

it is elementary to check that $\overline{y}$ satisfies (3.1) (and it is remarkable that the dependence in time is in some sense arbitrary). The boundary condition (3.3) translates into

$$\partial_n \theta(t, x) = 0 \text{ for all } (t, x) \in [0, T] \times (\partial \Omega \setminus \Sigma).$$

Main proposition

As one may expect, the solution $\overline{y}$ that one constructs is at the intersection of the two above constraints. An important part of the proof is devoted to showing the following existence result.

Proposition 3.2.6 (Coron). *There exists* $\overline{\theta} \in C^\infty([0, T] \times \overline{\Omega}; \mathbb{R})$, *compactly supported in time in* $(0, T)$, *such that*

$$\Delta_x \overline{\theta}(t, x) = 0 \text{ in } [0, T] \times \Omega, \ \partial_n \overline{\theta}(t, x) = 0 \text{ on } [0, T] \times (\partial \Omega \setminus \Sigma), \quad (3.15)$$

and such that the flow of $\nabla \overline{\theta}$ *makes all the points in* $\overline{\Omega}$ *leave the domain.*

Remark 3.2.7. The flow of $\nabla \overline{\theta}$ is not very well-defined, because $\nabla \overline{\theta}$ is not everywhere tangent on the boundary, and hence the flow is not "internal" to Ω. It follows that the flow $\Phi(t, s, x)$ solution to:

$$\partial_t \Phi(t, s, x) = \overline{y}(t, \Phi(t, s, x)) \text{ and } \Phi(t, t, x) = x,$$

is not defined for all time (t, s). An elementary way to define this flow properly is to extend $\overline{\theta}$ into a function of $C^\infty([0, T]; C_c^\infty(\mathbb{R}^n))$ (which of course is no longer harmonic outside Ω). This allows to define a flow globally and make the statement mathematically accurate.

Idea of the construction of $\overline{y}$

For the rest of Paragraph 3.2.3.2, we explain the steps to prove Proposition 3.2.6. As we will see, it is the consequence of the following one:

Proposition 3.2.8. *Given a curve* $\gamma \in C^k([0, 1]; \Omega \cup \Sigma)$, *there exists* $\theta \in C^k([0, 1] \times \overline{\Omega}; \mathbb{R})$ *satisfying* (3.15) *such that the flow* Φ *of* $\nabla \theta$ *satisfies:*

$$\Phi(t, 0, \gamma(0)) = \gamma(t).$$

The same holds for $\gamma \in C^k([0, 1]; \partial\Omega)$.

Idea of the proof of Proposition 3.2.6 assuming Proposition 3.2.8. This is mainly a matter of compactness of $\overline{\Omega}$. For each x in $\overline{\Omega}$, one can find a curve γ in $\overline{\Omega}$ driving x outside of $\overline{\Omega}$. To make this statement rigorous, extend a little bit Ω across Σ, to obtain a new smooth domain $\tilde{\Omega}$. Hence one can find a corresponding harmonic flow $\nabla\theta$ (depending on x), defined on $[0, 1] \times \overline{\tilde{\Omega}}$.

Now, by continuity of the flow, together with x, a small neighborhood of x in $\overline{\Omega}$ is sent outside $\overline{\Omega}$ by this flow, say $\mathcal{V}_x$. By compactness of $\overline{\Omega}$, we can find a finite number of points $x_1, \ldots x_n$ such that $\overline{\Omega} = \mathcal{V}_{x_1} \cup \cdots \cup \mathcal{V}_{x_n}$. Call $\nabla\theta_1, \ldots \nabla\theta_n$ the corresponding flows (Fig. 3.6).

Now, let us notice that the time T is not an issue here. If one is able to find a function $\theta \in C^\infty([0, \tilde{T}] \times \overline{\Omega}; \mathbb{R})$, harmonic in x, satisfying the homogeneous Neumann boundary condition on $\partial\Omega \setminus \Sigma$ for all $t \in [0, \tilde{T}]$ and whose flow satisfies (3.14), then it is just a matter of time-rescaling to prove Proposition 3.2.6.

Now since we do not care about the size of the time interval, the function θ is obtained by gluing in time several flows of this type. Precisely, we construct θ as follows:

- $\theta(t, x) := \theta_1(t, x)$ during $[0, 1]$.
- Then $\theta(t, x) := -\theta_1(2 - t, x)$ during $[1, 2]$. In this way, we know that the corresponding Φ satisfies $\Phi(2, 0, x) = x$ for all $x \in \overline{\Omega}$.
- And then we iterate: we set $\theta(t, x) := \theta_3(t + 2, x)$ during $[2, 3]$, and then $\theta(t, x) := -\theta_3(4 - t, x)$ during $[3, 4]$, etc.

It is then not difficult to see that the θ that we have constructed is convenient. $\square$

Following a curve

Now we have to establish Proposition 3.2.8. A somewhat close statement was noticed independently by T. Kato in another context. In [57], Kato proves that

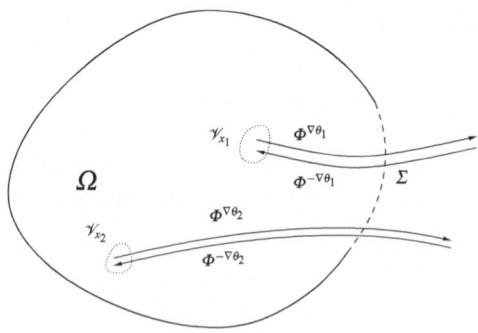

Fig. 3.6 The flow of θ

without control, the trajectories of the flow of solutions of the Euler equation are C^∞ with respect to time; with non-homogeneous boundary conditions, he notices that this is no longer true.

A way to prove Proposition 3.2.8 is to establish the following lemma.

Lemma 3.2.9. *For $x \in \overline{\Omega}$, one has the following:*

$$\left\{ \nabla\theta(x), \ \theta \in C^\infty(\overline{\Omega};\mathbb{R}) \text{ satisfying } (3.15) \right\} = \begin{cases} \mathbb{R}^n \text{ for } x \in \Omega \cup \Sigma, \\ T_x\partial\Omega \text{ for } x \in \partial\Omega \setminus \Sigma. \end{cases}$$

Proof of Proposition 3.2.8 assuming Lemma 3.2.9. We suppose that Lemma 3.2.9 is established. Then for each $t \in [0, 1]$, one can find a finite number of functions θ_1, $\ldots$, θ_N, satisfying (3.15), such that

$$\text{Span}\{\nabla\theta_1(\gamma(t)), \ldots, \nabla\theta_N(\gamma(t))\} = \mathbb{R}^n \ (\text{or } T_{\gamma(t)}\partial\Omega). \tag{3.16}$$

Using the continuity in time of the curve γ, the continuity in space of these functions $\nabla\theta_i$ and the openness of the condition (3.16), we see that we still have $\text{Span}\{\nabla\theta_1(\gamma(s)), \ldots, \nabla\theta_N(\gamma(s))\} = \mathbb{R}^n$ or $T_{\gamma(s)}\partial\Omega$ for s in a small neighborhood $\mathcal{U}_t$ of t. In particular, for each t, in such a neighborhood $\mathcal{U}_t$ we are able to describe $\dot{\gamma}(s)$ as follows:

$$\forall s \in \mathcal{U}_t, \ \dot{\gamma}(s) = \sum_{i=1}^{N} \lambda_i(s)\nabla\theta_i(\gamma(s)),$$
$$=: \nabla\theta^t(s, \gamma(s)),$$

for suitable functions $\lambda_i(s)$.

Now we use the compactness of $[0, 1]$, and extract a finite subcover of $[0, 1]$ by $\mathcal{U}_{t_1}, \ldots, \mathcal{U}_{t_n}$. Then one can construct $\theta(t, x)$ with the form

$$\theta(t, x) = \sum_{i=1}^{N} \rho_i(t)\theta^{t_i}(t, x).$$

where $\rho_1, \ldots, \rho_n$ is a partition of unity adapted to this covering of $[0, 1]$. Then one can check easily that this θ is convenient. $\qquad\qquad\square$

The possible directions of $\nabla\theta(x)$

Now it remains to prove Lemma 3.2.9. In 2D, this can be proved by using Runge's theorem (of approximation of holomorphic functions by rational functions). Indeed, as is very classical, complex analysis is very useful to construct such flows in dimension 2 because, setting $Vf := (\text{Re } f, -\text{Im } f)$, we have:

f satisfies the Cauchy–Riemann equations $\iff$ curl $Vf = \text{div } Vf = 0$.

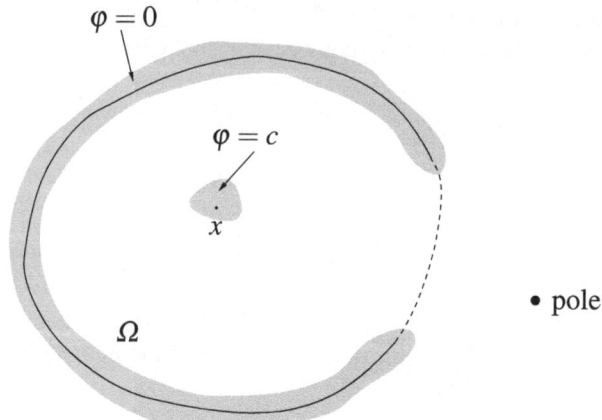

Fig. 3.7 Use of Runge's theorem

Of course, in a simply connected domain, there is no difference between a curl-free vector field and a gradient field. There are now two cases.

1. Let us first consider the case $x \in \Omega \cup \Sigma$. The idea is the following: define the holomorphic function φ as follows: in some neighborhood of $\partial\Omega \setminus \Sigma$ in the complex plane, $\varphi(z) = 0$, and in some neighborhood of x (disjoint from the latter), $\varphi(z) = c$ where c is an arbitrary complex constant, see Fig. 3.7. Now approximate φ by a rational function f whose only pole belongs to the unbounded component of $\mathbb{C} \setminus \overline{\Omega}$. (Recall that Σ meets all the connected components of $\partial\Omega$; hence we do not need more than one pole.)

   The resulting rational function is in particular holomorphic in a neighborhood of $\overline{\Omega}$, and even in a neighborhood of the complement of the unbounded component of $\mathbb{C} \setminus \overline{\Omega}$. And since the only pole is in the unbounded component of $\mathbb{C} \setminus \overline{\Omega}$, one can see that Vf is a gradient field. But it is not quite satisfactory yet, because this vector field $V(f) = \nabla\theta$ does not satisfy $\partial_n\theta = 0$ on $\partial\Omega \setminus \Sigma$ exactly, but merely $\partial_n\theta = \mathcal{O}(\varepsilon)$, where ε is the approximation parameter, in any C^k norm.
   But it suffices to subtract a solution of a Neumann problem to get $\partial_n\theta = 0$ on $\partial\Omega \setminus \Sigma$ exactly. To that purpose, choose g on the boundary such that $g = \partial_n\theta$ on $\partial\Omega \setminus \Sigma$, $\int_{\partial\Omega} g = 0$ and $\|g\|_{C^k(\Sigma)} = \mathcal{O}(\|g\|_{C^k(\partial\Omega\setminus\Sigma)})$ on Σ. Then solve

$$\Delta\psi = 0 \text{ in } \Omega, \quad \partial_n\psi = g \text{ on } \partial\Omega.$$

   Using standard elliptic estimates, we deduce that the size of ψ (in $C^{k,\alpha}(\Omega)$ norm for instance) is also of order ε, so $\theta - \psi$ is convenient.

2. If $x \in \partial\Omega \setminus \Sigma$, the situation is more difficult. Of course, we can no longer approximate 0 near $\partial\Omega \setminus \Sigma$ and c near x at the same time. Instead, we approximate the following function introduced for $a \in \Omega$:

Fig. 3.8 The boundary case

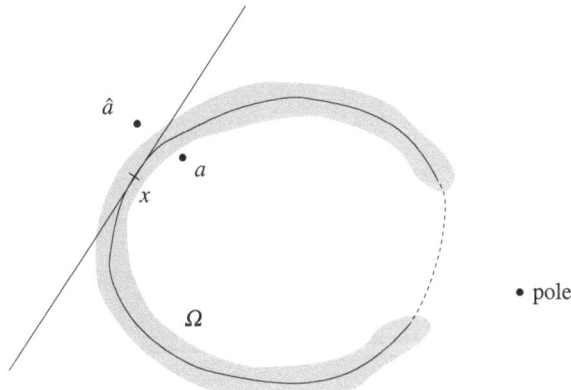

$$R_a(z) := N(a) \left[\frac{1}{z-a} - \frac{1}{z-\hat{a}} \right], \quad \text{as } a \to x, \tag{3.17}$$

where a belongs to Ω and $\hat{a}$ is as in Fig. 3.8, symmetrically disposed with respect to $T_x \partial\Omega$. Moreover, $N(a) = \mathcal{O}(d(x,a))$ is a normalization factor intended to get that $R_a(x)$ is of order 1, while on $\{z \in \partial\Omega \,/\, d(x,z) \geq c > 0\}$, one has $R_a(z) \to 0$ uniformly as $a \to x$. One can interpret $R_a(z)$ as a dipolar expansion.

Now we consider a sufficiently close to x (in a way that $x - a$ is not orthogonal to $T_x \partial\Omega$), and introduce a neighborhood $\mathscr{V}$ of $\partial\Omega$ which does not contain a nor $\hat{a}$, see again Fig. 3.8. Again we apply Runge's theorem to approximate R_a on $\mathscr{V}$, with a pole in the unbounded component of $\mathbb{C} \setminus \overline{\Omega}$. This results in a rational function f, which itself yields a vector field Vf. Again, this vector field is a gradient in Ω, but it does not necessarily satisfy $\partial_n \theta - 0$. Therefore, as previously, we have to subtract a function ψ defined as above. Then the result follows from asymptotic developments as $a \to x$, which allow to prove that the error between $V(R_a)$ and $\nabla\psi$ is small as $a \to x$. □

3.2.3.3 Using the Function $\overline{y}$

Now let us explain how we can use the function $\overline{y}$ constructed above to establish Theorem 3.2.1.

Local zero-controllability, rough idea

We first consider the case where v_0 is small enough (in some fixed, sufficiently strong norm), and $v_1 = 0$. The general case will be deduced from this particular one. Let us also suppose that Ω is simply connected; we will come back to the additional difficulty of a non-trivial topology of the domain later.

In the following construction, we will put aside the regularity/compatibility conditions issues in a first time. Given a small v_0, we try to construct a solution of the following system:

$$\begin{cases} \text{curl}\,(v) = \omega \text{ in } [0, T] \times \Omega, \\ \text{div}\,(v) = 0 \text{ in } [0, T] \times \Omega, \\ v.n = \overline{y}.n \text{ on } [0, T] \times \partial\Omega, \\ \partial_t \omega + (v.\nabla)\omega = 0 \text{ in } [0, T] \times \Omega, \\ \omega_{|t=0} = \omega_0 = \text{curl}\,(v_0) \text{ in } \Omega, \\ \omega = 0 \text{ on } \{(t, x) \in [0, T] \times \Sigma \ / \ \nabla\theta.n < 0\}. \end{cases} \tag{3.18}$$

Mainly, this is exactly (3.1) rewritten in vorticity form. This is very classical in fluid mechanics (and by the way this allows to get rid of the pressure): the vorticity follows a transport equation depending on the velocity field, and the velocity field can be recovered from the vorticity by using the div/curl elliptic system. The important fact here concerns the boundary conditions (that is, the control); in this form, Yudovich's boundary conditions become more natural. The normal velocity is directly inherited from $\overline{y}$, and the entering vorticity is 0; this clearly involves compatibility conditions issues on Σ and at $t = 0$, but this gives the main idea.

Hence we assume that by some procedure we managed to find a solution of (3.18). If there were no regularity issues due to the non-homogeneous boundary conditions, it would mainly be a matter of finding a fixed point of some operator defined as follows. First, one maps ω to v by the elliptic div-curl system. Then, to v a new vorticity, say $\tilde{\omega}$, by the transport system. Using the smallness of v_0 this would yield a fixed point of the operator.

Let us now explain why such a solution would drive the state of the system from v_0 at $t = 0$ to 0 at time T, provided that v_0 is sufficiently small.

• The main principle—this is the core of the application of the return method to the Euler equation—is the following. The vorticity is transported by the flow of the velocity; hence its value $\omega(t, x)$ comes either from the initial datum ω_0 (if in the flow of v the point x does not come from Σ), either from Σ and in that case it is 0 (since the entering vorticity is null). It follows that if $\|v_0\| \ll 1$ (for a norm stronger than Lipschitz), then the vorticity of the solution will also be small for all times. Hence the solution is close to the one obtained with no vorticity, that is to say with the solution associated to $v(0, \cdot) = 0$.

• But this solution corresponding to $v(0, \cdot) = 0$ is precisely $\overline{y}$.

• Consequently, for $\|v_0\|$ small enough, v stays close to $\overline{y}$ for all time. Therefore, using a Gronwall argument and (3.14), one can show that the flow of v makes all points in $\overline{\Omega}$ at $t = 0$ leave the domain.

• It follows that all points at time T in Ω come from Σ in the flow of v. Hence, since the vorticity follows the velocity flow and since the entering vorticity is null, we deduce that $\omega(T) = 0$.

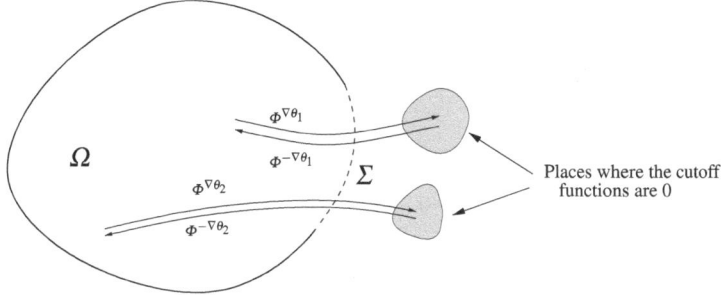

Fig. 3.9 The smooth process

- Now $v(T).n = \overline{y}(T).n = 0$. Since $\text{curl}\,(v(T)) = \text{div}\,(v(T)) = 0$ and since Ω is simply connected, we can affirm that $v(T) = 0$. (Not when Ω is multiply connected, as we will see later.)

How to make the construction smooth

Now, of course, we would like to do this in a smooth manner. Indeed putting the entering vorticity to 0 makes it unlikely to construct a regular solution. The main idea is the following. We have to elaborate a new fixed point strategy. Let us be given a velocity vector field v in $[0, T] \times \Omega$, starting from v_0 and close to $\overline{y}$ (remember that v_0 is small, so these two conditions are compatible). Then the construction is as follows:

- Extend the velocity field v to a velocity field $\tilde{v}$ defined on $\mathbb{R}^2$ and compactly supported in x.
- Transport the initial vorticity ω_0 (also extended on $\mathbb{R}^2$ with compact support) by $\tilde{v}$, and a finite number of times, transform the vorticity by

$$\omega(t^+, x) = \varphi(x)\omega(t^-, x), \tag{3.19}$$

where φ is a cutoff function such that $\varphi = 1$ on $\overline{\Omega}$, see Fig. 3.9. Here we use the particular form of $\overline{y}$: its flow brings points of Ω outside of the domain "one piece after another" (recall the covering of $\overline{\Omega}$ by $\mathscr{V}_{x_1}, \ldots, \mathscr{V}_{x_n}$ in the construction of $\overline{y}$). Hence the idea is to put the vorticity to zero on these "pieces" $\mathscr{V}_{x_i}$ of Ω, one after another, when the flow makes them go out of $\overline{\Omega}$.
- Then one associates a new velocity field in Ω by the div /curl elliptic system in Ω, where the normal velocity on the boundary has to be close to $\overline{y}$ and compatible with $v_0.n$ at $t = 0$.

Then one shows that:

- This operator has a fixed point when v_0 is small enough.

- This fixed point is regular. Of course we introduce discontinuities in time by the above process when applying a cutoff function between t^+ and t^- in (3.19). However, these discontinuities are only with respect to the variable t, and take place outside of $\overline{\Omega}$. It follows that *inside* $\overline{\Omega}$, the solution is smooth. . . .

Passage to the global controllability

The natural question now is: what if $v_1 \neq 0$ and v_0 is not small? The main point is to use the *time-scale invariance* of the equation: for $\lambda > 0$,

$v(t, x)$ is a solution of the equation defined in $[0, T] \times \Omega$

$$\Longleftrightarrow v^\lambda(t, x) := \lambda v(\lambda t, x) \text{ is a solution of the equation}$$

$$\text{defined in } [0, T/\lambda] \times \Omega. \quad (3.20)$$

The pressure associated to v_λ is

$$p_\lambda(t, x) = \lambda^2 p(t, x) \text{ in } [0, T/\lambda] \times \Omega.$$

It is remarkable that the Euler equation has this particular scale invariance, which concerns the time variable only and not the space one.

Now, the idea is the following:

- Using this scale invariance, we see that bringing v_0 to 0 in time T is equivalent to bringing λv_0 to 0 in time T/λ.
- We know how to bring any v_0 such that $\|v_0\| \leq \varepsilon$ to 0 in time T. Hence we know how to bring any v_0 with *larger* norm in *smaller* time (take λ large). In particular, we can bring a large initial condition very fast to 0. . . and then stay at 0 till the planned time of controllability.
- For what concerns $v_1 \neq 0$, use the *reversibility* of the equation, which corresponds to $\lambda = -1$ in (3.20). If a solution $v(t, x)$ goes from v_0 to 0 in time T, then $-v(T - t, x)$ is again a solution, going this time from 0 to $-v_0$. Hence to go from v_0 to v_1, apply the following recipe: bring v_0 to 0 in time $T/2$, and then 0 to v_1 in time $T/2$.

Multiply connected domains

Above we treated the case where Ω is simply connected. But in multiply connected domains,

$$\left.\begin{array}{l} \text{curl}\,(v(T)) = 0 \text{ in } \Omega, \\ \text{div}\,(v(T)) = 0 \text{ in } \Omega, \\ v.n = 0 \text{ on } \partial\Omega, \end{array}\right\} \nRightarrow v(T) = 0, \quad (3.21)$$

Fig. 3.10 Harmonic tangent
field in the annulus

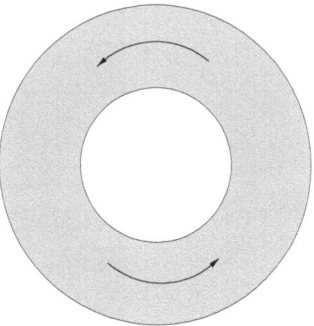

because there is a non-trivial finite-dimensional vector space of *harmonic tangent vector fields* (representing the first tangential de Rham cohomology space of the domain), that is, of solutions of the above system homogeneous div /curl elliptic system. Precisely, if $\partial\Omega$ has $g + 1$ connected components (recall that we are still in dimension 2), then the dimension of this space is g (Fig. 3.10).

Hence applying the strategy above, we can bring the vorticity of the solution to 0, but not the velocity field. We only know that $v(T)$ is a harmonic tangent vector field. Hence we have in fact controllability up to a finite dimensional space; but it turns out that solving this "finite-dimensional" problem has the same level of difficulty than bringing "the other directions to zero". (Even, as we will see, in 3D, this part is by far the most difficult part.)

Now, the space of harmonic tangent field is characterized by the g velocity circulations around the g inner boundary components (for instance), which we call $\Gamma_1, \ldots, \Gamma_g$. Instead of (3.21), we have then

$$\left.\begin{array}{r}\mathrm{curl}\,(v(T)) = 0 \text{ in } \Omega, \\ \mathrm{div}\,(v(T)) = 0 \text{ in } \Omega, \\ v.n = 0 \text{ on } \partial\Omega, \\ \int_{\Gamma_i} v(T).\tau\,d\sigma = 0 \text{ for } i = 1, \ldots, g,\end{array}\right\} \Rightarrow v(T) = 0. \qquad (3.22)$$

Hence our goal is to bring these g velocity circulations to zero. To do so, the idea consists in making some vorticity pass across the domain (from one component of $\mathbb{R}^2 \setminus \overline{\Omega}$ to another, typically from an inner component to the outer one) as in Fig. 3.11.

The rough idea is the following. We want to bring the velocity circulation around some Γ_i to 0. But by Kelvin's circulation theorem, this circulation is constant, when the curve follows the flow. If we make some vorticity cross the domain, by Stokes theorem the difference between the velocity circulation around Γ_i at time 0 (plain line in Fig. 3.11) and the one at time T (dotted line in Fig. 3.11), will be given by the total flux of the vorticity across $[0, T] \times \Gamma_i$. Hence by using this principle we can fix the circulation around Γ_i.

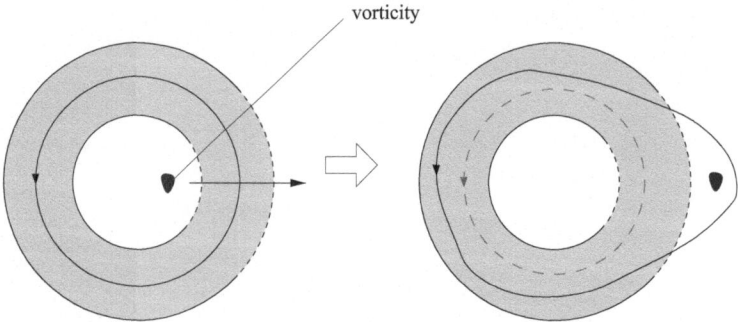

Fig. 3.11 Modifying the velocity circulation by making vorticity cross the domain

Of course, to do this properly, we have to construct another reference solution $\overline{y}$ making points going from an inner component to the outer one, and to construct the solution close to this new $\overline{y}$. This uses the principles showed above, we omit the details.

3.2.3.4   What About 3D?

Here we only give a few ideas about the 3D case.

There are four main differences for what concerns dimension 3.

1. *Construction of $\overline{y}$*. We have no longer access to the complex variable arguments to describe the potential flows. We will see however that there are similar tools that we can use in dimension 3.
2. *Transport of the vorticity*. In dimension 3, the vorticity is no longer transported by the flow but is affected by a "stretching" term:

$$\partial_t \omega + (v.\nabla\omega) = (\omega.\nabla)v. \tag{3.23}$$

However using (3.23), one can prove easily that the support of the vorticity is transported by the velocity flow. This property suffices to our purpose, when following the ideas above.
3. *Blow up*. The solution could blow up. Indeed, it is still unknown whether regular solutions of the 3D Euler equation, which exist locally in time, are global in time or can blow up in finite time. But mainly, as we follow the lines of the proof described above, we see that the main part of the work is done with initial states v_0 satisfying $\|v_0\| \leq \varepsilon$. Even if it is not know that the solutions of the 3D Euler equation remain regular for all time, we know that they have a time of existence at least of $1/\|v_0\|$. Hence we can work with solution which will not be singular before the final time T.

After the time-rescaling procedure, this means that we act *sufficiently fast* to avoid the blow up, and bring the solution to 0 exactly for some small time.

4. *Topology.* This is by far the most difficult issue. Of course, the topology of regular open sets in 3D is different and more complex that in 2D.

Let us say a few words about the new ingredients needed for the 3D case.

Construction of $\overline{y}$

We cannot use the complex variable argument here, but in 3-D, there exist Runge-type theorems of approximations of *harmonic functions* by harmonic functions defined on a larger set. For instance, the following result of Walsh can be used (see e.g. [43, 82]).

Theorem 3.2.10 (Walsh (1929)). *Let K a compact set in $\mathbb{R}^n$ such that $\mathbb{R}^n \setminus K$ is connected. Then for each function u harmonic in an open set containing K, for each $\varepsilon > 0$, there exists a harmonic polynomial v such that*

$$\|u - v\|_\infty \leq \varepsilon.$$

This can replace the use of Runge's theorem in the above steps. For what concerns (3.17), we can replace the functions $z \mapsto \frac{1}{z-a}$ by the fundamental solution of the Laplacian in $\mathbb{R}^3$, and one can make the same type of dipolar developments. The projection of the direction of $a - x$ on $T_{\gamma(t)}\partial\Omega$ will give the direction of the resulting $\nabla\theta$ up to small errors. See [44] for more details.

The difficulty coming from the topology

Of course, the topology of smooth bounded open of $\mathbb{R}^3$ is by far more complex that in the 2D case. Note in particular that in 3D, multiply connected domains can have a connected boundary, and simply connected domains can have several boundary components. The difficulty concerns as before multiply connected domains (whether $\partial\Omega$ is connected or not.)

In dimension 3, to get rid of tangential harmonic vector fields, one uses *vortex filaments* (or regularization of vortex filaments), that has to cross the domain, as described in Fig. 3.12.

We recall that given a Jordan curve J, a vortex filament located at J is the distribution of vorticity given by

$$\omega = \alpha\delta_J(x)\tau(x),$$

where δ_J is the linear measure on J, τ the unit tangent on J, and α is a real parameter. This distribution is naturally divergence-free and is an important object in three-dimensional fluid mechanics. Of course, to get a smooth solution, one has to mollify it at some stage.

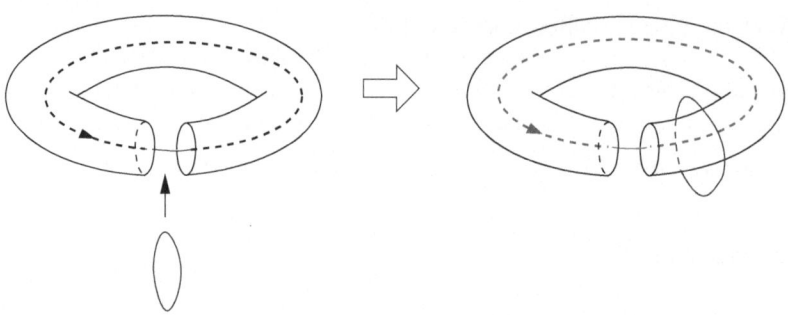

Fig. 3.12 Making vortex filaments cross the domain

One can check that the "change of velocity circulation" around the curve represented in dotted lines in Fig. 3.12, is given by the intensity of the vortex filament. The main part consists in finding $\overline{y}$ whose flow makes the vortex filament cross the domain, using the previous tools. We refer again to [44] for more details.

3.2.4 References

An important majority of the arguments described above come from the seminal work of Coron [25] (see also [24]). Extensions of this work concerning the controllability of the Euler equation can be found in [44,45].

For what concerns the connected problem of asymptotic stabilization by the boundary control, we refer to Coron [26,27] and to [46].

A general reference concerning this problem and the use of the return method is Coron's book [29].

Let us finally give some other references for what concerns the Navier–Stokes equations, which is closely related to Euler equation (let us underline that this bibliography is far from being complete). For Navier–Stokes, due to the regularizing effect of the equation, one would like to show the controllability to trajectories. Several results on this direction:

- Fursikov–Imanuvilov [41], Imanuvilov [55], Fernandez-Cara–Guerrero–Imanuvilov–Puel [40], have obtained results of *local* controllability to trajectories.
- Coron [28], Coron–Fursikov [30], Chapouly [18]: obtained *global* approximate controllability, with Navier boundary conditions. (These results rely on the controllability of the Euler equation!).

However the global controllability to zero for the Navier–Stokes equation with Dirichlet boundary conditions leaves us the following problem.

Open problem 3 *The problem of global controllability to trajectories of the Navier–Stokes equations with $v = 0$ on $\partial\Omega \setminus \Sigma$ is still open.*

Let us finally give some references concerning the controllability of the Navier–Stokes equation by means of low modes: Agrachev–Sarychev [1, 2] and Shirikyan [80]. A related technique was used recently by Nersesian [72] for the compressible Euler equation.

3.3 Approximate Lagrangian Controllability of the Euler Equation

In this section, we describe how the techniques developed in Sect. 3.2, can be used to fulfill other purposes for the fluid, namely, to control the displacement of the fluid during the time interval $[0, T]$ rather than its velocity field at final time T.

3.3.1 The Question of Lagrangian Controllability

3.3.1.1 Controlling the Displacement of a Fluid

Again we consider a smooth bounded domain $\Omega \subset \mathbb{R}^2$ (we consider only $n = 2$ here), and Σ a nonempty open set of $\partial\Omega$ and the control system

$$\begin{cases} \partial_t v + (v.\nabla)v + \nabla p = 0 \text{ in } [0, T] \times \Omega, \\ \text{div } v = 0 \text{ in } [0, T] \times \Omega, \\ v.n = 0 \text{ on } [0, T] \times [\partial\Omega \setminus \Sigma]. \end{cases} \tag{3.24}$$

As previously the control is the boundary data on Σ, e.g.

$$\begin{cases} v(t, x).n(x) \text{ on } [0, T] \times \Sigma, \\ \text{curl } v(t, x) \text{ on } \Sigma_T^- := \{(t, x) \in [0, T] \times \Sigma \ / \ v(t, x).n(x) < 0\}. \end{cases} \tag{3.25}$$

But here, we will be interested in another type of controllability, which is natural for equations from fluid mechanics: is possible *to drive a zone of fluid* from a given place to another by using the control? This question is based on a suggestion by J.-P. Puel. The first study on the subject is due to Horsin [54] where the Burgers equation is considered. One can think for instance to a polluted zone in the fluid, which we would like to transfer to a zone where it can be treated (Fig. 3.13).

First definition

Now before giving the precise definition of the problem under view, let us make a few remarks to motivate it:

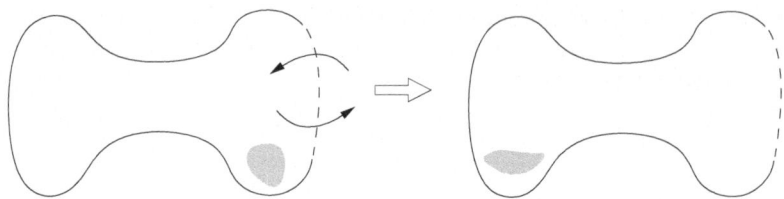

Fig. 3.13 Controlling the displacement of a fluid zone

- First, it is natural, in order to control the fluid zone during the whole displacement to ask that it remains inside the domain Ω during the whole time interval. This is not straightforward, since the condition $v.n = 0$ is not imposed on the whole boundary.
- In the sequel, we will consider only fluid zones given by the interior (inside Ω) of smooth (C^∞) Jordan curves. This seems a natural class of domains; of course generalizations could be considered.
- Due to the incompressibility of the fluid, the starting zone and the target zone must have the same area, if one wants to be able to drive one to another by the flow of the velocity field.
- We have also to require that there is no topological obstruction to move a zone to the other one. In other words, one should be able to deform continuously a curve on the other one. Hence we will suppose that the two curves are *homotopic* in Ω.

Definition 3.3.1. We will say that the system satisfies the exact *Lagrangian* controllability property, if given two smooth Jordan curves γ_0, γ_1 in Ω, homotopic in Ω and surrounding the same area, a time $T > 0$ and an initial datum v_0, there exists a control such that the flow given by the velocity field drives γ_0 to γ_1, by staying inside the domain.

An objection. But one can see that the exact Lagrangian controllability does not hold in general. As way to see this is the following. Denote $\Phi^v(t, s, x)$ the flow associated to the velocity field v; see (3.13). Now we observe the following.

- Let us suppose that $\omega_0 := \operatorname{curl} v_0 = 0$. In that case if the flow $\Phi^v(t, 0, x)$ maintains γ_0 inside the domain, then for all t we have that in the neighborhood of $\Phi^v(t, 0, \gamma_0)$.
$$\omega(t, \cdot) = \operatorname{curl} v(t, \cdot) = 0,$$
  since, due to (3.11), the vorticity satisfies $\omega(t, \Phi^v(t, 0, x)) = \omega_0(x)$.
- Since $\operatorname{curl} v = \operatorname{div} v = 0$, locally around the points of γ_0, v is the gradient of a harmonic function; v is therefore real-analytic in a neighborhood $\Phi^v(t, 0, \gamma_0)$.
- Hence if γ_0 is real-analytic, its real-analyticity is propagated over time.
- Now if γ_1 is smooth but non real-analytic, we see that we cannot drive γ_0 to γ_1 by keeping the curve inside Ω. Hence the exact Lagrangian controllability cannot hold.

Approximate Lagrangian controllability

Since the exact Lagrangian controllability does not hold, this leads us, as for the controllability in the usual sense, to soften the question and wonder if it is possible, at least, to drive the initial fluid zone *arbitrarily close* to the target.

This motivates the following definition.

Definition 3.3.2. We will say that the system satisfies the property of *approximate Lagrangian controllability* in C^k, if given two smooth Jordan curves γ_0, γ_1 in Ω, homotopic in Ω and surrounding the same area, a time $T > 0$, an initial datum v_0 *and a real number* $\varepsilon > 0$, we can find a control such that the flow of the velocity field maintains γ_0 inside Ω for all time $t \in [0, T]$ and satisfies, up to reparameterization of the curves:

$$\|\Phi^v(T, 0, \gamma_0) - \gamma_1\|_{C^k} \leq \varepsilon.$$

Here, $(t, x) \mapsto \Phi^v(t, 0, x)$ is again the flow of the vector field v. One parameterizes Jordan curves by the circle $\mathbb{S}^1$.

Main result

The main result that we describe in this section is the following. It can be found in [48].

Theorem 3.3.3 (G.-Horsin). *Provided that $\Sigma \neq \emptyset$, the approximate Lagrangian controllability holds in all C^k.*

In other words, consider two smooth Jordan curves γ_0, γ_1 in Ω, homotopic in Ω and surrounding the same area. Let $k \in \mathbb{N}$. We consider $v_0 \in C^\infty(\overline{\Omega}; \mathbb{R}^2)$ satisfying

$$div(v_0) = 0 \text{ in } \Omega \text{ and } v_0.n = 0 \text{ on } [0, T] \times (\partial\Omega \setminus \Sigma).$$

For any $T > 0$, $\varepsilon > 0$, there exists a solution v of the Euler equation in $C^\infty([0, T] \times \overline{\Omega}; \mathbb{R}^2)$ with

$$v.n = 0 \text{ on } [0, T] \times (\partial\Omega \setminus \Sigma) \text{ and } v_{|t=0} = v_0 \text{ in } \Omega,$$

and whose flow satisfies

$$\forall t \in [0, T], \ \Phi^v(t, 0, \gamma_0) \subset \Omega,$$

and up to reparameterization

$$\|\gamma_1 - \Phi^v(T, 0, \gamma_0)\|_{C^k} \leq \varepsilon.$$

3.3.1.2 A Connected Result: Vortex Patches

We now discuss a closely related problem. Indeed, it turns out that the techniques used to prove Theorem 3.3.3 can be used to answer the following question: is it possible to control the shape of vortex patches?

Let us first explain what vortex patches are. The starting point is the following, see [86].

Theorem 3.3.4 (Yudovich (1963)). *For any $v_0 \in C^0(\overline{\Omega}; \mathbb{R}^2)$ such that div $(v_0) = 0$ in Ω, $v_0.n = 0$ on $\partial\Omega$ and* **curl** $v_0 \in L^\infty$, *there exists a unique (weak) global solution of the Euler equation starting from v_0 and satisfying $v.n = 0$ on the boundary.*

A particular case of initial data with vorticity in $L^\infty(\Omega; \mathbb{R})$ is the one of *vortex patches*.

Definition 3.3.5. A vortex patch is a solution of the Euler equation whose initial datum is the characteristic function of the interior of a smooth Jordan curve (at least $C^{1,\alpha}$).

An important result in the theory of vortex patches is the following, see [19, 20].

Theorem 3.3.6 ([19]). *In $\mathbb{R}^2$, the regularity of the boundary of the vortex patch is propagated globally in time.*

There are many other references on the subject of vortex patches, see also for instance: Bertozzi–Constantin [10], Danchin [33], Depauw [34], Dutrifoy [38], Gamblin and Saint-Raymond [42], Serfati [78], Sueur [81],...

Hence one can wonder whether it is possible, in the framework of the control system (3.24)–(3.25), to control the shape of a vortex patch, that is, to control the evolution of its boundary. Let us underline that this problem is different from the one considered above, because in the context of vortex patches, the solutions are not regular, while in Theorem 3.3.3 the solutions are smooth.

A result that one can prove is the following.

Theorem 3.3.7 (G.-Horsin). *Consider two smooth Jordan curves γ_0, γ_1 in Ω, homotopic in Ω and surrounding the same area. Suppose also that the control zone Σ is in the exterior of these curves. Let $v_0 \in \mathscr{L}ip(\overline{\Omega}; \mathbb{R}^2)$ with $v_0.n \in C^\infty(\partial\Omega)$ a vortex patch initial condition corresponding to γ_0, i.e. such that*

$$curl\,(v_0) = \mathbf{1}_{Int(\gamma_0)} \text{ in } \Omega, \quad div\,(v_0) = 0 \text{ in } \Omega, \quad v_0.n = 0 \text{ on } \partial\Omega \setminus \Sigma.$$

Then for any $T > 0$, any $k \in \mathbb{N}$, any $\varepsilon > 0$, there exists $u \in L^\infty([0, T]; \mathscr{L}ip(\overline{\Omega}))$ a solution of the Euler equation such that

$$curl\,v = 0 \text{ on } [0, T] \times \Sigma,$$

$$v.n = 0 \text{ on } [0, T] \times (\partial\Omega \setminus \Sigma) \text{ and } v_{|t=0} = v_0 \text{ in } \Omega,$$

that $\Phi^v(T, 0, \gamma_0)$ does not leave the domain and that, up to reparameterization, one has

$$\|\gamma_1 - \Phi^v(T, 0, \gamma_0)\|_{C^k} \leq \varepsilon.$$

Note that in the above result, we impose the entering vorticity to be zero. The reason for this is that we want the vortex patch to stay a vortex patch; hence we do not want to add vorticity inside the domain.

Remark 3.3.8. Let us focus on the regularity of the velocity field:

- As long as the patch stays regular, one has $v(t, \cdot) \in \mathscr{L}ip(\Omega)$ (see for instance [20]).
- Without the regularity of the patch, the velocity field $v(t, \cdot)$ is merely log-Lipschitz:

$$|v(t, x) - v(t, y)| \lesssim |x - y| \max(1, -\log(|x - y|)).$$

This estimate is a central argument in [85, 86].

   Hence we obtain a result on the shape of the patch in C^k, despite the fact that the velocity field is Lipschitz only. This is connected to the fact that this velocity field is in fact more regular in several "good directions", see the references above.

3.3.2   Ideas of Proof

Let us now give a few ideas of the proofs of Theorems 3.3.3 and 3.3.7.

3.3.2.1   The Main Proposition

We exploit the same idea to use *potential flows* and complex analysis as in Sect. 3.2. If we follow the ideas of Sect. 3.2, we would like to find a potential flow which makes the fluid approximately go from one zone to another. In particular, this will answer to the problem in the particular case where $v_0 = 0$.

   Precisely, the core of the proof is to show the following proposition.

Proposition 3.3.9. *Consider two smooth Jordan curves γ_0, γ_1 in Ω, homotopic in Ω and surrounding the same area. For any $k \in \mathbb{N}$, $\varepsilon > 0$, there exists $\theta \in C_0^\infty([0, 1]; C^\infty(\overline{\Omega}; \mathbb{R}))$ such that*

$$\Delta_x \theta(t, \cdot) = 0 \text{ in } \Omega, \text{ for all } t \in [0, 1],$$

$$\frac{\partial \theta}{\partial n} = 0 \text{ on } [0, 1] \times (\partial\Omega \setminus \Sigma),$$

whose flow satisfies

$$\forall t \in [0, 1], \ \Phi^{\nabla\theta}(t, 0, \gamma_0) \subset \Omega,$$

and, up to reparameterization of the curves,

$$\|\gamma_1 - \Phi^{\nabla\theta}(1, 0, \gamma_0)\|_{C^k} \le \varepsilon.$$

In other words, there exists a potential flow driving γ_0 to γ_1 (approximately in C^k) and fulfilling the boundary condition on $\partial\Omega \setminus \Sigma$. The time interval here is fixed to be $[0, 1]$; one can change the parameterization in time to transform it into any $[0, T]$.

A large part of the proof consists in establishing this proposition. This is proven in two steps:

- *Part 1:* Find a solenoidal (divergence-free) vector field driving γ_0 to γ_1.
- *Part 2:* Approximate (at each time) the above vector field on the curve (or to be more precise, its normal part), by the gradient of a harmonic function defined on $\overline{\Omega}$ and satisfying the constraint on $\partial\Omega \setminus \Sigma$.

3.3.2.2   Part 1: Finding a Solenoidal Vector Field Driving Exactly γ_0 to γ_1

In this paragraph, we consider the problem of driving γ_0 to γ_1 (exactly), by a divergence-free vector field. Of course, this constraint on the vector field is significantly weaker than the constraint to be a potential flow. In return, one can obtain an exact result. In more precise form, one can prove the following proposition. We denote by $\text{Int}(\gamma)$ the interior of a Jordan curve γ in the sense of Jordan's theorem, that is, the (unique) bounded component of $\mathbb{R}^2 \setminus \gamma$. We also denote by $|A|$ the Lebesgue measure of a measurable subset $A \subset \mathbb{R}^2$.

Proposition 3.3.10. *Consider γ_0 and γ_1 two smooth (C^∞) Jordan curves which are homotopic in Ω and satisfy*

$$|Int(\gamma_0)| = |Int(\gamma_1)|. \tag{3.26}$$

Then there exists $v \in C_0^\infty((0, 1) \times \Omega; \mathbb{R}^2)$ such that

$$div\, v = 0 \text{ in } (0, 1) \times \Omega,$$

$$\Phi^v(1, 0, \gamma_0) = \gamma_1.$$

Note that, even without the divergence constraint, the result is not trivial (but it is known for a long time in that case). Indeed, the two Jordan curves being homotopic means that one can find a continuous function $\Gamma : [0, 1] \times \mathbb{S}^1 \to \Omega$ such that $\Gamma(0, \cdot) = \gamma_0$ and $\Gamma(1, \cdot) = \gamma_1$. But it does not say that the deformation is regular, nor the fact that for $t \in (0, 1)$, $\Gamma(t, \cdot)$ is still a Jordan curve...

Ideas of proof of Proposition 3.3.10

As a matter of fact, the case where $\text{Int}(\gamma_0)$ and $\text{Int}(\gamma_1)$ do not intersect can be treated rather simply. An idea in this case would be for instance to draw a "pipe" between the two domains, and to "blow" the first domain into the second one, through the pipe.

But this is less clear if the two domains intersect. In that case, one has in particular to be sure that the deformation of γ_0 does not self-intersect. And one should keep in mind that we have the constraint that the fluid zone under view should stay inside Ω: there could be very few room left inside Ω (in particular, it may be impossible to separate the two zones in order to apply the strategy described above) Now we reason as follows.

1. A way to treat the general case is to get in the opposite case where the two zones intersect:
$$\text{Int}(\gamma_0) \cap \text{Int}(\gamma_1) \neq \emptyset. \tag{3.27}$$

   To prove that one can reduce the study to the case described by (3.27), it is enough to find a solenoidal vector field v driving some point of γ_0 inside $\text{Int}(\gamma_1)$ (while letting some other point outside), and to consider $\Phi^v(1, 0, \gamma_0)$ as a new initial curve. Note that one cannot have $\gamma_0 \subset \text{Int}(\gamma_1)$ due to (3.26).

   Constructing such a vector field is not difficult. Indeed, we have much flexibility to construct a solenoidal vector field: any vector field of the form

$$v(t, x) = \nabla^\perp \psi(t, x) = (-\partial_{x_2} \psi, \partial_{x_1} \psi), \tag{3.28}$$

   is automatically solenoidal. Hence a possible procedure is the following:

   - Draw a smooth curve $\mathscr{C}$ in Ω from some point of γ_0 to some point inside $\text{Int}(\gamma_1)$.
   - Introduce the velocity vector $\dot{\mathscr{C}}$ on the graph $(t, \mathscr{C}(t))$ of the curve.
   - Extend this field on $[0, 1] \times \Omega$ with the form (3.28); using a cutoff function (applied to ψ) if necessary to ensure that this field is compactly supported in Ω.

   One can check that this procedure allows to construct v as claimed.
   Let us add, that, using a small translation if necessary, we can moreover suppose that γ_0 and γ_1 intersect transversally. Of course, a translation is obtained by the flow of a solenoidal vector field, and again we can make the corresponding vector field compactly supported in Ω. And by a small translation of γ_0, one can make the curves transverse (by using the parametric form of Thom's transversality Theorem for instance).
2. We are now in the situation described by Fig. 3.14. As a matter fact, things can be way messier, but let us give the idea of the proof when Ω is simply connected, so that $\partial\Omega$ is connected and no connected components of $\partial\Omega$ can be enclosed by γ_0 and γ_1. The goal is to deform γ_0 on γ_1 in an area-preserving way.

   Now to make the construction, as described above, we first define the vector field on the curve itself as it evolves through time, and then to extend it on the whole $[0, T] \times \Omega$. We work only inside the symmetric difference of the two interiors (colored on Fig. 3.14) to deform one curve to another (see the arrows in Fig. 3.14). The goal is, on each component of this symmetric difference, to find a vector field which drives the segment of γ_0 to the one of γ_1. This can be done

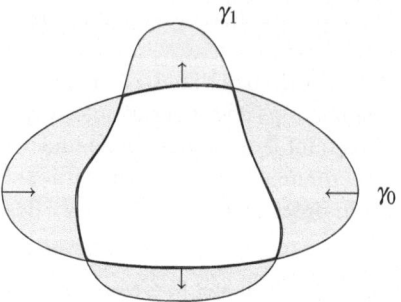

Fig. 3.14 Deforming one curve on another

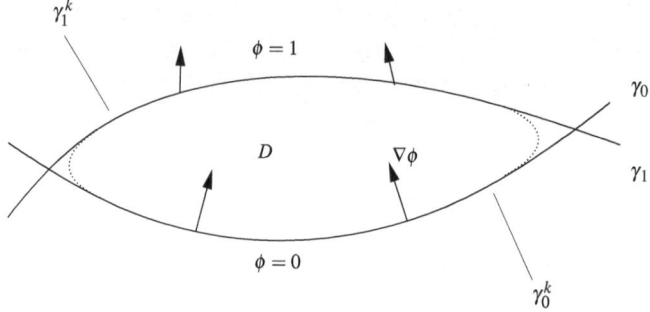

Fig. 3.15 Construction of a vector field on S^k

inside $\text{Int}(\gamma_0)$ (zones on the left and on the right in Fig. 3.14) or inside $\text{Int}(\gamma_1)$ (zones on the top and on the bottom in Fig. 3.14). To do so, there are several steps:

- Denote S_k the components of the symmetric differences. Each S_k is circumscribed by a segment of γ_0 that we denote γ_0^k and a segment of γ_1 that we denote γ_1^k. We aim at constructing a vector field driving the interval γ_0^k of γ_0 on the interval γ_1^k of γ_1.
- A way to do this (inspired from [24]) is to consider $\nabla \phi$ where ϕ is the harmonic extension of a function g equal to 0 (respectively 1) on the interval γ_0^k (resp. on the interval γ_1^k) and "regularized near the intersections" $\gamma_0^k \cap \gamma_1^k$. The regularization near the intersections consists in introducing near the two intersection points a small curve joining γ_0^k and γ_1^k so that the resulting domain D is smooth (see Fig. 3.15), and considering g going smoothly from 0 to 1 on these small curves. We extend ϕ in a harmonic way in D, and then the extension of the vector field to the whole domain S_k can be done by using local coordinates near the two intersection points (which is made easy due to the transversality of γ_0 and γ_1).
- Next we normalize the vector fields in order that the corresponding flow, satisfies in each S_k and for all $t \in [0, 1]$:

$$\text{Area}(\gamma_0^k, \Phi(t, 0, \gamma_0^k)) = t \, \text{Area}(\gamma_0^k, \gamma_1^k), \tag{3.29}$$

where $\text{Area}(\gamma_0^k, \Phi(t, 0, \gamma_0^k))$ denote the area enclosed between γ_0^k and $\Phi(t, 0, \gamma_0^k)$.

- Then we have to glue these vector fields defined in each S_k together, in a way that it is smooth at the points of $\gamma_0 \cap \gamma_1$. Again, we can use local coordinates to make an explicit construction here.
- Then, thanks to (3.29), the vector field restricted to $\{(t, \Phi(t, 0, \gamma_0))\}$ can be extended to a global solenoidal vector field.

3.3.2.3   Part 2: Approximating the Flow of the Reference Vector Field by a Potential Flow

Now that we have a reference vector field driving γ_0 to γ_1, we have to explain how we can approximate its flow on γ_0 by the action of a potential flow. This is given by the following proposition.

Proposition 3.3.11. *Let γ_0 a smooth (C^∞) Jordan curve; let $X \in C^0([0, 1]; C^\infty (\overline{\Omega}))$ a smooth solenoidal vector field, with $X.n = 0$ on $[0, 1] \times \partial\Omega$. Then for all $k \in \mathbb{N}$ and $\varepsilon > 0$ there exists $\theta \in C^\infty([0, 1] \times \overline{\Omega}; \mathbb{R})$ such that*

$$\Delta_x \theta(t, \cdot) = 0 \text{ in } \Omega, \text{ for all } t \in [0, 1],$$

$$\frac{\partial \theta}{\partial n} = 0 \text{ on } [0, 1] \times (\partial\Omega \setminus \Sigma),$$

and whose flow satisfies

$$\forall t \in [0, 1], \ \Phi^{\nabla\theta}(t, 0, \gamma_0) \subset \Omega,$$

and, up to reparameterization,

$$\|\Phi^X(t, 0, \gamma_0) - \Phi^{\nabla\theta}(t, 0, \gamma_0)\|_{C^k} \leq \varepsilon, \ \forall t \in [0, 1].$$

Ideas of proof for Part 2

The proof follows three successive steps of growing generality, namely:

- We first treat the case when all the data, that is, both γ_0 and X are real-analytic (in the x-variable for the latter).
- Then we relax the assumption by assuming only X to be real-analytic (while γ_0 is merely C^∞).
- And finally we relax the assumption by assuming only C^∞ smoothness of the data.

Again, for simplicity, we assume that Ω is simply connected (but this is not as crucial as for the controllability in the usual sense.)

First step: when the data are real-analytic:

$$\gamma_0 \in C^\omega(\mathbb{S}^1; \mathbb{R}^2) \text{ and } X \in C^0([0, 1]; C^\omega(\overline{\Omega})).$$

Let $\gamma(t) := \Phi^X(t, 0, \gamma_0)$. For any t, this is a real-analytic curve. The main principle is the following. If we want the action of the potential flow $\nabla\theta$ on γ_0 to generate exactly $\gamma(t)$ (up to reparameterization), we only have to mimic the normal part of X on $\gamma(t)$ (the tangential part is "absorbed by the reparameterization".)

Hence it is natural to consider for each time the solution of the following elliptic problem:

$$\begin{cases} \Delta_x \psi(t, \cdot) = 0 \text{ in } \mathrm{Int}(\gamma(t)), \\ \dfrac{\partial \psi}{\partial n}(t, \cdot) = X(t, \cdot).n(\cdot) \text{ on } \gamma(t), \\ \int_{\gamma(t)} \psi(t, \cdot)\, d\sigma = 0. \end{cases}$$

This is the only harmonic function defined in $\mathrm{Int}(\gamma(t))$ which has exactly the "correct" normal part on $\gamma(t)$. Unfortunately, this function cannot be extended as a harmonic function on Ω in general, nor a fortiori in a way that satisfies $\partial_n \psi = 0$ on $[0, 1] \times (\partial\Omega \setminus \Sigma)$.

But here is the place where the real-analyticity plays a crucial role: as $\gamma(t)$ and $X.n$ on $\gamma(t)$ are analytic, we can extend the solution ψ across the boundary $\gamma(t)$ (this is a classical Cauchy–Kowalewsky-style result, see for instance [71]).

Moreover, using the continuity in time of X and γ with values in C^ω (see e.g. [59] for more details on the topology of C^ω), we see that the size of the neighborhood of $\gamma(t)$ where this solution can be extended can be estimated from below.

Now this vector field is still not globally defined on Ω, but we can use Runge's theorem in a similar way as in Sect. 3.2. Proceeding in the same way, we can obtain approximations defined on $\overline{\Omega}$, and which satisfy

$$\nabla\tilde{\psi}(t, \cdot).n = 0 \text{ on } \partial\Omega \setminus \Sigma. \tag{3.30}$$

As previously, (3.30) is not obtained exactly in a first time, but one can remove the solution of a suitable Neumann problem to get this relation exactly.

Finally, we obtain the function θ as:

$$\theta(t, x) = \sum_{k=1}^{n} \rho_i(t)\tilde{\psi}(t_i, \cdot),$$

with ρ_i a certain partition of unity of $[0, 1]$. (We use that Runge's approximation obtained at time t, is still an acceptable approximation in some neighborhood of t, so that by compactness of $[0, 1]$ we can consider only a finite number of $\tilde{\psi}(t_i, \cdot)$.)

The rest of the proof consists in explaining why the cost of changing X by $\nabla\theta$ is small: this is mainly a Gronwall's lemma and the use of reparameterization to compensate the discrepancy of the tangential components of the vector fields. By this process, by choosing a sufficiently small parameter in Runge's theorem, and using the fact that the size of the neighborhood of $\gamma(t)$ on which ψ can be extended is uniform, one can also obtain estimates of $\|\nabla\theta\|_{C^k}$ on $\Phi^{\nabla\theta}(t, 0, \gamma_0)$ in terms of $\|\nabla\psi(t)\|_{C^k}$ on $\gamma(t)$ only.

Second step: when only the vector field is real analytic:

$$X \in C^0([0, 1]; C^\omega(\overline{\Omega})) \text{ but } \gamma_0 \in C^\infty(\mathbb{S}^1; \mathbb{R}^2).$$

The idea is of course to use the previous step. We can approach γ_0 by real analytic curves, from the outside. This comes from a general result by H. Whitney [84], or in a simpler way in our case:

- We consider C_0 the complement of $\mathrm{Int}(\gamma_0)$ in the Riemann sphere. By Riemann's conformal mapping theorem, there exists φ a conformal transformation from C_0 to $\overline{B}_\mathbb{C}(0, 1)$.
- Then, since such a conformal transformation is regular up to the boundary when γ_0 is regular (say, C^∞) (this is Kellogg–Warschawski's theorem, see e.g. [74]), the curve $\varphi(S(0, 1 - \nu))$ is an appropriate approximation as $\nu \to 0^+$.
- Next, we apply the process of Part 1 on the ν-approximation γ_0^ν of γ_0. We obtain a function θ^ν. Call $\gamma_\nu(t) := \Phi^{\nabla\theta^\nu}(t, 0, \gamma_0^\nu)$.
- The central point is to show that, on $\gamma^\nu(t)$, we have uniform estimates on $\nabla\theta^\nu$ as $\nu \to 0^+$.
- Due to the construction in the Step 1, we have only to prove uniform estimates on the $\nabla\psi^\nu(t)$ constructed on $\gamma^\nu(t)$ as $\nu \to 0^+$.
- This is obtained by noting that the constants in elliptic estimates in $\mathrm{Int}(\gamma^\nu(t))$ are bounded independently from ν. Indeed, $\gamma^\nu(t)$ converges to $\gamma(t)$ for the C^∞ topology. It follows that we have uniform estimates on θ *inside* $\mathrm{Int}(\gamma^\nu(t))$, in all the C^k-norms in terms of X, γ_0 and k only. In particular, these estimates do not blow up as $\nu \to 0^+$.
- We are then able to conclude by Gronwall's lemma, because $\Phi^{\nabla\theta^\nu}(t, 0, \gamma_0)$ is precisely included in $\mathrm{Int}(\gamma^\nu(t))$ since γ_0 is in the inside of γ_0^ν.

Third step: when both data are merely C^∞:

$$\gamma_0 \in C^\infty(\mathbb{S}^1; \mathbb{R}^2) \text{ and } X \in C^0([0, 1]; C^\infty(\overline{\Omega})).$$

Again this is a consequence of the previous step. We use Whitney's analytic approximation theorem [83]: X can be approached arbitrarily for the $C^0([0, 1]; C^\infty(\overline{\Omega}))$-topology by $X^n \in C^0([0, 1]; C^\omega(\overline{\Omega}))$.
Hence we construct by the step above potential flows corresponding to X^n. Then we can prove by using the previous step and Gronwall's lemma that for n sufficiently large, we have a good approximation of the flow on γ_0.

3.3.2.4   How to Deduce the Results from the Main Proposition

The idea uses here the same argument of time-scale invariance of the Euler equation.

1. As before, we first consider the case when $\|v_0\|_{C^{k+1,\alpha}} \ll 1$. In that case, proceeding as previously, one can construct a solution of the Euler equation, starting from v_0, such that the normal velocity on the boundary is mainly $\nabla\theta$ (if fact we have to take $v_0.n$ into account), and such that $\|v(t,\cdot)\|_{C^{k+1,\alpha}}$ is of the same order as $v_0\|_{C^{k+1,\alpha}}$. To make the construction, one can use an analogous fixed point scheme as in Sect. 3.2, using an extension operator (but here we do not need the cutoff functions). Then standard perturbation arguments show that one has

$$\|\Phi^v(T,0,\gamma_0) - \gamma_1\|_{C^k} \le \|\Phi^v(T,0,\gamma_0) - \Phi^{\nabla\theta}(T,0,\gamma_0)\|_{C^k}$$
$$+ \|\Phi^{\nabla\theta}(T,0,\gamma_0) - \gamma_1\|_{C^k} \lesssim \|v_0\|_{C^{k+1,\alpha}} + \varepsilon.$$

2. Then one uses again the time scale invariance of the equation as follows. We cut the time interval in two parts: for $v > 0$ small, there are two phases, namely, during the time intervals $[0, T - v]$ and $[T - v, T]$, such as described in Fig. 3.16. The control is performed as follows:

   - In a first time, during the time interval $[0, T - v]$, we "do nothing", that is we mainly wait. In fact, we have to take $v_0.n$ into account, and to preserve the regularity of the solution. But we have no other purpose during this time interval than to wait the second phase and to let the size of the solution v stay of the same order as v_0. This can be done by introducing the same type of fixed point scheme as previously. We can drive the normal part of the velocity on Σ to 0 during this phase.
   - At the very end of the time interval, that is, during $[T - v, T]$, we act fast and violently to drive $\tilde\gamma_0 := \Phi(T - v, 0, \gamma_0)$ to γ_1. The control is given by the normal part of $\frac{1}{v}\nabla\theta(t - T + v, \cdot)$ for what concerns the normal velocity. The part of the control concerning the vorticity is used just in order not to ruin the regularity and that the size of the vorticity stays of the same order as ω_0.

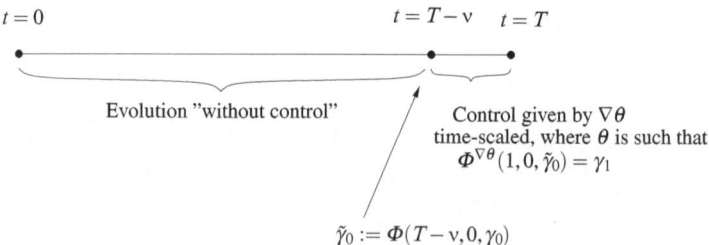

Fig. 3.16 The two phases of the control

3. Now let v be the resulting solution in $[0, T] \times \Omega$. If we change back the time scale to get back the dynamics of the time interval $[T - v, T]$ to the time interval $[0, 1]$, the evolution is driven by the Euler equation, with:

- As boundary condition (on the normal trace) the same as $\nabla \theta$.
- As initial condition $vu(T - v, \cdot)$, which is clearly small as $v \to 0^+$.

Hence as in Sect. 3.2, we are in the same situation as if the initial datum was small! And we can show that the solution that we constructed on $[0, T]$ satisfies:

$$\|\Phi^v(T, 0, \gamma_0) - \gamma_1\|_{C^k} \lesssim v + \varepsilon.$$

This allows to prove Theorem 3.3.3.

The case of vortex patches

Let us now say a few words concerning the proof of Theorem 3.3.7. The construction is similar, but we can no longer use

$$\|vu(T - v, \cdot)\|_{C^{k+1,\alpha}} \lesssim v,$$

because v is Lipschitz only! But we use instead arguments due to:

- Depauw [34], which has studied vortex patches in a domain, and showed that the regularity propagates as in Chemin's theorem; hence the "first phase" $[0, T - v]$ can be done in the same way.
- Bertozzi–Constantin [10], who tackled the problem of the regularity of vortex patches by using the integro-differential equation satisfied by their boundary γ:

$$\frac{d}{dt}\gamma(t, s) = -\frac{1}{2\pi} \int_0^{2\pi} \log|x - \gamma(\sigma)| \, \tau(\sigma) \, d\sigma$$

$+$ here, terms due to the presence of $\partial \Omega$ and of the control.

Using this approach, we can see that, despite the fact that the flow is merely Lipschitz, it propagates the regularity of the boundary of the patch. And including the terms due to the boundary and to the control is not a real issue, since these terms are regular.

3.3.3 *Comments*

The main reference concerning this section is [48], where the technical details are written. This lets several problems open, though.

What about 3D?

Several problems appear when considering the dimension 3:

- How to deform (in a smooth, volume-preserving way) a domain to another one?
- How to prevent the solution from potentially blowing up?
- The others parts of the proof do not depend on the dimension...

Results in that direction are partial:

Proposition 3.3.12 (G.-Horsin, in progress). *If B_1 and B_2 are two smooth open sets in Ω, diffeomorphic to a ball, with same volume, at positive distance from $\partial\Omega$ and disjoint, then one can smoothly deform B_1 to B_2 inside Ω in a volume-preserving manner.*

This yields to

Corollary 3.3.13 (G.-Horsin, in progress). *Let B_1 and B_2 as previously, and S_1, S_2 their boundary. Let $k \in \mathbb{N}$. We consider $v_0 \in C^\infty(\overline{\Omega}; \mathbb{R}^3)$ satisfying*

$$div\,(v_0) = 0 \text{ in } \Omega \text{ and } v_0.n = 0 \text{ on } [0, T] \times (\partial\Omega \setminus \Sigma).$$

For any $\varepsilon > 0$, there exists $T > 0$ and a solution v of the Euler equation in $C^\infty([0, T] \times \overline{\Omega}; \mathbb{R}^3)$ with

$$v.n = 0 \text{ on } [0, T] \times (\partial\Omega \setminus \Sigma) \text{ and } v_{|t=0} = v_0 \text{ in } \Omega,$$

and whose flow satisfies

$$\forall t \in [0, T], \ \Phi^v(t, 0, B_1) \subset \Omega,$$

and up to reparameterization

$$\|S_2 - \Phi^v(T, 0, S_1)\|_{C^k} \leq \varepsilon.$$

The fact that the result is valid *for short control times* comes as a way to avoid blow-up.

Open problems

Other open problems can be raised in this field:

- *More complex domains.* What can be said if the fluid zone to be displaced is no longer a Jordan domain, or about more general situations in 3D?
- *Numerics.* Can we find an efficient algorithm to compute the control?
- *Navier–Stokes equations.* Can we obtain a similar result for incompressible Navier–Stokes equations?

$$\begin{cases} \partial_t v + (v.\nabla)v - \Delta v + \nabla p = 0 \text{ in } [0, T] \times \Omega, \\ \text{div } v = 0 \text{ in } [0, T] \times \Omega. \end{cases}$$

This question can be raised both in the cases of Dirichlet's boundary conditions and with Navier's (for which one could try to use the techniques of Coron [28] and Chapouly [18]).

- *Stabilization.* Can we find a feedback control:

$$\text{control(t)} = f(\gamma(t), v(t)),$$

stabilizing a fluid zone at a fixed place?

3.4 Controllability of the 1D Isentropic (Compressible) Euler Equation

In this last section, we consider a different model, namely the 1D isentropic compressible Euler equation. Despite the fact that this equation, like the incompressible Euler equation, models an inviscid fluid evolving under the influence of pressure, the mathematical properties of the two equations are rather different. However, for what concerns the controllability of this equation, the basic principle of using the return method is common, even if it takes different forms.

3.4.1 Introduction

The models that we consider here are the following. There are two versions of the one-dimensional isentropic Euler equations: in Eulerian coordinates or in Lagrangian coordinates (that is, when following the flow). These equations read:

- In Eulerian coordinates:

$$\begin{cases} \partial_t \rho + \partial_x (m) = 0, \\ \partial_t (m) + \partial_x (\frac{m^2}{\rho} + \kappa \rho^\gamma) = 0. \end{cases} \tag{EI}$$

- In Lagrangian coordinates (the equation is also known as the *p*-system):

$$\begin{cases} \partial_t \tau - \partial_x v = 0, \\ \partial_t v + \partial_x (\kappa \tau^{-\gamma}) = 0. \end{cases} \tag{P}$$

Above, the various notations are:

- $t \in \mathbb{R}^+$ is the time, $x \in \mathbb{R}$ is the position.
- $\rho = \rho(t, x) \geq 0$ is the density of the fluid.

- $m(t, x)$ is the momentum ($v(t, x) = \frac{m(t,x)}{\rho(t,x)}$ is the velocity of the fluid).
- $\tau := 1/\rho$ is the specific volume.
- The pressure law is $p(\rho) = \kappa\rho^\gamma = \kappa\tau^{-\gamma}$, $\gamma \in (1, 3]$, $\kappa > 0$.

As is well-know, these two equations stand respectively for the conservation of mass and momentum in the fluid.

The controllability problem

What we consider in this chapter is the study of the above equations from the point of view of controllability. The equation will be posed on a bounded domain of the real line, say $[0, 1]$; hence (t, x) belongs to $[0, T] \times [0, 1]$.

The *state* of the system will be the couple of both unknowns, that is:

$$\text{Case (EI): } u = (\rho, m), \quad \text{Case (P): } u = (\tau, v). \tag{3.31}$$

The *control* will be the "boundary data", which is a very delicate matter for this type of equations (see for instance [4, 5, 37]). As before, in order to avoid dealing with this issue, we will not look for the control itself, but rather for the solution itself. Hence we will not focus on this aspect.

Finally the *controllability problem* is the following: given $u_0 = (\rho_0, m_0)$ (or $u_0 = (\tau_0, v_0)$) and $u_1 = (\rho_1, m_1)$ (or $u_1 = (\tau_1, v_1)$), can we find *a solution* of the system driving u_0 at initial time to u_1 at time T? For which T?

Class of solutions

Both (EI) and (P) are classical examples of *hyperbolic systems of conservation laws:*

$$u_t + f(u)_x = 0,$$

where $u : \mathbb{R}^+ \times \mathbb{R} \to \mathbb{R}^n$ and $f : \mathbb{R}^n \to \mathbb{R}^n$ (see the next section for more details on this class of equations).

Hyperbolic systems of conservation laws are known to develop singularities in finite time. This is due to the mechanism of *formation of shocks*, which are easy to see for instance for what concerns the Burgers equation (for which $n = 1$):

$$u_t + (u^2)_x = 0.$$

One can use the method of characteristics to show that u is constant along the characteristics associated to u (that is, $u(t, \Phi^u(t, 0, x)) = u(0, x)$). As a consequence, these characteristics are straight lines. For many u_0, the straight lines can cross. These leads to shock waves appearing in the solution, such as described in Fig. 3.17.

Fig. 3.17 Formation of shock

When considering control problems associated to equations such as (EI) or
(P), a possibility is to consider regular solutions (say C^1), whose existence for a
relevant interval of time is ensured by the smallness of the data. See for instance the
references given in Sect. 3.4.4 for such studies.

But from both mathematical and physical viewpoints, one should also consider
the case of *discontinuous* weak solutions in which *shock waves* may appear, which
are to be understood in the sense of distributions (his makes sense for $u \in L^\infty$
for instance). But a classical issue for what concerns weak solutions containing
discontinuities, is that in general in this context *uniqueness is lost*. Hence it is
natural to consider weak solutions which satisfy additional requirements, aimed
at selecting among all weak solutions, the physically relevant one. These will be
called *entropy conditions*. Here we will consider a special class of *entropy solutions*.
These solutions will be of bounded variation in the variable x uniformly in t, that
is will belong to $L^\infty(0, T; BV(\mathbb{R}))$. Moreover these solutions will be of small total
variation (mainly) and will avoid vacuum; as we will see they are constructed by
a particular technique known as the *wave front tracking algorithm*. We discuss this
more precisely in the next section.

Let us underline that it is very important to specify which class of solutions are
considered (regular solutions or weak entropy solutions), because the properties
of the equation in the two contexts are really different. For instance, the system is
reversible in the case of C^1 solutions, not in the context of weak entropy solutions.
As the reader knows (or guesses), this is not a detail when it comes to controllability
questions.

3.4.2 Basic Facts on Systems of Conservation Laws

In this section, we recall some basic facts about (one-dimensional) systems of
conservation laws and a particular way to construct solutions of these systems
known as the *wave front tracking algorithm*. The reader familiar with this is
encouraged to skip the section; the one who would like to know more precisely the
theory (and to see the proofs) is referred to Bressan [13], Dafermos [32], Holden
and Risebro [52] or LeFloch [61].

3.4.2.1 Systems of Conservations Laws

Equations (EI) and (P) are PDEs of a particular class, known as *systems of conservation laws*. Here, we consider only one-dimensional problems, and these are written as follows

$$u_t + f(u)_x = 0, \quad f : \Omega \subset \mathbb{R}^n \to \mathbb{R}^n, \tag{3.32}$$

where f is a smooth *flux* function (let us say, of class C^2 to fix the ideas) satisfying the following *strict hyperbolicity condition*:

for all $u \in \Omega$, $A(u) := df(u)$ has n real distinct eigenvalues $\lambda_1(u) < \cdots < \lambda_n(u)$.

These scalar functions $\lambda_1, \ldots, \lambda_n : \Omega \to \mathbb{R}$ are the *characteristic speeds of the system*. Denote by $r_i(u)$, $i = 1 \ldots n$, some corresponding eigenvectors.

The theory concerning such systems is simplified when the characteristic fields (λ_i, r_i) satisfy a condition called the *genuine non-linearity* the sense of Lax [60]:

$$\nabla \lambda_i . r_i \neq 0 \quad \text{for all } u \text{ in } \Omega.$$

When this condition is satisfied—in particular this is the case for what concerns (EI) and (P)—, we normalize the vector fields r_i so that

$$\nabla \lambda_i . r_i = 1 \quad \text{in } \Omega. \tag{3.33}$$

For what concerns the two systems that we consider above, by standard computations one can show that (EI) and (P) satisfy the strict hyperbolicity condition (away from the vacuum $\rho = 0$) and that both characteristic fields are genuinely nonlinear. The characteristic speeds are as follows:

- Case (EI): $u = (\rho, m) \in \mathbb{R}^+ \times \mathbb{R}$:

$$\lambda_1 = \frac{m}{\rho} - \sqrt{\kappa \gamma} \rho^{\frac{\gamma-1}{2}} \quad \text{and} \quad \lambda_2 = \frac{m}{\rho} + \sqrt{\kappa \gamma} \rho^{\frac{\gamma-1}{2}},$$

- Case (P): $u = (\tau, v) \in \mathbb{R}^+ \times \mathbb{R}$:

$$\lambda_1 = -\sqrt{\kappa \gamma \tau^{-\gamma-1}} \quad \text{and} \quad \lambda_2 = \sqrt{\kappa \gamma \tau^{-\gamma-1}}.$$

One can see immediately an important difference between the two cases: for what concerns (P), the characteristic speeds have a constant sign, while this is not the case for (EI). This is very important for our problem, since the sign of the characteristic speed indicates the direction in which the solution propagates; and in particular the way the boundary control propagates inside the domain.

Entropy solutions

Now, as we indicated above, we will consider weak solutions which may contain discontinuities. Since in general uniqueness does not hold in this context, it is natural to introduce *entropy solutions*, which are weak solutions which fulfill additional admissibility conditions, aimed at selecting among the set of weak solutions, the physically acceptable one. A way to introduce the *entropy criterion* is the following.

One defines an *entropy/entropy flux couple* as a couple of functions $(\eta, q) : \Omega \to \mathbb{R}^2$ such that

$$\forall u \in \Omega, \quad D\eta(u).Df(u) = Dq(u).$$

Then one defines an *entropy solution*: as a (weak) solution of (3.32) such that for any entropy couple (η, q) with η *convex*, one has:

$$\eta(u)_t + q(u)_x \le 0, \tag{3.34}$$

in the sense of measures. Of course, if the solution u is regular, then (3.34) takes place as an equality, by the chain rule. This is no longer necessarily true for discontinuous solutions.

A way to justify the conditions (3.34) is the following. One can show that the solutions obtained by *vanishing viscosity*, i.e. as limits of solutions of the system where a small viscosity term has been added:

$$u_t^\varepsilon + (f(u^\varepsilon))_x - \varepsilon u_{xx}^\varepsilon = 0,$$

are entropy solutions. This explains the physical meaning of entropy solutions: in some sense, entropy solutions are solutions from which viscosity has disappeared, except for what concerns *the selection of admissible discontinuities*. We will see later another formulation of this selection at the level of a single discontinuity.

A celebrated result concerning hyperbolic systems of conservation laws with genuinely nonlinear fields is due to Glimm [50]. In this paper is shown the existence of global in time entropy solutions for such systems with the assumption that the initial data is of small total variation. The resulting entropy solution is then of small total variation uniformly in time.

There is now a huge literature on the subject, and it is virtually impossible to refer to all the works of the field in this course; see for instance the books [13, 32, 52, 61, 79] and references therein. Let us however underline that the situation is now well understood in the context of solutions with small total variation in the general case (not limited to the genuine nonlinearity assumption) for what concerns existence as well as uniqueness, stability issues, etc. See in particular Bianchini–Bressan [11].

Riemann problem

Now let us explain a way to construct solutions of (3.32). We will restrict ourselves to the case when $n = 2$ ("2×2 systems") and when both the fields are genuinely nonlinear. This is sufficient to treat (EI) and (P). The wave front tracking method uses as an elementary brick the solutions of the so-called *Riemann problem*, which consists in finding self-similar solutions $u = \bar{u}(x/t)$ to

$$\begin{cases} u_t + (f(u))_x = 0 \\ u_{|\mathbb{R}^-} = u_l \text{ and } u_{|\mathbb{R}^+} = u_r, \end{cases} \tag{3.35}$$

where u_l and u_r are constants of Ω. The fact that, given such initial data, one should look for self-similar solutions of (3.32) is due to the scale invariance of the equation under the change of variables $(t, x) \mapsto (\lambda t, \lambda x)$.

Of course, solutions of (3.35) are very particular cases of solutions of (3.32); however we will see that more general solutions can be constructed by "gluing together" pieces of solutions obtained as solutions to the Riemann problem.

Now in the particular case under view (the genuine nonlinearity is essential here), the Riemann problem can be solved by introducing *Lax's wave curves*. These are curves inside Ω which consist of all points $u_r \in \Omega$ that can be connected to a fixed u_l by particular solutions of (3.35), which are *shock waves* or *rarefaction waves*, which we now describe.

Elementary waves

Let us now describe these elementary waves:

* *Shock waves* are admissible discontinuous solutions joining u_l and u_r, as in Fig. 3.18. More precisely, a shock is a simple discontinuity between the states u_l and u_r (on the left and the right, respectively), traveling at speed s satisfying *Rankine–Hugoniot relations*:

$$[f(u)] = s[u] \quad \text{(jump condition)}, \tag{3.36}$$

and *Lax's inequalities*:

$$\lambda_i(u_r) < s < \lambda_i(u_l) \text{ and } \lambda_{i-1}(u_l) < s < \lambda_{i+1}(u_r). \tag{3.37}$$

Lax's inequalities are associated to each characteristic family ($i = 1, \dots, n$), and each shock satisfies exactly one of them. As a consequence, there is a family of shocks associated to each characteristic family ($i = 1, \dots, n$).

In (3.36), the brackets denote the jump of the quantity across the discontinuity: $[f(u)] := f(u_r) - f(u_l)$ and $[u] := u_r - u_l$.

The Rankine–Hugoniot relation (3.36) ensures that this is a solution in the sense of distributions, Lax's inequalities (3.37) (associated to each characteristic family) give the entropy criterion.

One can show that, for each $i \in \{1, \dots, n\}$, there is a curve $S-i$, passing through u_l and tangent to $r_i(u_l)$ at u_l, corresponding to the points u_r that fulfill (3.36).

Fig. 3.18 A shock wave

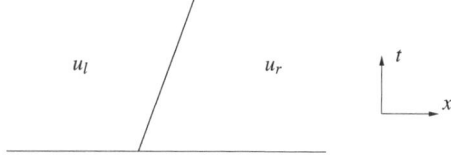

Fig. 3.19 A rarefaction wave

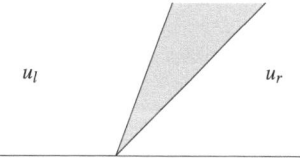

Only *half* of this curve satisfies (3.37). It is elementary to see that, as u_r tends to u_l along the i-th shock curve, one has $s \to \lambda_i(u_l)$.

- *Rarefaction waves* are *regular* (self-similar) solutions joining u_l to u_r, as described in Fig. 3.19. They are obtained with the help integral curves of r_i as follows. We introduce the orbits of the vector fields r_i

$$
\begin{cases}
\frac{d}{d\sigma} R_i(\sigma) = r_i(R_i(\sigma)), \\
R_i(0) = u_l.
\end{cases}
\tag{3.38}
$$

Now, for $\sigma \geq 0$, if $u_r = R_i(\sigma, u_l)$, then one can construct the following function:

$$
u(t,x) = \begin{cases}
u_l & \text{if } \frac{x}{t} < \lambda_i(u_l), \\
R_i(\sigma)(u_l) & \text{if } \frac{x}{t} = \lambda_i(R_i(\sigma)(u_l)), \\
u_r & \text{if } \frac{x}{t} > \lambda_i(u_r).
\end{cases}
\tag{3.39}
$$

Using (3.33) one sees that this gives a solution of (3.32).

Again, for each $i \in \{1, \ldots, n\}$, there is a curve, passing through u_l and tangent to $r_i(u_l)$ at u_l, corresponding to the orbit of the vector field r_i. But only *half* of this curve satisfies that the characteristic speed progresses across the wave (so that Fig. 3.19 is valid). Due to (3.33), this corresponds indeed to $\sigma \geq 0$.

Lax's wave curves

Now *wave curves* are constructed as follows. Given u_l, we associate:

- The curves of i-shocks (or to be more precise, the half curves of i-shocks), given by all states u_r which can be connected by u_l through a shock of the i-th family.
- The curves of i-rarefactions (or again, the half curves of i-rarefactions), given by all states u_r which can be connected by u_l through a rarefaction waves of the i-th family.

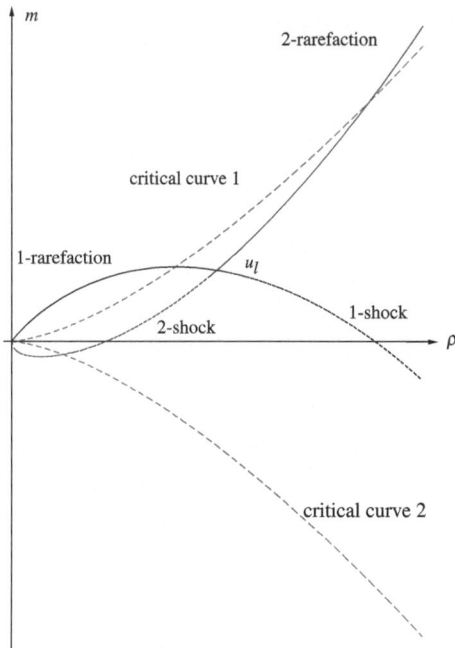

Fig. 3.20 Lax's curves for (EI) in (ρ, m) coordinates

- Lax's curves, which we will denote by Φ_i, obtained by gluing together these two half curves S_i and R_i.

One can show that Lax's curves are regular, because the i-shock curves and the i-rarefaction curves have a second-order tangency at u_l (with suitable parameterization). Figure 3.20 gives an example in the case of system (EI). The main point is that when u_r belongs to the i-th curve associated to u_l, that is to say, when $u_r = \Phi_i(\sigma, u_l)$, then there is an elementary wave joining u_l on the left to u_r on the right and giving an entropy solution to the Riemann problem.

In Fig. 3.20, we have also represented the critical curves defined as the locus where one of the characteristic speeds vanishes.

Remark 3.4.1. The curves that we describe above are *right* shock, rarefaction or wave curves, because they describe the states that can be connected on the right to some fixed left state u_l. We could define in the same way *left* shock, rarefaction or wave curves describing the states that can be connected on the left to some fixed right state u_r.

These curves allow to solve the Riemann problem (at least, when u_l and u_r are sufficiently close one to another, which is sufficient to our purpose, since we will consider small BV solutions). Indeed, Lax [60] proved that one can solve (at least locally) the Riemann problem by first following the 1-curve then

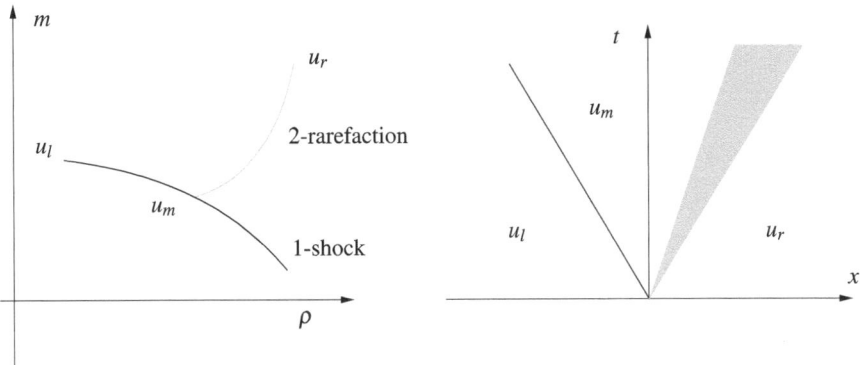

Fig. 3.21 Resolution of the Riemann problem

the 2-curve. In other words, one can connect any u_r sufficiently close to u_l by, first, a shock/rarefaction of the first family and, then, a shock/rarefaction of the second family, as in Fig. 3.21. Another way to express this result is to see that $(\sigma_1, \varsigma_2) \mapsto \Phi_2(\sigma_2, \Phi_1(\sigma_1, u_l))$ is locally onto near $(0,0)$; this is a consequence of the inverse mapping theorem. Moreover one can estimate (σ_1, σ_2) in terms of $\Phi_2(\sigma_2, \Phi_1(\sigma_1, u_l)) - u_l$ and vice versa, with constants independent of u_l, that is :

$$c(|\sigma_1| + |\sigma_2|) \leq |\Phi_2(\sigma_2, \Phi_1(\sigma_1, u_l)) - u_l| \leq (|\sigma_1| + |\sigma_2|). \qquad (3.40)$$

Riemann invariants

Let us finally introduce the *Riemann invariants*. We will say that $w^i : \Omega \to \mathbb{R}$ is a *i-Riemann invariant* when we have

$$r_i.\nabla w^i = 0 \text{ in } \Omega. \qquad (3.41)$$

It is elementary to determine for (EI) and (P) new coordinates given by a 1-Riemann invariant and a 2-Riemann invariant:

• Case (EI):

$$w^1(u) = \frac{m}{\rho} + \frac{2\sqrt{\kappa\gamma}}{\gamma - 1}\rho^{\frac{\gamma-1}{2}} \text{ and } w^2(u) = \frac{m}{\rho} - \frac{2\sqrt{\kappa\gamma}}{\gamma - 1}\rho^{\frac{\gamma-1}{2}},$$

• Case (P):

$$w^1(u) = v + \frac{2\sqrt{\kappa\gamma}}{\gamma - 1}\tau^{-\frac{\gamma-1}{2}} \text{ and } w^2(u) = v - \frac{2\sqrt{\kappa\gamma}}{\gamma - 1}\tau^{-\frac{\gamma-1}{2}}.$$

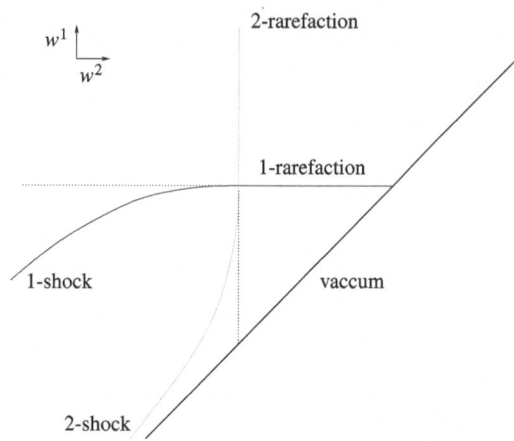

Fig. 3.22 Lax's curves in (w^1, w^2) coordinates

It is particularly interesting to parameterize the wave curves by these Riemann invariants, because in these coordinates, naturally, rarefaction curves become half straight lines, see Fig. 3.22.

Front-tracking algorithm

Now let us discuss a particular way to construct entropy solutions to systems of conservation laws, known as the *wave-front tracking algorithm*. This algorithm was introduced by Dafermos [31] in the scalar case ($n = 1$), and extended by Di Perna [36] for 2×2 systems, and then extended by Bressan [12], Risebro [75], Ancona–Marson [9], G.-LeFloch [49], etc.

Let us underline that there are other ways to construct entropy solutions of systems of conservation laws, such as Glimm's random choice method [50], the vanishing viscosity approach [11], etc.

The basic principle is as follows:

– Construct a suitable sequence of *piecewise constant functions over a polygonal subdivision* of $\mathbb{R}^+ \times \mathbb{R}$. These approximations are called *front-tracking approximations*.

– Prove estimates in $L_t^\infty(BV_x)$ for the approximations.

– Extract by compactness a converging subsequence. Then prove that the limit is an entropy solution.

To fulfill this purpose, an algorithm is the following (we more or less follow Di Perna [36], who considers 2×2 genuinely nonlinear systems):

• Let $\nu > 0$ (which will go to 0^+).
• Approximate the initial condition on $\mathbb{R}$ by piecewise constant functions: $u_0^\nu \to u_0$ in L_{loc}^1 as $\nu \to 0^+$.

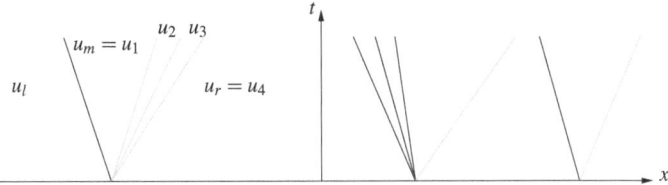

Fig. 3.23 A front-tracking algorithm, phase 1

- At each discontinuity of u_0^ν, let us say x_0:

  - Solve the corresponding Riemann problem (where the discontinuity is placed at x_0 rather than 0).
  - Replace rarefaction waves by *rarefaction fans*. These are piecewise constant functions according to the variable $\frac{x-x_0}{t}$, approximating the solution given by (3.39) (recentered to (t_0, x_0) instead of $(0, 0)$). To be more precise, let us consider as in Fig. 3.23 a rarefaction wave at $x = 0$, separating u_m and u_r, let us say $u_r = R_i(\sigma, u_m)$, $\sigma > 0$. Then introduce states $u_1 := u_m, u_2, \ldots,$ $u_k = u_r$ in a way that $u_{j+1} = R_i(s_j, u_j)$ with $0 < s_j \le \nu$ (and let us say, all s_j but s_{k-1} are equal to ν). Then the rarefaction fan is given by (for (t, x) close to $(0, x_0)$):

$$
u^\nu(t, x) = \begin{cases} u_m & \text{if } \frac{x-x_0}{t} < \lambda_i(u_m), \\ u_{j+1} & \text{if } \lambda_i(u_j) \le \frac{x-x_0}{t} < \lambda_i(u_{j+1}), \text{ for } j \le k - 1, \quad (3.42) \\ u_r & \text{if } \frac{x-x_0}{t} \ge \lambda_i(u_r). \end{cases}
$$

At this stage, we can hence construct front-tracking approximations for small times, by extending the discontinuities along straight lines, see Fig. 3.23. We have to explain how to extend them for all $t \ge 0$, precisely, to explain how we define the approximation after two such discontinuities meet. All discontinuities (representing a shock or approximating a rarefaction) are called *fronts*. We call an *interaction point* a point where to fronts meet.

- To extend the approximation across an interaction point, iterate the procedure without splitting again rarefactions (this is specific to 2×2 system). In other words, when two fronts meet, we solve the Riemann problem between the leftmost and the rightmost states, and for what concerns the rarefaction waves, we cut them into pieces as previously if there was no rarefaction front of the same family among the incoming fronts, or we approximate it by a single front otherwise. See Fig. 3.24.

One can show than this algorithm defines a piecewise constant function for all $t \ge 0$, with a finite number of fronts and discrete interaction points. (As matter of fact, to prove this, one uses estimates that are described below.)

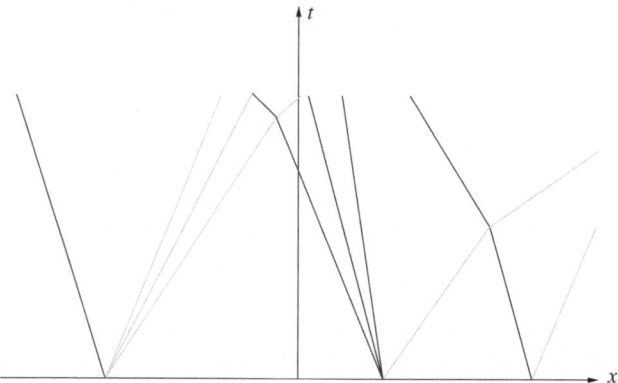

Fig. 3.24 A front-tracking algorithm, phase 2

Estimates for front tracking approximations

Now to complete the program, one has to prove estimates on front-tracking approximations, in order to get compactness and to be able to pass to the limit. (Actually, one already needs estimates to prove the above claim of well-defined approximations.)

A central argument is due to Glimm [50], and allows to obtain a bound on the total variation of the approximations, uniformly in time.

Consider the approximation u^ν obtained by the above process, defined on $\mathbb{R}^+ \times \mathbb{R}$. First, it is easy to see that the total variation of u^ν does not change except at interaction times. Hence one has only to analyze what happens at the interaction points. To that purpose, a first step is to decide a way to measure the size of a front in a front-tracking approximation. We will call σ_i the *strength* of a front, the real number such that $u_r = \Phi_i(\sigma_i, u_l)$ (so that $\sigma_i > 0$ for rarefactions, $\sigma_i < 0$ for shocks). The value $|\sigma_i|$ measures the size of the discontinuity (remember (3.40)); the sign of σ_i encodes the nature of the wave.

Now, at an interaction point where a i-wave meets a j-wave, one proves that, whether $i = j$ or $i \neq j$, one has the following relations between the strengths of the incoming waves, and the strengths of the outgoing ones (Fig. 3.25):

$$\sigma_i'' = \sigma_i + \sigma_i' + \mathcal{O}(1)|\sigma_i \sigma_j'|. \tag{3.43}$$

Estimates (3.43) are known as *Glimm's interaction estimates*.
In other words, what (3.43) proves is that:

- If $i \neq j$, the strength of the i- and j-outgoing waves are almost the same as the i- and j-incoming ones, up to a quadratic error.

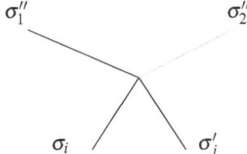

Fig. 3.25 Analysis of an interaction

- If $i = j$, the strength of the outgoing wave of family $i = j$ is almost the sum of the strengths of the incoming waves (up to a quadratic error), and the strength of the outgoing wave in the other family is of quadratic order.

Now consider the functionals

$$V(\tau) = \sum_{\substack{\alpha \text{ front at time } t}} |\sigma_\alpha| \; ; \quad Q(\tau) = \sum_{\substack{\alpha.\beta \\ \text{approaching fronts}}} |\sigma_\alpha|.|\sigma_\beta|,$$

By *approaching fronts*, we mean fronts of different families where the leftmost front is of a faster family (that is, having a higher index), or fronts of the same family (provided that one of the two at least is a shock).

An important feature of the functional V is that, due to (3.40), there exists $C_1, C_2 > 0$ such that on a front-tracking approximation u^ν, one has

$$C_1 TV(u^\nu(\tau)) \leq V(\tau) \leq C_2 TV(u^\nu(\tau)).$$

Using the above interaction estimates, we see that, at an interaction time,

$$\sum_{\substack{\alpha \text{ outgoing fronts} \\ \text{after interaction at time } t}} |\sigma_\alpha| \; \leq \sum_{\substack{\alpha \text{ incoming fronts} \\ \text{interacting at time } t}} |\sigma_\alpha| \; + \mathcal{O}(1)[Q(t^-) - Q(t^+)].$$

It follows that for some $C > 0$, if $TV(u_0)$ is small enough, then the functional,

$$V(t) + CQ(t),$$

known as *Glimm's functional*, is non-increasing over time.

From this we obtain a bound in $L^\infty(BV)$ of the sequence. Now one uses the *finite speed of propagation*: the slope of the fronts is bounded. This gives a $\mathscr{L}ip(L^1_{loc})$ bound.

Passage to the limit

Hence, with the help of these two bounds and of Helly's theorem, one obtains the *compactness* of the front-tracking sequence in L^1_{loc}.

Finally, one proves that a limit point of this sequence is indeed an entropy solution. To do so, given an entropy couple (η, q) with η convex, we have to estimate

$$\int_0^T \int_{\mathbb{R}} \varphi(t, x)(\eta(u^{\nu})_t + q(u^{\nu})_x), \tag{3.44}$$

for $\varphi \in C_c^{\infty}((0, T) \times \mathbb{R})$ with $\varphi \geq 0$. We only need to see the contributions of the fronts in the integral (3.44). More precisely, by Stokes' theorem, one can transform this integral into:

$$\int_0^T \sum_{\alpha \text{ front at time } t} \varphi(t, \alpha(t)) \Big\{ \dot{\alpha}(t) \big[\eta(u^{\nu}(\alpha(t)^+)) - \eta(u^{\nu}(\alpha(t)^-)) \big]$$

$$- \big[q(u^{\nu}(\alpha(t)^+)) - q(u^{\nu}(\alpha(t)^-)) \big] \Big\} \, dt,$$

where $\alpha(t)$ denotes the position of the front α at time t, and $\dot{\alpha}(t)$ is speed. Then the analysis is as follows:

- Shock fronts give a negative contribution (this comes from the admissibility of shocks—they satisfy the entropy inequality).

- Rarefaction fronts are not exact entropy solutions. Hence each rarefaction front gives a contribution to (3.44). One can see that this contribution is of order $\mathcal{O}(\nu^2)$; this is due to the fact that they are of strength at most ν, and travel to the correct velocity up to an error of size ν. Since using the bound on the total variation, the total strength of rarefaction fronts is at most $\mathcal{O}(1)TV(u_0)$, the total contribution of rarefaction fronts in (3.44) is at most $\mathcal{O}(1)TV(u_0)\nu \ldots$

That the equation is satisfied in the sense of distributions corresponds to the particular case $(\eta(u), q(u)) = (u, f(u))$. This completes our description of the existence theory by the front-tracking algorithm.

A remark

For the isentropic Euler system (in Eulerian or Lagrangian coordinates), existence theory of entropy solutions has been shown for much more general solutions [68, 69]:

Theorem 3.4.2 (Lions, Perthame, Souganidis, Tadmor). *Let $(\rho_0, v_0) \in L^{\infty}(\mathbb{R})$, $\rho_0 \geq 0$. Then for all $\gamma > 1$, there exists a global entropy solution of* (EI) *with initial data* (ρ_0, v_0).

3.4.3 The Controllability Problem

The problem

Let us now be more precise on the controllability problem that we consider. As explained above, we will not focus on finding the control on the boundary, but rather the solution itself; this allows to avoid the difficulties of the initial-boundary value problem. Next, in order to be able to use the front-tracking method, we will consider states with small total variation.

Hence the problem becomes the following: given u_0 an initial state (remember that the state is given by (3.31)) and u_1 a target, both supposed to have a small total variation, is it possible to find an entropy solution u, defined on $[0, T] \times [0, 1]$ and driving u_0 to u_1, for *some* time $T > 0$? Note that one does not necessarily expect the controllability here to hold for any time $T > 0$. This is mainly a consequence of the finite speed of propagation of the equation.

But as we will see, the main problem here is the final state u_1. This is due to the fact that a nonlinear effect of genuinely nonlinear systems, known as the decay of positive waves (see [13]), probably prevents all u_1 to be reachable. On another side, describing exactly the set of u_1 that can be attained starting from u_0 seems out of reach for the moment. What we can prove is that, under *sufficient conditions*, u_1 can be reached (for some time T) starting from u_0.

Results

Precisely, here is what one can prove, see [47]. We begin with the Eulerian case.

Theorem 3.4.3 (G.). *There exists $c > 0$ depending on γ such that the following holds. Consider $\bar{u}_0$ and $\bar{u}_1$ two states in $\mathbb{R}^{+*} \times \mathbb{R}$. Set $\overline{\lambda_1} := \lambda_1(\bar{u}_1)$ and $\overline{\lambda_2} := \lambda_2(\bar{u}_1)$. There exist $\varepsilon_1 = \varepsilon_1(\bar{u}_0) > 0$, $\varepsilon_2 = \varepsilon_2(\bar{u}_1) > 0$, and $T = T(\bar{u}_0, \bar{u}_1) > 0$, such that, for any $u_0, u_1 \in BV([0, 1])$ satisfying:*

$$\|u_0 - \bar{u}_0\| \le \varepsilon_1 \text{ and } TV(u_0) \le \varepsilon_1,$$

$$\|u_1 - \bar{u}_1\| \le \varepsilon_2 \text{ and } TV(u_1) \le \varepsilon_2,$$

and $\forall x, y \in [0, 1]$ such that $x < y$,

$$\begin{cases} \dfrac{w^2(u_1(x)) - w^2(u_1(y))}{x - y} \le c \max\left(\dfrac{\overline{\lambda_2} - \overline{\lambda_1}}{1 - y}, \dfrac{\overline{\lambda_1}}{x}, \dfrac{-\overline{\lambda_1}}{1 - y}\right), \\ \dfrac{w^1(u_1(x)) - w^1(u_1(y))}{x - y} \le c \max\left(\dfrac{\overline{\lambda_2} - \overline{\lambda_1}}{x}, \dfrac{-\overline{\lambda_2}}{1 - y}, \dfrac{\overline{\lambda_2}}{x}\right), \end{cases} \qquad (3.45)$$

there is an entropy solution u of (EI) in $[0, T] \times [0, 1]$ such that

$$u_{|t=0} = u_0 \ \text{and} \ u_{|t=T} = u_1.$$

The statement concerning the Lagrangian system is the following.

Theorem 3.4.4 (G.). *Consider $\bar{u}_0$ and $\bar{u}_1$ two states in $\mathbb{R}^{+*} \times \mathbb{R}$. There exists $c = c(\gamma, \bar{u}_1) > 0$ such that the following holds. Set $\overline{\lambda_1} := \lambda_1(\bar{u}_1)$ and $\overline{\lambda_2} := \lambda_1(\bar{u}_2)$. There exist $\varepsilon_1 = \varepsilon_1(\bar{u}_0) > 0$, $\varepsilon_2 = \varepsilon_2(\bar{u}_1) > 0$, and $T = T(\bar{u}_0, \bar{u}_1) > 0$, such that, for any $u_0, u_1 \in BV([0,1])$ satisfying:*

$$\|u_0 - \bar{u}_0\| \leq \varepsilon_1 \ \text{and} \ TV(u_0) \leq \varepsilon_1,$$

$$\|u_1 - \bar{u}_1\| \leq \varepsilon_2 \ \text{and} \ TV(u_1) \leq \varepsilon_2,$$

and $\forall x, y \in [0,1]$ such that $x < y$,

$$\begin{cases} \dfrac{w^2(u_1(x)) - w^2(u_1(y))}{x - y} \leq c \dfrac{\overline{\lambda_2} - \overline{\lambda_1}}{1 - y}, \\ \dfrac{w^1(u_1(x)) - w^1(u_1(y))}{x - y} \leq c \dfrac{\overline{\lambda_2} - \overline{\lambda_1}}{x}, \end{cases} \tag{3.46}$$

there is an entropy solution u of (P) in $[0, T] \times [0, 1]$ such that

$$u_{|t=0} = u_0 \ \text{and} \ u_{|t=T} = u_1.$$

In other words, for both systems, we consider u_0 and u_1 that have small total variation, more precisely which are close in the sense of the BV norm to two constant states $\bar{u}_0$ and $\bar{u}_1$. Provided that u_1 satisfy these "semi-Lipschitz" inequalities (3.45) or (3.46) (written in the coordinates given by the Riemann invariants), where the constant depends on $\bar{u}_1$ and can degenerate on the boundary, then one can drive u_0 to u_1.

The semi-Lipschitz inequalities

Let us comment a little bit these semi-Lipschitz inequalities that we require on the final state. These are close to *Oleinik's inequality*, which is valid for entropy solutions of uniformly convex scalar conservation laws. This inequality states that if $f : \mathbb{R} \to \mathbb{R}$ is such that $f'' \geq c > 0$, then the entropy solutions of

$$u_t + (f(u))_x = 0,$$

satisfy

$$\forall t > 0, \ \forall x < y, \quad \frac{u(t, y) - u(t, x)}{y - x} \leq \frac{1}{ct}.$$

Fig. 3.26 A trajectory
violating (3.45) or (3.46)

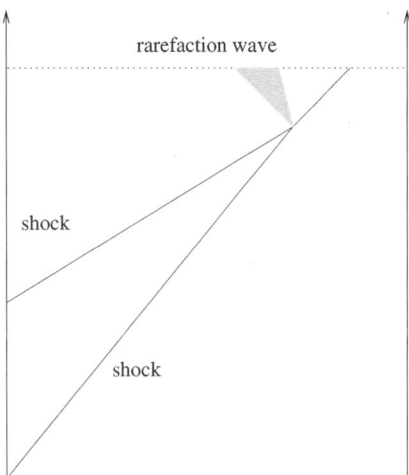

(See for instance [32]). The Oleinik inequality describes the spreading of rarefaction
waves: shock waves yield a negative left hand side, while rarefaction waves given
by formula (3.39) naturally spread and satisfy an inequality of this type.

Now, for what concerns the trajectories of systems (EI) or (P), the Oleinik-type
conditions on the Riemann invariants are not satisfied in general. (See however
Bressan–Colombo [15].)

In particular, it is not difficult to construct solutions of (EI) or (P) which violate
this condition: the meeting of two shocks of the same family can generate a
rarefaction wave in the other family, in contradiction with these inequalities, if the
time T considered is very close after the interaction time, as in Fig. 3.26. But as we
explained earlier, these are *sufficient conditions* for the final state to be reachable.

3.4.4 Some References

Before giving ideas of the proof, let us give several references concerning the control
of systems of conservation laws.

Classical solutions

As we explained earlier, the theory for the control of systems of conservation laws
highly depends on whether you consider classical solutions (let us say, of class C^1),
or entropy solutions (with discontinuities). Concerning the former, a very important
result is the following [63].

Theorem 3.4.5 (Li–Rao (2003)). *Consider*

$$\partial_t u + A(u)u_x = F(u),$$

*such that $A(u)$ has n distinct real eigenvalues $\lambda_1(u) < \cdots < \lambda_k(u) \le -c < 0$ and
$0 \le c < \lambda_{k+1}(u) < \cdots < \lambda_n(u)$ and $F(0) = 0$. Then there exists $\varepsilon > 0$ such
that for all $\phi, \psi \in C^1([0, 1])$ such that $\|\phi\|_{C^1} + \|\psi\|_{C^1} < \varepsilon$, there exists a solution
$u \in C^1([0, T] \times [0, 1])$ such that*

$$u_{|t=0} = \phi, \ and \ u_{|t=T} = \psi.$$

Note that in this context of classical solutions, Theorem 3.4.5 is an extremely
general result. Not only this theorem considers general systems (not limited to
$n = 2$ or to genuinely nonlinear fields), but it even considers the case where A
is not the jacobian of some f (non-conservative systems) and a right-hand side
(balance laws). Nothing so general is known in the context of entropy solutions.
Note however that the condition of strict separation of the characteristic speeds from
0 is not required in Theorems 3.4.3 and 3.4.4.

Since the above result, *many* other developments and generalizations have
appeared. For this, we refer in particular the recent book by Li Ta-Tsien [64] and
references therein.

Entropy solutions

In this context, one does not expect to have a result with such a wide range as
Theorem 3.4.5. In fact, new phenomena appear, proving that such a general result is
not true in general. Let us list several results in the field:

- Ancona and Marson [7]: For the scalar equation $u_t + (f(u))_x = 0$ with $f'' \ge c > 0$, they give a complete description of the attainable set starting from 0.
- Horsin [53] has studied the controllability problem for the Burgers equation $u_t + (u^2/2)_x = 0$ with general $u_0 \in BV$ using Coron's return method.
- Bressan and Coclite [14]: For systems with genuinely nonlinear fields and satisfying $\lambda_1(\cdot) < \cdots < \lambda_k(\cdot) \le -c < 0$ and $0 < c \le \lambda_{k+1}(\cdot) < \cdots < \lambda_n(\cdot)$, for any constant state ω, one can find u such that

$$u(t, \cdot) \to \omega \ as \ t \to +\infty.$$

- Ancona and Coclite [6]: Temple systems satisfying $\lambda_1(u) < \cdots < \lambda_k(u) \le -c < 0$ and $0 < c \le \lambda_{k+1}(u) < \cdots < \lambda_n(u)$, are controllable in L^∞ provided the final state satisfies the Oleinik-type condition.
- Bressan and Coclite [14]: For a class of systems containing Di Perna's system [35]:

$$\begin{cases} \partial_t \rho + \partial_x(\rho u) = 0, \\ \partial_t u + \partial_x \left(\frac{u^2}{2} + \frac{K^2}{\gamma-1}\rho^{\gamma-1} \right) = 0, \end{cases} \tag{3.47}$$

there are initial conditions $\varphi \in BV([0, 1])$ of arbitrary small total variation such
that any entropy solution u *remaining of small total variation* satisfies:

for any $t, u(t, \cdot)$ is not constant.

This is particularly striking, when comparing to Theorem 3.4.5: in the C^1 framework, any small C^1 data can be driven to a constant in finite time.

- Ancona–Marson [8]: In this paper, they consider the asymptotic stabilization by controlling one side only.
- Perrollaz [73]: In this paper, the author considers the controllability of scalar conservation laws with an additional control on the left hand side:

$$u_t + f(u)_x = v(t),$$

and proves that this control can help in a very important way. This follows a study by Chapouly [17] in the C^1 framework.

3.4.5 Sketch of Proof

The proof relies again on the *return method*: the idea is to connect u_0 and u_1 via a solution which goes far away from u_0 and u_1. It is worth noticing that we will not use a *linearization* technique here; this is due to the low level of regularity.

The proof also uses a central difference between Euler system and DiPerna's one (3.47): for the Euler system, the interaction of two shocks of the same family generate *a rarefaction wave in the other family*. For DiPerna's system, it generates a shock. And this is central in Bressan and Coclite's negative result cited above.

The proof is split in three steps:

- Drive u_0 to a constant state.
- Drive the previous state to any constant state.
- Drive a constant state to u_1 or, in other words, find a solution from *some* constant state to u_1.

In the sequel, the argument is performed at the level of front-tracking approximations, which we almost consider as genuine solutions.

3.4.5.1 Driving u_0 to a Constant State

A first idea

In the Eulerian case, an idea is the following: to make a (*very*) strong 2-shock enter the domain through the left side.

More precisely, one considers a state U_l such that the Riemann problem $(U_l, \bar{u}_0)$ is solved by a 2-shock. One computes easily that the set of $U_l = (\rho_l, m_l)$ that can be connected *from the left* to $\bar{u}_0$ by a 2-shock can be parameterized by ρ_l as follows:

$$[\overline{\rho_0}, +\infty) \ni \rho_l \mapsto (\rho_l, m_l) \text{ with } \frac{m_l}{\rho_l} = \frac{\overline{m_0}}{\overline{\rho_0}} + \sqrt{\kappa \frac{1}{\rho_l \overline{\rho_0}} \frac{\rho_l^\gamma - \overline{\rho_0}^\gamma}{\rho_l - \overline{\rho_0}}} (\rho_l - \overline{\rho_0}). \quad (3.48)$$

The corresponding shock speed is given by:

$$s = \frac{\overline{m_0}}{\overline{\rho_0}} + \sqrt{\kappa \frac{\rho_l}{\overline{\rho_0}} \frac{\rho_l^\gamma - \overline{\rho_0}^\gamma}{\rho_l - \overline{\rho_0}}}, \tag{3.49}$$

and the 1-characteristic speed on the left (that is, at U_l) is:

$$\lambda_1(\rho_l, m_l) = \frac{\overline{m_0}}{\overline{\rho_0}} + \sqrt{\kappa \frac{\rho_l}{\overline{\rho_0}} \frac{\rho_l^\gamma - \overline{\rho_0}^\gamma}{\rho_l - \overline{\rho_0}}} (1 - \frac{\overline{\rho_0}}{\rho_l}) - \sqrt{\kappa \gamma} \rho_l^{\frac{\gamma-1}{2}}. \tag{3.50}$$

It follows that one can choose U_l so that:

$$s \geq 2 \text{ and } \lambda_1(U_l) \geq 2.$$

Now, one constructs a solution on the whole real line $\mathbb{R}$ with initial condition:

$$\begin{cases} U_l \text{ on } (-\infty, 0), \\ u_0 \text{ on } [0, 1], \\ \overline{u_0} \text{ on } (1, +\infty). \end{cases} \tag{3.51}$$

Several authors (Alber [3], Schochet [77], Corli and Sablé-Tougeron [22], Chern [21], Lewicka–Trivisa [62], Bressan–Colombo [16],...) have studied the existence of BV solutions in the neighborhood of a strong shock, under Majda's stability condition [70]:

i. s is not an eigenvalue of $A(u^\pm)$,
ii. $\{r_j(u^+) / \lambda_j(u^+) > s\} \cup \{u^+ - u^-\} \cup \{r_j(u^-) / \lambda_j(u^-) < s\}$
    is a basis of $\mathbb{R}^2$,

which is satisfied for any shock here. According to these studies, one can construct a global in time solution on $\mathbb{R}$ associated to the initial condition (3.51). As we will see, restricting this solution to $[0, T] \times [0, 1]$ will give a solution steering u_0 to a constant state (in the Eulerian case).

Let us give more details about the way to construct a solution "near a strong shock". Schochet proved in this context that the Riemann problem is solvable in a neighborhood of the strong shock and gave interaction estimates on the interactions $\Gamma + \gamma \rightarrow \Gamma' + \gamma'$, where Γ represented the large shock. That is, the interaction of the large shock with a small wave yields again a large shock (whose strength has been a little bit modified, but which stays strong) plus small waves. Moreover we have estimates on the strengths of the outgoing waves in terms of the incoming ones, replacing (3.43) which is valid for the interaction of small waves. Let us say the strong shock is of the family j and interacts with a small wave of the family k, then we have:

$$\sigma_i' = \mathcal{O}(1)|\sigma_k| \text{ for } i \neq j \text{ and } \sigma_j' = \sigma_j + \mathcal{O}(1)|\sigma_k|. \tag{3.52}$$

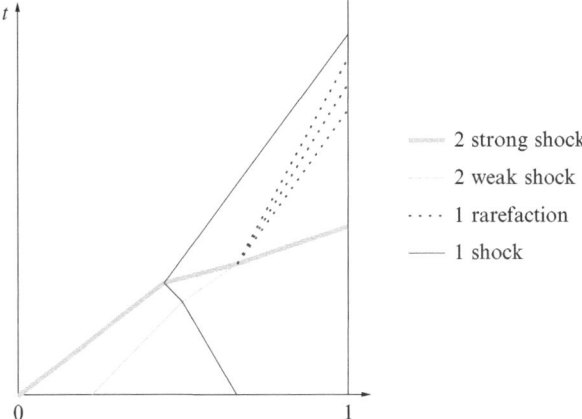

Fig. 3.27 The use of a (*very*) strong shock

As opposed to Glimm's estimates (3.43), estimates (3.52) are linear with respect to
the strength of incoming small waves; but this is compensated by the fact that there
is only one strong shock in our solution. Using this tool, one can construct a global
solution with the initial condition described above (because a standard wave crosses
the strong shock at most once).

Now on the left of the 2-strong shock, all characteristic speeds (whether of the
first or the second family) are positive and bounded away from 0, hence fronts leave
the domain, see Fig. 3.27.

Remark 3.4.6. In this context of the perturbation of a large shock, we call the shock
with large amplitude a strong (or large) shock. By contrast, we call the other waves
weak.

Drawbacks of the previous construction:

- A first problem is that even for a small perturbation of a constant, the solution
  constructed above is *huge*. One would like a control reasonably small when the
  perturbation is small.
- The previous strategy fails in the case of the *p*-system, for which λ_1 is always
  negative. One could have the impression that in a first time, the strong 2-
  shock "filters" the 2-waves from the initial datum, so that even in this case the
  above strategy allows to reach a constant state. But it should be noted that the
  interactions of 1-fronts do generate new 2-waves, see Fig. 3.28...

A better strategy

This leads us to invent another strategy. The starting point is the following. If
above the 2-strong shock and within the first characteristic family, there were only

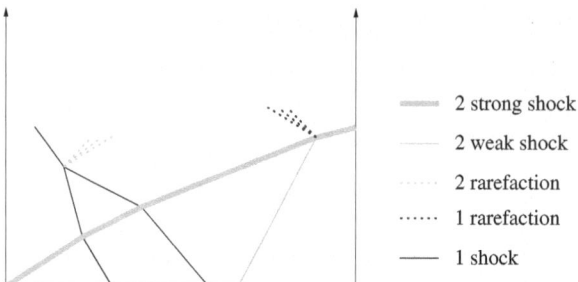

Fig. 3.28 A strong shock in the Lagrangian case

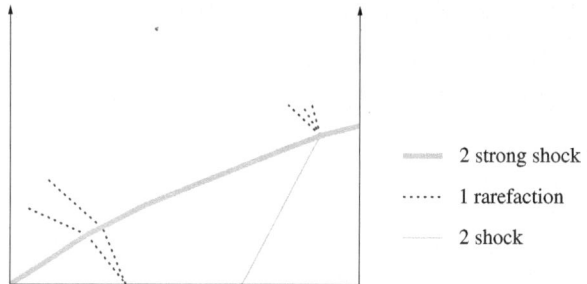

Fig. 3.29 Situation with only 1-rarefaction waves above the 2-strong shock

1-rarefaction waves, then the problem would be solved, because there would be no interaction above the strong shock. Let us explain why. The situation is described in Fig. 3.29. Rarefaction fronts of the first family above the strong shock will not interact, since their behavior consists in going away one from another (this is due to genuine nonlinearity). But since the 2-waves have been absorbed by the strong 2-shock, there are no interaction at all above the 2-strong shock. Hence fronts travel without crossing above the 2-strong shock, and eventually leave the domain provided that the 2-strong shock has been chosen in a way that avoids null characteristic speeds on its left. This is made possible by the above formulas concerning the strong shock, for both the Eulerian and the Lagrangian system.

Consequently, one would like to understand how to prevent 1-shock waves to emerge from the 2-strong shock. There are two situations that can make a 1-shock enter the domain above the 2-strong shock:

• The meeting of the strong shock with a 1-shock.
• The meeting of the strong shock with a 2-rarefaction front.

See Fig. 3.30.

The main idea is the following. One can prove that it is possible to construct additional small 2-shocks that—provided that they arrive from the left at the right interaction time with the right intensity—kill the outgoing shock in the manner described in Fig. 3.31. This is possible thanks to the fact that normally, the

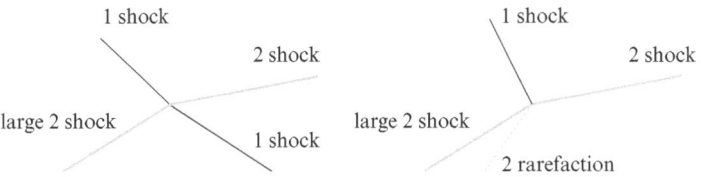

Fig. 3.30 The two situations generating a 1-shock above the strong shock

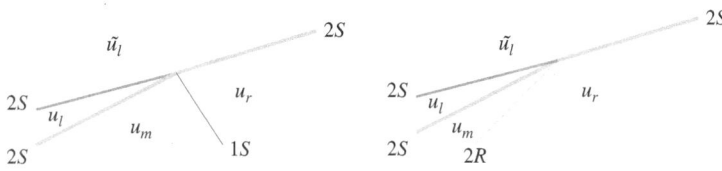

Fig. 3.31 Additional 2-shocks killing the emerging 1-shock wave

interaction of two shocks of the same family generate a *rarefaction* in the other family. Hence we use a *cancellation* effect. Indeed the interaction of the large shock with the incoming 1-shock or 2-rarefaction wave normally generates a 1-rarefaction wave, while the interaction of the large shock with the additional 2-shock normally generates a 1-shock. This is where we use the central difference with respect to DiPerna's model.

Together with this construction, we have an estimate on the size of these 2-shocks in terms of the incoming 1-shock or 2-rarefaction (as long as the strong shock is strong...). The proof of the existence of these two shocks and the corresponding estimates is obtained by the inverse mapping theorem and a precised version of (3.52).

An important problem remains: how to construct an approximation in which these 2-*shocks come at the right time and with the right strength?*

The construction

The idea in order to construct such an approximation is the following. First we construct the solution *under* the strong 2-shock, taking the additional 2-shocks described above in to account, as in Fig. 3.32. In other words, we imagine that we have succeeded to send our additional 2-shocks exactly as we wish. Then taking this information into account (the additional 2-shocks influence the strong one), we can construct the front-tracking approximation under the strong 2-shock.

In a second time, we construct the approximations beyond the strong 2-shock. To that purpose, we have to extend:

- The 1-rarefaction waves *forward* in time.
- The 2-shocks *backward* in time.

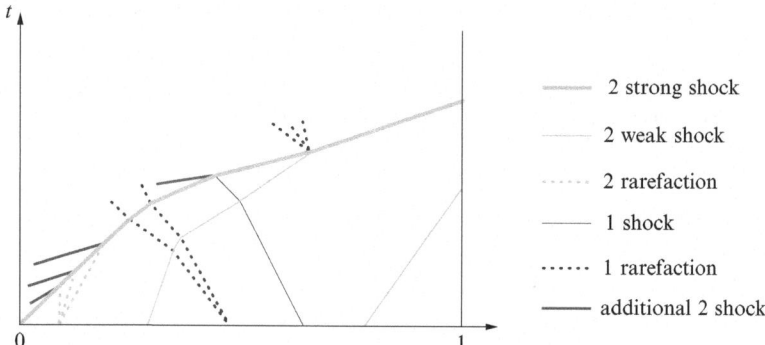

Fig. 3.32 First part of the construction

Fig. 3.33 Second part of the
construction

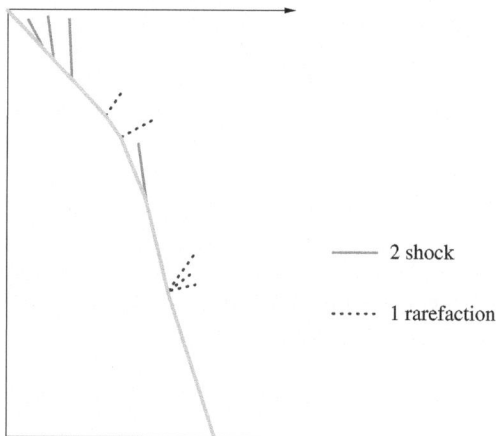

We construct this approximation by using $1 - x$ as the time variable. This is classical for what concerns C^1 solutions. But this raises more difficulties when it comes to entropy weak solutions. Indeed, the direction of time is very important in the selection of admissible discontinuities. Now, once we consider $1 - x$ as the time (that is, after "rotating" the figure), we are led to an initial-boundary value problem, with a moving boundary (the strong 2-shock constructed in the previous step), see Fig. 3.33. We have to describe how we complete the approximation.

The idea is to extend the front-tracking approximation by expanding the fronts emerging from the moving left boundary, that is, from the strong 2-shock; see Fig. 3.33. These fronts are either 1-rarefaction fronts of 2-shocks. Then we have to solve the "interactions" that we meet in this situation. Let us see how we can "treat" these interactions. There are two possible types of interactions. Either the two incoming fronts are of the same family, or they are of opposite family.

But there are no interactions of fronts of the same family because:

- Rarefaction fronts go forward in time, and hence do not meet because of the genuine nonlinearity (they spread).

Fig. 3.34 "Side" interactions

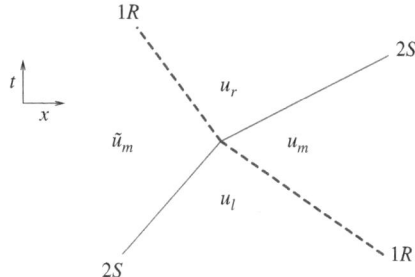

Fig. 3.35 The approximation

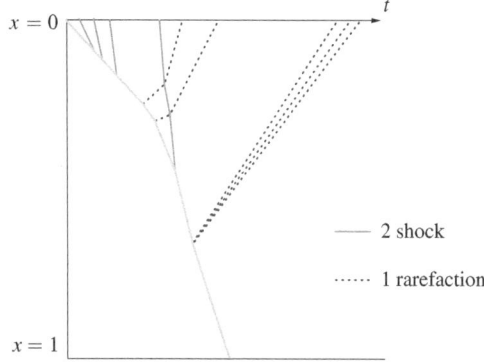

— 2 shock

······ 1 rarefaction

- Shocks go backward in time, and hence do not meet thanks to Lax's inequalities (in the usual direction of time they tend to run one into another).

And this is fortunate, because it is not clear how we could have solved these interactions in a way that would result in an entropy solution in the limit.

For what concerns interactions of front of opposite families, it is not difficult to see that this can be solved as in Lax's Theorem. In other words, one can extend the approximation above the interaction of a 2-shock going backward in time and a 1-rarefaction front going forward in time, as fronts of the same nature, see Fig. 3.34. In the same way, we can obtain Glimm-type estimates (the strength of the waves is conserved across the interaction up to a quadratic error).

Finally we get an approximation as described in Fig. 3.35.

After this construction, it remains then to prove $L_t^\infty(BV_x)$ estimates by adapting Glimm's functionals to the situation, and then to use arguments comparable to the ones used for the Cauchy problem. The main idea for this is to consider functionals measuring the strength of the waves along curves which are not time slices. One can compare the total strength of the waves "under the strong shock" with their initial total strength, and then compare the total strength above the strong shock to the total strength under it. Using the above interaction estimates allows to bound the total variation of the solution uniformly in time.

Fig. 3.36 Moving between
constant states inside a zone

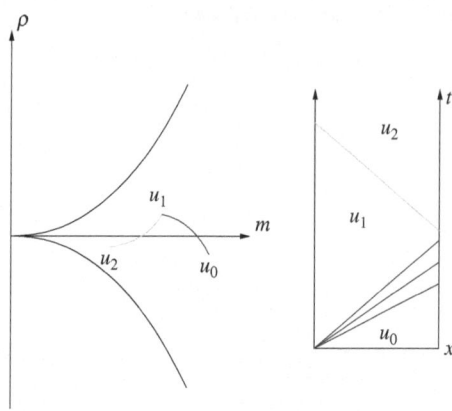

3.4.5.2   Travelling Between Constant States

This step is actually very elementary. There are three different zones in the case of
(EI):

$$\mathscr{D}_1 := \{u, \ 0 < \lambda_1 < \lambda_2\},$$

$$\mathscr{D}_2 := \{u, \ \lambda_1 < 0 < \lambda_2\},$$

$$\mathscr{D}_3 := \{u, \ \lambda_1 < \lambda_2 < 0\}.$$

In the case of equation (P), the situation is even simpler since there is only one zone,
that is $\mathscr{D}_2$.

Now, inside each zone, one can move along (right or left) wave curves, using
simple waves, such as described in Fig. 3.36. This corresponds to simple waves
(shock or rarefaction), that we make cross the domain one after another, from the
left or from the right, according to the sign of the speed of the wave. This works
inside each $\mathscr{D}_i$, because inside each zone the zero characteristic speed is not met.

In the case of equation (EI), it remains to explain how to move from a zone
to another. A way to do this is to use strong shocks as in Fig. 3.37. Recall also
formulas (3.48)–(3.50) in the first approach to treat the initial condition.

3.4.5.3   Driving a Constant State to u_1

Let us finally explain how one can reach u_1 from some constant state.

The construction consists in starting from u_1 and to build approximation of a
solution by a *backward in time* front-tracking algorithm. For the usual front-tracking
algorithm, the standard elementary brick is the Riemann problem. But the equivalent
in the backward setting (in a way that will yield an entropy solution in the usual
sense of time) is not clear.

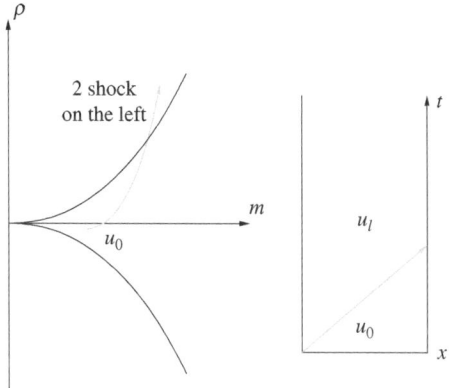

Fig. 3.37 Traveling between zones

Fig. 3.38 A case with non existence for the backward Riemann problem

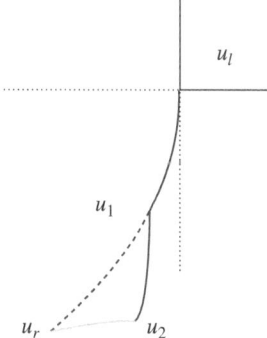

Fig. 3.39 Construction of two solutions of the backward Riemann problem

A backward Riemann problem?

The Riemann problem for negative times is ill-posed. There are two reasons for that. First, existence is not granted in general: typically, a rarefaction wave as in Fig. 3.38 cannot be extended backward in a way that respects entropy criterions.

But even when one has existence, in general one does not have uniqueness. A simple example using wave curves is presented in Figs. 3.39 and 3.40.

The difficulty with the well-posedness of the backward Riemann problem is the raison d'être of the semi-Lipschitz inequalities (3.45)–(3.46).

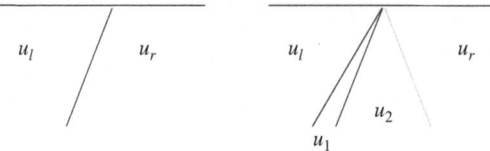

Fig. 3.40 Two solutions of the backward Riemann problem

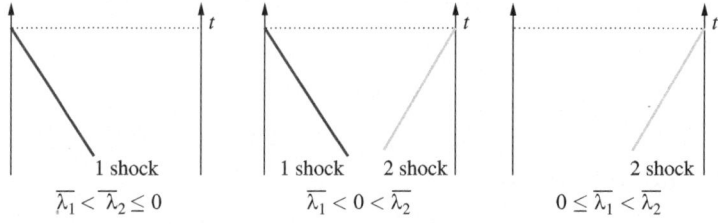

Fig. 3.41 Additional strong shocks

The construction

Let us now describe the construction. A first idea is again to construct a solution which includes strong shocks (backward in time), see Fig. 3.41. The fact that we can use one or two strong shocks depends on the sign of the characteristic speeds in the case (EI) (and this explains the complex form of the semi-Lipschitz inequalities (3.45) in this case by the way).

The presence of these strong shocks will help for both the question of non-existence for the backward Riemann problem (together of course with (3.45) and (3.46)) and for the following important issue. Another problem can indeed be posed by characteristic speeds which can be close to 0. Let us recall indeed that we have no assumption of separation of the characteristic speeds from 0 in Theorem 3.4.3. Of course, fronts having a velocity close to zero do not leave the domain; hence they make it impossible to reach a constant state. But we can manage to have non-zero characteristic speeds "under the strong shocks" constructed above, which excludes this difficulty.

Now, let us begin the construction of the backward front-tracking algorithm:

1. *Final state approximation.* Using the assumptions on u_1, we find *particular* piecewise constant approximations of u_1. These approximations are selected in order that at each discontinuity point, we can "approximately" solve the backward Riemann problem as in Fig. 3.42 using:

   • Either "shock fans", that is a succession of small shocks focusing at the same point.

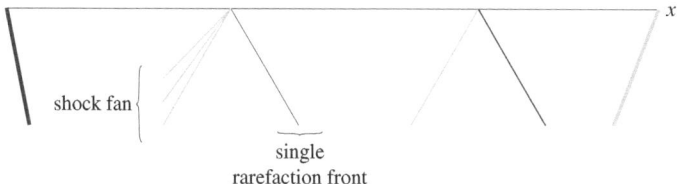

Fig. 3.42 Approximation of the final state

- Either single rarefaction fronts with small amplitude, with this additional constraint that the distance between two successive rarefaction fronts of the same family is estimated from below.
- Either the (strong) shocks from the boundary.

That this is actually possible is a consequence of (3.45) and (3.46). Let us describe this a little bit. A simple approximation of u_1 consists in constructing piecewise constant functions having only horizontal or vertical jumps in Riemann coordinates (that is, where w^2 or w^1 remains constant, respectively). But in fact, the negative jumps (for which w^i decreases) would not yield a backward Riemann problem easily solvable in terms of simple waves (see Fig. 3.22), because even in Riemann coordinates, shock curves are not straight lines.

Instead, one constructs an approximation of u_1 having the following features. There are two possible jumps:

- Either a positive jump in w^i (with the other Riemann invariant w^j constant) of size at most ν (the approximation parameter), which gives naturally a rarefaction front, since rarefaction curves are straight lines in Riemann coordinates.
- Either a negative jump in w^i, and in that case we approximate a horizontal/vertical segment by a succession of small shocks in the family $j \neq i$. Consequently in that case, w^j for $j \neq i$ is not constant across the discontinuity. However a succession of small shocks gives an accurate approximation of a horizontal/vertical negative jump in Riemann coordinates.

Using (3.45) and (3.46), we can moreover make sure that the successive positive jumps in w^i are distant one from another of at least $C\nu$.

2. *Extending the backward front-tracking approximations.* We extend the resulting fronts till two of these fronts meet at a *backward interaction point*. The backward interactions are treated as follows. Whether the two incoming fronts are of the same family or not is very important here.
Interactions inside a family. This depends on the nature of the fronts that are meeting:

- Shock/shock interactions: such interactions do not occur inside a characteristic family, as a consequence of Lax's inequalities.

- Rarefaction front/shock interactions: these do not occur either as a consequence of Lax's inequalities and estimates on the sizes of the rarefaction fronts, which stay small, hence with a speed close to the characteristic speed.
- Rarefaction/rarefaction interactions: these are likely to happen. These must be avoided, because if we allow many rarefaction fronts to merge, this will result in an non-entropic solution in the limit. As we will see later, the additional strong shock that we let enter the domain will be useful to kill the rarefaction fronts before this can happen.

Interactions of fronts of different families. There are two types of these interactions, depending on whether a strong shock is involved or not. We call weak waves the fronts that are not one of the strong shocks

- Weak waves: if the two incoming fronts are weak, one can "solve" the interactions, just as in Lax's Theorem, see Fig. 3.43.
- Strong shock/weak shock interaction: again, one can extend the solution by a strong shock and a weak shock, satisfying Schochet's interaction estimates, see Fig. 3.44.
- Strong shock/weak rarefaction interaction: we solve the backward interaction in terms of two incoming shocks of the same family (one strong, one weak), see Fig. 3.45. In other terms, we use the opportunity of this meeting to kill the rarefaction fronts which are the main obstacle to get an entropy solution in the limit. Let us underline that we *choose* to do this, since there is no uniqueness in the backward Riemann problem.

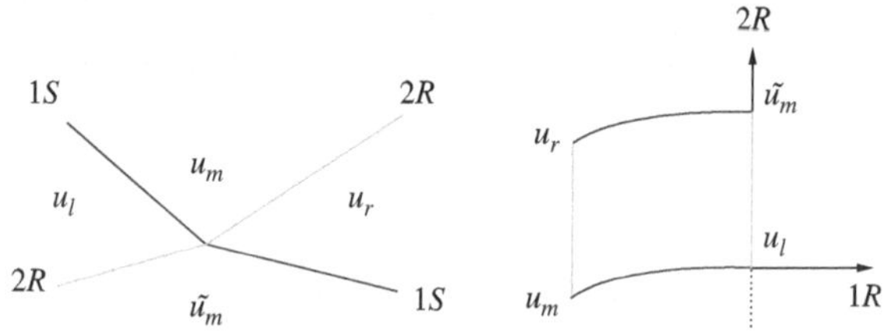

Fig. 3.43 A backward interaction between weak waves of opposite family

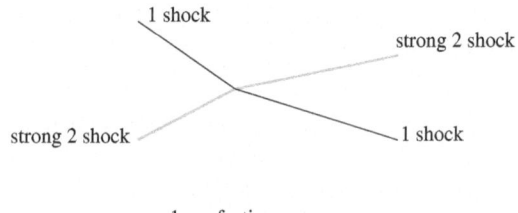

Fig. 3.44 A backward interaction between a strong shock and a weak shock of opposite family

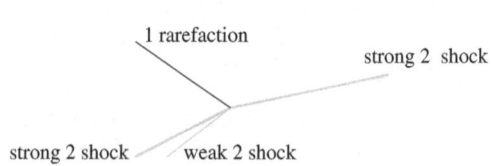

Fig. 3.45 A backward interaction between a strong shock and a weak rarefaction of opposite family

Focusing of rarefaction waves

Now that we have given the main construction, we have to check that indeed this prevents rarefaction/rarefaction interactions to occur.

The main point is to estimate the distance between consecutive rarefaction fronts of the same family in order to prove that before possibly meeting, they must:

- Either leave the domain.
- Either be killed by the meeting of a strong shock of the opposite family.

For that, one has to estimate the (backward-in time) focusing of rarefaction waves. This is done by using Glimm and Lax's estimates [51] on the spreading of rarefaction waves (forward-in-time).

Let us say a few words of this. We measure the distance between two successive 1-rarefaction fronts $\mathscr{C}_1$ and $\mathscr{C}_2$ "in the direction" of the second characteristic family, see Fig. 3.46. Supposing that there has been no merging of rarefaction fronts yet, the strength of these rarefaction fronts are of order ν.

Now, roughly speaking: there are two "sources" in the difference of speed between $\mathscr{C}_1$ and $\mathscr{C}_2$:

- The strength of these fronts $\mathscr{C}_1$ and $\mathscr{C}_2$ themselves, of order $\mathscr{O}(\nu)$. More precisely if σ_1 and σ_2 are the strengths of the fronts (measured through w_1), this adds approximately $\frac{1}{2}\frac{\partial\lambda_1}{\partial w_1}[\sigma_1 + \sigma_2] + \mathscr{O}(\nu^2)$ to $\dot{\mathscr{C}}_1 - \dot{\mathscr{C}}_2$.
- The fronts of the second family that cross the two curves. When these fronts are between $\mathscr{C}_1$ and $\mathscr{C}_2$, they add an error between $\dot{\mathscr{C}}_1$ and $\dot{\mathscr{C}}_2$. The corresponding "additional deviation" is of order $\mathscr{O}(1)|\mathscr{C}_2(t) - \mathscr{C}_1(t)|$. Indeed, the fronts of the other family "do not stay long" between $\mathscr{C}_1$ and $\mathscr{C}_2$, see again Fig. 3.46.

Using the construction of the approximations of u_1 and Oleinik-type semi-Lipschitz conditions on u_1, we can give a lower bound on $\mathscr{C}_1(T) - \mathscr{C}_2(T)$ in terms of ν. Then one is able by using a Gronwall argument to estimate $\mathscr{C}_1(t) - \mathscr{C}_2(t)$ from below for $t \leq T$. If the constants in these semi-Lipschitz assumptions on u_1 are small enough,

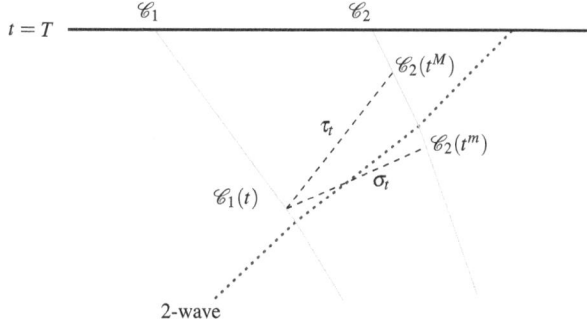

Fig. 3.46 Non crossing of characteristics

Fig. 3.47 The approximation
of the backward problem

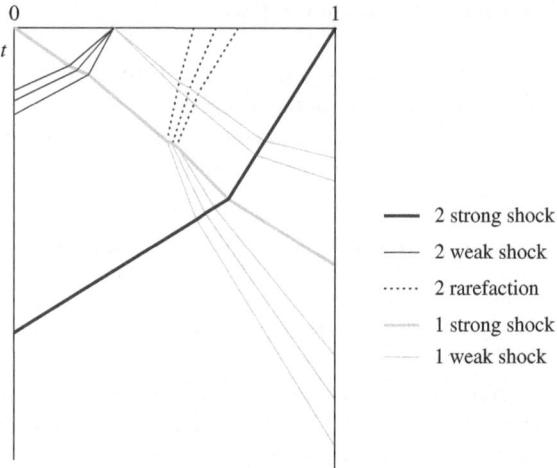

we are thus capable to affirm that the backward rarefaction fronts do not meet before either leaving the domain, or meeting a strong shock of the opposite family, in which case one kills rarefaction fronts.

Conclusion

Taking into account that backward rarefaction fronts do not merge, we get an approximation as described in Fig. 3.47.

The rest of the proof consists in establishing estimates in $L_t^\infty(BV_x)$ for these approximations, as in the standard case. The main difference with the usual method is that one considers modified Glimm functionals V and Q to take into account the strong shocks and the construction, and that the estimates go backward in time (but are not of different nature).

Hence again, we can obtain the compactness of the sequence of approximations, and hence one can obtain a limit point of this sequence. It remains to prove that it is an entropy solution. The main point is that, since the rarefaction fronts never merge, they are always of order $\mathscr{O}(\nu)$. Hence we can be sure as for the usual front-tracking algorithm to obtain an entropy solution in the limit, and this ends the proof.

3.4.6 Comments

There is a *huge gap* between what is known in the framework of C^1 solutions and what is known for entropy solutions. For instance, the controllability of the *full compressible Euler equation*, with the equation of energy is a completely open problem.

Open problem 4 *What can be said on the controllability of the* 1D *full compressible Euler equation:*

$$\begin{cases} \partial_t \rho + \partial_x(\rho u) = 0, \\ \partial_t(\rho u) + \partial_x(\rho u^2 2 + \rho\theta) = 0, \\ \partial_t(\frac{\rho u^2}{2} + \frac{\rho\theta}{\gamma-1}) + \partial_x(\frac{\rho u^3}{2} + \frac{\gamma\rho\theta u}{\gamma-1}) = 0, \end{cases}$$

where θ is the temperature, by means of boundary controls? More generally, can we widen the class of systems of conservation laws where one can prove the reachability of constant states?

As we underlined earlier, the situation is necessarily more complex in the context of entropy solutions than in the class of classical ones: in the case of C^1 solutions, both Euler and Di Perna's systems are locally controllable...

Acknowledgements The author is very thankful to Professors P. Cannarsa and J.-M. Coron, as well as to the CIME staff, for organizing this wonderful summer school in Cetraro.

References

1. A.A. Agrachev, A.V. Sarychev, Navier–Stokes equations: Controllability by means of low modes forcing. J. Math. Fluid Mech. **7**(1), 108–152 (2005)
2. A.A. Agrachev, A.V. Sarychev, Controllability of 2D Euler and Navier–Stokes equations by degenerate forcing. Comm. Math. Phys. **265**(3), 673–697 (2006)
3. H.D. Alber, Local existence of weak solutions to the quasilinear wave equation for large initial values. Math. Z. **190**(2), 249–276 (1985)
4. D. Amadori, Initial-boundary value problems for nonlinear systems of conservation laws. NoDEA Nonlinear Differ. Equat. Appl. **4**(1), 1–42 (1997)
5. D. Amadori, R.M. Colombo, Continuous dependence for 2×2 conservation laws with boundary. J. Differ. Equat. **138**(2), 229–266 (1997)
6. F. Ancona, G.M. Coclite, On the attainable set for Temple class systems with boundary controls. SIAM J. Contr. Optim. **43**(6), 2166–2190 (2005)
7. F. Ancona, A. Marson, On the attainable set for scalar nonlinear conservation laws with boundary control. SIAM J. Contr. Optim. **36**(1), 290–312 (1998)
8. F. Ancona, A. Marson, Asymptotic stabilization of systems of conservation laws by controls acting at a single boundary point. Control methods in PDE-dynamical systems. Contemp. Math. **426**, 1–43 (2007)
9. F. Ancona, A. Marson, Existence theory by front tracking for general nonlinear hyperbolic systems. Arch. Ration. Mech. Anal. **185**, 287–340 (2007)
10. A.L. Bertozzi, P. Constantin, Global regularity for vortex patches. Comm. Math. Phys. **152**(1), 19–28 (1993)
11. S. Bianchini, A. Bressan, Vanishing viscosity solutions of nonlinear hyperbolic systems. Ann. Math. **161**(1), 223–342 (2005)
12. A. Bressan, Global solutions of systems of conservation laws by wave front tracking. J. Math. Anal. Appl. **170**, 414–432 (1992)
13. A. Bressan, *Hyperbolic Systems of Conservation Laws, the One-Dimensional Problem*, Oxford Lecture Series in Mathematics and its Applications 20, (Oxford University Press, Oxford, 2000)

14. A. Bressan, G.M. Coclite, On the boundary control of systems of conservation laws. SIAM J. Contr. Optim. **41**(2), 607–622 (2002)

15. A. Bressan, R.M. Colombo, Decay of positive waves in nonlinear systems of conservation laws. Ann. Sc. Norm. Sup. Pisa **IV-26**, 133–160 (1998)

16. A. Bressan, R.M. Colombo, Unique solutions of 2×2 conservation laws with large data. Indiana Univ. Math. J. **44**(3), 677–725 (1995)

17. M. Chapouly, Global controllability of nonviscous and viscous Burgers-type equations. SIAM J. Contr. Optim. **48**(3), 1567–1599 (2009)

18. M. Chapouly, On the global null controllability of a Navier–Stokes system with Navier slip boundary conditions. J. Differ. Equat. **247**(7), 2094–2123 (2009)

19. J.-Y. Chemin, Persistance de structures géométriques dans les fluides incompressibles bidimensionnels. Ann. Sci. École Norm. Sup. **26**(4), 517–542 (1993)

20. J.-Y. Chemin, *Fluides Parfaits Incompressibles*, (Astérisque 230, 1995).

21. I-L. Chern, Stability theorem and truncation error analysis for the Glimm scheme and for a front tracking method for flows with strong discontinuities. Comm. Pure Appl. Math. **42**, 815–844 (1989)

22. A. Corli, M. Sablé-Tougeron, Perturbations of bounded variation of a strong shock wave. J. Differ. Equat. **138**(2), 195–228 (1997)

23. J.-M. Coron, Global Asymptotic Stabilization for controllable systems without drift. Math. Contr. Signal Syst. **5**, 295–312 (1992)

24. J.-M. Coron, Contrôlabilité exacte frontière de l'équation d'Euler des fluides parfaits incompressibles bidimensionnels. C. R. Acad. Sci. Paris Sér. I Math. **317**(3), 271–276 (1993)

25. J.-M. Coron, On the controllability of 2-D incompressible perfect fluids. J. Math. Pures Appl. **75**, 155–188 (1996)

26. J.-M. Coron, *Sur la stabilization des fluides parfaits incompressibles bidimensionnels*, séminaire : équations aux dérivées partielles, 1998–1999, Exp. No. VII, (École Polytechnique, Palaiseau, 1999)

27. J.-M. Coron, On the null asymptotic stabilization of 2-D incompressible Euler equation in a simply connected domain. SIAM J. Control Optim. **37**(6), 1874–1896 (1999)

28. J.-M. Coron, On the controllability of the 2-D incompressible Navier–Stokes equations with the Navier slip boundary conditions. ESAIM Contrôle Optim. Calc. Var. **1**, 35–75 (1995/1996)

29. J.-M. Coron, *Control and Nonlinearity*, Mathematical Surveys and Monographs 136, (American Mathematical Society, Providence, RI, 2007)

30. J.-M. Coron, A.V. Fursikov, Global exact controllability of the 2D Navier–Stokes equations on a manifold without boundary. Russ. J. Math. Phys. **4**(4), 429–448 (1996)

31. C.M. Dafermos, Polygonal approximations of solutions of the initial value problem for a conservation law. J. Math. Anal. Appl. **38**, 33–41 (1972)

32. C.M. Dafermos, *Hyperbolic Conservation Laws in Continuum Physics*, vol 325 Grundlehren Math. Wissenschaften Series, (Springer, Berlin, 2000)

33. R. Danchin, Évolution d'une singularité de type cusp dans une poche de tourbillon. Rev. Mat. Iberoamericana **16**(2), 281–329 (2000)

34. N. Depauw, Poche de tourbillon pour Euler 2D dans un ouvert à bord. J. Math. Pures Appl. **78**(3), 313–351 (1999)

35. R.J. DiPerna, Global solutions to a class of nonlinear hyperbolic systems of equations. Comm. Pure Appl. Math. **26**, 1–28 (1973)

36. R.J. DiPerna, Global existence of solutions to nonlinear hyperbolic systems of conservation laws. J. Differ. Equat. **20**(1), 187–212 (1976)

37. F. Dubois, P.G. LeFloch, Boundary conditions for nonlinear hyperbolic systems of conservation laws. J. Differ. Equat. **71**(1), 93–122 (1988)

38. A. Dutrifoy, On 3-D vortex patches in bounded domains. Comm. Part. Differ. Equat. **28**(7–8), 1237–1263 (2003)

39. D. Ebin, J. Marsden, Groups of diffeomorphisms and the motion of an incompressible fluid. Ann. of Math. **92**, 102–163 (1970)

40. E. Fernàndez-Cara, S. Guerrero, O.Y. Imanuvilov, J.-P. Puel, Some controllability results for the N-dimensional Navier–Stokes and Boussinesq systems with $N - 1$ scalar controls. SIAM J. Contr. Optim. **45**(1), 146–173 (2006)

41. A.V. Fursikov, O.Y. Imanuvilov, Exact local controllability of two-dimensional Navier–Stokes equations. Russ. Mat. Sb. **187**(9), 103–138 (1996); translation in Sb. Math. **187**(9), 1355–1390 (1996)

42. P. Gamblin, X. Saint Raymond, On three-dimensional vortex patches. Bull. Soc. Math. France **123**(3), 375–424 (1995)

43. S.J. Gardiner, *Harmonic approximation*, London Mathematical Society Lecture Note Series, 221 (Cambridge University Press, Cambridge, 1995)

44. O. Glass, Exact boundary controllability of 3-D Euler equation. ESAIM Contr. Optim. Calc. Var. **5**, 1–44 (2000)

45. O. Glass, An addendum to a J.-M. Coron theorem concerning the controllability of the Euler system for 2D incompressible inviscid fluids. J. Math. Pures Appl. **80**(8), 845–877 (2001)

46. O. Glass, Asymptotic stabilizability by stationary feedback of the two-dimensional Euler equation: The multiconnected case. SIAM J. Contr. Optim. **44**(3), 1105–1147 (2005)

47. O. Glass, On the controllability of the 1-D isentropic Euler equation. J. Eur. Math. Soc. **9**(3), 427–486 (2007)

48. O. Glass, T. Horsin, Approximate Lagrangian controllability for the 2-D Euler equation. Application to the control of the shape of vortex patches. J. Math. Pures Appl. **93**(1), 61–90 (2010)

49. O. Glass, P.G. LeFloch, Nonlinear Hyperbolic Systems: Nondegenerate Flux, Inner Speed Variation, and Graph Solutions. Arch. Ration. Mech. Anal. **185**, 409–480 (2007)

50. J. Glimm, Solutions in the large for nonlinear hyperbolic systems of equations. Comm. Pure Appl. Math. **18**, 697–715 (1965)

51. J. Glimm, P.D. Lax, Decay of solutions of systems of nonlinear hyperbolic conservation laws. Amer. Math. Soc., Providence, RI. **101**, (1970)

52. H. Holden, N.H. Risebro, *Front Tracking for Hyperbolic Conservation Laws*, Applied Mathematical Sciences, 152, (Springer, New York, 2002)

53. T. Horsin, On the controllability of the Burgers equation. ESAIM: Control Opt. Calc. Var. **3**, 83–95 (1998)

54. T. Horsin, Local exact Lagrangian controllability of the Burgers viscous equation. Ann. Inst. H. Poincaré Anal. Non Linéaire **25**(2), 219–230 (2008)

55. O.Y. Imanuvilov, Remarks on exact controllability for the Navier–Stokes equations. ESAIM Control Optim. Calc. Var. **6**, 39–72 (2001)

56. T. Kato, Nonstationary flows of viscous and ideal fluids in $\mathbf{R}^3$. J. Funct. Anal. **9**, 296–305 (1972)

57. T. Kato, *On the Smoothness of Trajectories in Incompressible Perfect Fluids*, Nonlinear wave equations, (Providence, RI, 1998); Contemp. Math. **263**, 109–130. (2000), Amer. Math. Soc., Providence, RI.

58. A.V. Kazhikov, Note on the formulation of the problem of flow through a bounded region using equations of perfect fluid. PMM USSR, **44**, 672–674 (1981)

59. S.G. Krantz, H.R. Parks, *A primer of Real Analytic Functions*, 2nd edn. Basler Lehrbücher, (Birkhäuser, Basel, 1992)

60. P.D. Lax, Hyperbolic systems of conservation laws II. Comm. Pure Appl. Math. **10**, 537–566 (1957)

61. P.G. LeFloch, *Hyperbolic Systems of Conservation Laws: The Theory of Classical and Nonclassical Shock Waves*, Lectures in Mathematics, (ETH Zürich, Birkäuser, 2002)

62. M. Lewicka, K. Trivisa, On the L^1 well posedness of systems of conservation laws near solutions containing two large shocks. J. Differ. Equat. **179**(1), 133–177 (2002)

63. T.-T. Li, B.-P. Rao, Exact boundary controllability for quasi-linear hyperbolic systems. SIAM J. Contr. Optim. **41**(6), 1748–1755 (2003)

64. T.-T. Li, *Controllability and Observability for Quasilinear Hyperbolic Systems*, AIMS Series on Applied Mathematics, 3. American Institute of Mathematical Sciences (AIMS), (Springfield, MO, Higher Education Press, Beijing, 2010)

65. L. Lichtenstein, Über einige existenzprobleme der hydrodynamic. Math. Z. **28**(1), 387–415 (1928)
66. J.-L. Lions, *Contrôlabilité exacte, perturbations et stabilisation de systèmes distribués*, Tomes 1 & 2. (Masson, RMA 8 & 9, Paris, 1988)
67. J.-L. Lions, in *Are There Connections Between Turbulence and Controllability?*, *Analysis and Optimization of Systems*, ed. by A. Bensoussan, J.-L. Lions (Springer, Berlin, 1990); Lecture Notes Control and Inform. Sci. 144
68. P.-L. Lions, B. Perthame, P.E. Souganidis, Existence and stability of entropy solutions for the hyperbolic systems of isentropic gas dynamics in Eulerian and Lagrangian coordinates. Comm. Pure Appl. Math. **49**(6), 599–638 (1996)
69. P.-L. Lions, B. Perthame, E. Tadmor, Existence and stability of entropy solutions to isentropic gas dynamics in Eulerian and Lagrangian variables. Comm. Math. Phys. **163**, 415–431 (1994)
70. A. Majda, *Compressible Fluid Flow and Systems of Conservation Laws in Several Space Variables*, (Springer, New York, 1984)
71. C.B. Morrey, *Multiple Integrals in the Calculus of Variations*, Die Grundlehren der mathematischen Wissenschaften, Band 130, (Springer, New York, 1966)
72. H. Nersisyan, Controllability of the 3D compressible Euler system. Comm. Partial Differential Equations **36**(9), 1544–1564 (2011)
73. V. Perrollaz, Exact controllability of scalar conservation laws with an additional control in the context of entropy solutions, Preprint (2010)
74. Ch. Pommerenke, *Boundary Behaviour of Conformal Maps*, Grundlehren der Mathematischen Wissenschaften, 299. (Springer, Berlin, 1992)
75. N.H. Risebro, A front-tracking alternative to the random choice method. Proc. Amer. Math. Soc. **117**(4), 1125–1139 (1993)
76. D.L. Russell, Controllability and stabilizability theory for linear partial differential equations. Recent progress and open questions. SIAM Rev. **20**, 639–739 (1978)
77. S. Schochet, Sufficient conditions for local existence via Glimm's scheme for large BV data. J. Differ. Equat. **89**(2), 317–354 (1991)
78. P. Serfati, Une preuve directe d'existence globale des vortex patches 2D. C. R. Acad. Sci. Paris Sér. I Math. **318**(6), 515–518 (1994)
79. D. Serre, *Systèmes de lois de conservation. I. Hyperbolicité, entropies, ondes de choc, & II. Structures géométriques, oscillation et problèmes mixtes.* Fondations. (Diderot Editeur, Paris, 1996)
80. A. Shirikyan, Approximate controllability of three-dimensional Navier–Stokes equations. Comm. Math. Phys. **266**(1), 123–151 (2006)
81. F. Sueur, Vorticity internal transition layers for the Navier–Stokes equations, preprint, arXiv:0812.2145 (2008)
82. J. L. Walsh, The approximation of harmonic functions by harmonic polynomials and by harmonic rational functions. Bull. Amer. Math. Soc. 35 (1929), no. 4, 499-544
83. H. Whitney, Analytic extensions of differentiable functions defined in closed sets. Trans. Amer. Math. Soc. **36**(1), 63–89 (1934)
84. H. Whitney, The imbedding of manifolds in families of analytic manifolds. Ann. of Math. (2) **37**(4), 865–878 (1936)
85. W. Wolibner, Un théorème sur l'existence du mouvement plan d'un fluide parfait, homogène, incompressible, pendant un temps infiniment long. Math. Z. **37**, 698–726 (1933)
86. V.I. Yudovich, Non-stationary flows of an ideal incompressible fluid. Ž. Vyčisl. Mat. i Mat. Fiz. **3**, 1032–1066 (1963); *(Russian). English translation in* USSR Comput. Math. Math. Physics **3**, 1407–1456 (1963)
87. V.I. Yudovich, The flow of a perfect, incompressible liquid through a given region. Dokl. Akad. Nauk SSSR **146**, 561–564 (1962); *(Russian). English translation in* Soviet Physics Dokl. **7**, 789–791 (1962)
88. E. Zuazua, Exact controllability for the semilinear wave equation. J. Math. Pures Appl. (9) **69**(1), 1–31 (1990)

Chapter 4
Carleman Estimates and Some Applications to Control Theory

Jérôme Le Rousseau

Abstract We prove Carleman estimates for elliptic and parabolic operators, using several methods: a microlocal approach where the main tool is the Gårding inequality and a more computational direct approach. Carleman estimates are proven locally and we describe how they can be patched together to form a global estimate. We expose how they can be used to provide unique continuation properties, as well as approximate and null controllability results.

4.1 Introduction

Carleman estimates are weighted energy estimates for the solutions of partial differential equations (PDEs). The weights are of exponential types. First introduced for the quantification of unique continuation, going back to the early work of T. Carleman himself [3], in recent years the field of applications of Carleman estimates has gone beyond this original domain: now they are also used in the study of inverse problems and control theory for PDEs. Here, we shall mainly survey the application to control theory in the case of parabolic equations, for which Carleman estimates have now become an essential technique.

These notes originate in part from an expository article by G. Lebeau and the author [12], where the fundamental ideas and mechanisms of Carleman estimates for elliptic and parabolic operators are presented through a microlocal point of view. Here, we also show those ideas but we also put a large emphasis on the exposition of the method A. Fursikov and O. Yu. Imanuvilov [5] for the derivation of such Carleman estimates. One of our goals is to compare the two approaches, distinguishing

J.L. Rousseau (✉)
Université d'Orléans, Laboratoire de Mathématiques, Analyse, Probabilités, Modélisation, Orléans (MAPMO), CNRS UMR 6628, Fédération Denis Poisson, FR CNRS 2964, B.P. 6759, 45067 Orléans cedex 2, France
e-mail: jlr@univ-orleans.fr

F. Alabau-Boussouira et al., *Control of Partial Differential Equations*,
Lecture Notes in Mathematics 2048, DOI 10.1007/978-3-642-27893-8_4,
© Springer-Verlag Berlin Heidelberg 2012

advantages and disadvantages, and to make connections between the two. In fact the Fursikov–Imanuvilov method has been widely used in the recent years in particular because of the simplicity of the argumentation. It however hides some of the positivity mechanisms that are at the heart of the derivation of Carleman estimates.

In Sect. 4.2 we start by presenting basic facts on differential operators with a large parameter, with an extension to pseudo-differential operators. The framework is that of semi-classical analysis. The key aspects will be the formulae for the composition and adjoint of such operators and the Gårding inequality.

We next consider second-order elliptic operators on a bounded domain and give proofs of:

- Carleman estimates away from the boundary (Sect. 4.3).
- Carleman estimates in the neighborhood of the boundary (Sect. 4.5).
- How to patch such estimates together to form a global estimate on the whole open set (Sect. 4.6).

For each case (away from the boundary or at the boundary) we present different proof strategies, microlocal techniques or a direct computational method, and we aim to pin down the nature of the positivity arguments they contain.

One of our goals is also to insist on the fact that the types of Carleman estimates we study here, that is, estimates with the loss of a half derivative, are of *local* nature. We believe that this simple observation is sometimes missed, which can lead to complications in proofs.

In Sect. 4.4 we exploit the local estimates of Sect. 4.3 to show how Carleman estimates are meaningful tools for questions such as unique continuation.

For the application to the controllability of parabolic equation we choose here to follow the approach of [5]. We thus pursue our analysis of Carleman estimates for parabolic operators in Sect. 4.7. Here also, we present different point of views for the proofs, and we insist again on the local nature of the estimates: they are proven away from the boundary and at the boundary, and as in the elliptic case, they can be patched together to form a global estimate. In Sect. 4.8 we use the parabolic estimates to prove:

- The approximate controllability of the heat equation through a unique continuation argument.
- The null-controllability of the heat equation by proving an observability inequality that follows from a global parabolic Carleman estimate.

Finally we gather some technical points in the appendix.

4.2 Differential and Pseudo-Differential Operators with a Large Parameter

If $p(x, \xi)$ is a polynomial in ξ of order less than or equal to m, with $x, \xi \in \mathbb{R}^n$, and $p(x, \xi) = \sum_{|\alpha| \leq m} a_\alpha(x) \xi^\alpha$, we set

$$p(x, D)u = \sum_{|\alpha| \le m} a_\alpha(x) D^\alpha u, \quad \text{with } D = \frac{\partial}{i}.$$

Here $\alpha = (\alpha_1, \dots, \alpha_n)$ is a multi-index:

$$\xi^\alpha = \xi_1^{\alpha_1} \cdots \xi_n^{\alpha_n}, \qquad D^\alpha = D_{x_1}^{\alpha_1} \cdots D_{x_n}^{\alpha_n}.$$

We observe that

$$p(x, D)u(x) = (2\pi)^{-n} \int_{\mathbb{R}^n} e^{i\langle x, \xi\rangle} p(x, \xi) \hat{u}(\xi) \, d\xi, \tag{4.1}$$

where $\hat{u}$ is the Fourier transform of u.

The polynomial function $p(x, \xi)$ is called the symbol of $P(x, D)$ and classically $p_m(x, \xi) = \sum_{|\alpha|=m} a_\alpha(x)\xi^\alpha$ is its principal symbol. For example the symbol of $-\Delta$ is $|\xi|^2$. Observe that the symbol of $-\Delta + \tau^2 V$, is $|\xi|^2 + \tau^2 V$, while its principal symbol is simply $|\xi|^2$. For τ large, we would want the contribution of $\tau^2 V$ to be taken into account in the principal symbol, i.e., we may want to give the same "strength" to ξ, a derivation, as to τ. For this reason, we introduce the following classes of symbols.

Definition 4.2.1. Let $\tau \in \mathbb{R}$ with $\tau \ge \tau_0 > 0$. We set $\lambda^2 = |(\tau, \xi)|^2 = \tau^2 + |\xi|^2$. Let $a(x, \xi, \tau) \in \mathscr{C}^\infty(\mathbb{R}^n \times \mathbb{R}^n \times \mathbb{R}^+)$, be such that for all multi-indices α, β we have

$$|\partial_x^\alpha \partial_\xi^\beta a(x, \xi, \tau)| \le C_{\alpha,\beta} \lambda^{m-|\beta|}, \quad x \in \mathbb{R}^n, \ \xi \in \mathbb{R}^n, \ \tau \ge \tau_0.$$

We write $a \in S(\lambda^m)$.

For $a \in S(\lambda^m)$ we call principal symbol the equivalence class of a in $S(\lambda^m)/S(\lambda^{m-1})$.

Lemma 4.2.2. *Let* $m \in \mathbb{R}$ *and* $a_j \in S(\lambda^{m-j})$ *with* $j \in \mathbb{N}$. *Then, there exists* $a \in S(\lambda^m)$ *such that*

$$\forall N \in \mathbb{N}, \quad a - \sum_{j<N} a_j \in S(\lambda^{m-N}).$$

We write $a \sim \sum_j a_j$. *The symbol* a *is unique up to* $S(\lambda^{-\infty})$, *in the sense that the difference of two such symbols is in* $S(\lambda^{-N})$ *for all* $N \in \mathbb{N}$.

We usually identify a_0 with the principal symbol of a.

With these symbol classes we can define pseudo-differential operators (ψDOs).

Definition 4.2.3. If $a \in S(\lambda^m)$, we set

$$\operatorname{Op}(a)u(x) := (2\pi)^{-n} \int_{\mathbb{R}^n} e^{i\langle x, \xi\rangle} a(x, \xi, \tau) \hat{u}(\xi) \, d\xi.$$

We denote by $\Psi(\lambda^m)$ the set of these ψDOs. For $A \in \Psi(\lambda^m)$, $\sigma(A)$ will denote its principal symbol.

We have $\mathrm{Op}(a) : \mathscr{S}(\mathbb{R}^n) \to \mathscr{S}(\mathbb{R}^n)$ continuously and $\mathrm{Op}(a)$ can be uniquely extended to $\mathscr{S}'(\mathbb{R}^n)$. Then $\mathrm{Op}(a) : \mathscr{S}'(\mathbb{R}^n) \to \mathscr{S}'(\mathbb{R}^n)$ continuously.

We now introduce Sobolev spaces and Sobolev norms which are adapted to the scaling (large) parameter τ. The natural norm on $L^2(\mathbb{R}^n)$ will be written as $\|u\|_0^2 := (\int |u(x)|^2 \, dx)^{\frac{1}{2}}$. Let $s \in \mathbb{R}$; we then set

$$\|u\|_s := \|\Lambda^s u\|_0, \quad \text{with } \Lambda^s := \mathrm{Op}(\lambda^s) \quad \text{and} \quad \mathscr{H}^s(\mathbb{R}^n) := \{u \in \mathscr{S}'(\mathbb{R}^n); \|u\|_s < \infty\}.$$

The space $\mathscr{H}^s(\mathbb{R}^n)$ is algebraically equal to the classical Sobolev space $H^s(\mathbb{R}^n)$. For a fixed value of τ, the norm $\|.\|_s$ is equivalent to the classical Sobolev norm that we write $\|.\|_{H^s}$. However, these norms are not uniformly equivalent as τ goes to ∞. In fact we only have

$$\|u\|_s \leq \|u\|_{H^s}, \quad \text{if } s \leq 0, \quad \text{and} \quad \|u\|_{H^s} \leq C \|u\|_s, \quad \text{if } s \geq 0.$$

For $s \geq 0$ note that we have

$$\|u\|_s \sim \tau^s \|u\|_0 + \|u\|_{H^s}.$$

The spaces $\mathscr{H}^s$ and $\mathscr{H}^{-s}$ are in duality, i.e. $\mathscr{H}^{-s} = (\mathscr{H}^s)'$ in the sense of distributional duality with $L^2 = \mathscr{H}^0$ as a pivot space.

The following continuity result holds.

Theorem 4.2.4. *If $a(x, \xi, \tau) \in S(\lambda^m)$ and $s \in \mathbb{R}$, we then have $\mathrm{Op}(a) : \mathscr{H}^s \to \mathscr{H}^{s-m}$ continuously, uniformly in τ.*

We shall compose ψDOs in the sequel. Such compositions yield a calculus at the level of operator symbols.

Theorem 4.2.5 (Symbol calculus). *Let $a \in S(\lambda^m)$ and $b \in S(\lambda^{m'})$. Then $\mathrm{Op}(a) \circ \mathrm{Op}(b) = \mathrm{Op}(c)$ for a certain $c \in S(\lambda^{m+m'})$ that admits the following asymptotic expansion*

$$c(x, \xi, \tau) = (a \, \sharp \, b)(x, \xi, \tau) \sim \sum_\alpha \frac{1}{i^{|\alpha|}\alpha!} \, \partial_\xi^\alpha a(x, \xi, \tau) \, \partial_x^\alpha b(x, \xi, \tau),$$

where $\alpha! = \alpha_1! \cdots \alpha_n!$. The first term in the expansion, the principal symbol, is ab; the second term is $\frac{1}{i} \sum_j \partial_{\xi_j} a(x, \xi, \tau) \, \partial_{x_j} b(x, \xi, \tau)$. It follows that the principal symbol of the commutator $[\mathrm{Op}(a), \mathrm{Op}(b)]$ is

$$\sigma([\mathrm{Op}(a), \mathrm{Op}(b)]) = \frac{1}{i}\{a, b\} \in S(\lambda^{m+m'-1}).$$

The symbol of the adjoint operator can be obtained as follows.

Theorem 4.2.6. *Let* $a \in S(\lambda^m)$. *Then* $\mathrm{Op}(a)^* = \mathrm{Op}(b)$ *for a certain* $b \in S(\lambda^m)$ *that admits the following asymptotic expansion*

$$b(x, \xi, \tau) \sim \sum_{\alpha} \frac{1}{i^{|\alpha|}\alpha!} \partial_\xi^\alpha \partial_x^\alpha \overline{a}(x, \xi, \tau).$$

The principal symbol of b *is simply* $\overline{a}$.

The following Gårding inequality is the important result we shall be interested in here.

Theorem 4.2.7 (Gårding inequality). *Let* K *be a compact set of* $\mathbb{R}^n$. *If* $a(x, \xi, \tau) \in S(\lambda^m)$, *with principal part* a_m, *if there exists* $C > 0$ *such that*

$$\mathfrak{Re}\, a_m(x, \xi, \tau) \geq C\lambda^m, \quad x \in K, \, \xi \in \mathbb{R}^n, \tau \geq \tau_0,$$

then for $0 < C' < C$ *and* $\tau_1 > 0$ *sufficiently large we have*

$$\mathfrak{Re}(\mathrm{Op}(a)u, u) \geq C' \|u\|_{m/2}^2, \quad u \in \mathscr{C}_c^\infty(K), \, \tau \geq \tau_1.$$

The positivity of the principal symbol of a thus implies a certain positivity for the operator $\mathrm{Op}(a)$. The value of τ_1 depends on C, C' and a finite number of constants $C_{\alpha,\beta}$ associated to the symbol $a(x, \xi, \tau)$ (see Definition 4.2.1). A proof of the Gårding inequality is provided in the appendix. It is also a consequence of the following so-called sharp Gårding inequality (see [8] for a proof).

Theorem 4.2.8 (Sharp Gårding inequality). *If* $a(x, \xi, \tau) \in S(\lambda^m)$, *with principal part* a_m, *is such that*

$$\mathfrak{Re}\, a_m(x, \xi, \tau) \geq 0, \quad x \in \mathbb{R}^n, \, \xi \in \mathbb{R}^n, \tau \geq \tau_0,$$

then there exist $C > 0$ *and* $\tau_1 > 0$ *such that*

$$\mathfrak{Re}(\mathrm{Op}(a)u, u) \geq -C \|u\|_{\frac{m-1}{2}}^2, \quad u \in \mathscr{S}(\mathbb{R}^n), \quad \tau \geq \tau_1.$$

For references on ψDOs the reader can consult [1, 6, 8, 18, 19].

4.3 Local Carleman Estimates for Elliptic Operators

We shall prove a local Carleman estimates for a second-order elliptic operator. To simplify notation we consider the Laplace operator $A = -\Delta = D \cdot D$ but the method we expose extends to more general second-order elliptic operators with a principal part of the form $\sum_{i,j} \partial_j (a_{ij}(x)\partial_i)$ with $a_{ij} \in \mathscr{C}^\infty(\mathbb{R}^n, \mathbb{R})$, $1 \leq i, j \leq n$ and $\sum_{i,j} a_{ij}(x)\xi_i\xi_j \geq C|\xi|^2$, with $C > 0$, for all $x, \xi \in \mathbb{R}^n$. In particular, we note that the Carleman estimates we prove below are insensitive[1] to changes in the operator by zero- or first-order terms.

Let $\varphi(x)$ be a real-valued function. We define the following conjugated operator $A_\varphi = e^{\tau\varphi} A e^{-\tau\varphi}$ to be considered as a differential operator with large parameter τ as introduced in Sect. 4.2. We have

$$
\begin{aligned}
A_\varphi &= -\Delta - |\tau\varphi'|^2 + \langle \tau\varphi', \nabla \rangle + \langle \nabla, \tau\varphi' \rangle \\
&= D \cdot D - |\tau\varphi'|^2 + i\big(\langle \tau\varphi', D \rangle + \langle D, \tau\varphi' \rangle \big) \\
&= D \cdot D - |\tau\varphi'|^2 + 2i \langle \tau\varphi', D \rangle + \tau\Delta\varphi.
\end{aligned}
$$

Its full symbol is given by $|\xi|^2 - |\tau\varphi'|^2 + 2i \langle \tau\varphi', \xi \rangle + \tau\Delta\varphi$. Its principal symbol is given by

$$
a_\varphi = \sigma(A_\varphi) = |\xi|^2 - |\tau\varphi'|^2 + 2i \langle \tau\varphi', \xi \rangle = \sum_j (\xi_j + i\tau\varphi'_{x_j})^2,
$$

i.e., we have "replaced" ξ_j by $\xi_j + i\tau\varphi'_{x_j}$. In fact we note that the symbol of $e^{\tau\varphi} D_j e^{-\tau\varphi}$ is $\xi_j + i\tau\varphi'_{x_j}$. We define the following symmetric operators

$$
\begin{aligned}
A_2 &= (A_\varphi + A_\varphi^*)/2 = D \cdot D - |\tau\varphi'|^2, \\
A_1 &= (A_\varphi - A_\varphi^*)/(2i) = \langle \tau\varphi', D \rangle + \langle D, \tau\varphi' \rangle,
\end{aligned}
$$

with respective principal symbols

$$
a_2 = |\xi|^2 - |\tau\varphi'|^2 \in S(\lambda^2), \qquad a_1 = 2\langle \xi, \tau\varphi' \rangle \in \tau S(\lambda).
$$

We have $a_\varphi = a_2 + i a_1$ and $A_\varphi = A_2 + i A_1$.

We choose φ that satisfies the following assumption.

Assumption 4.3.1 (L. Hörmander [7,9]) *Let V be a bounded open set in $\mathbb{R}^n$. We say that the weight function $\varphi \in \mathscr{C}^\infty(\mathbb{R}^n, \mathbb{R})$ satisfies the* sub-ellipticity *assumption in $\overline{V}$ if $|\varphi'| > 0$ in $\overline{V}$ and*

$$
\forall(x,\xi) \in \overline{V} \times \mathbb{R}^n, \quad a_\varphi(x,\xi) = 0 \quad \Rightarrow \quad \{a_2, a_1\}(x,\xi) \geq C > 0.
$$

Assumption 4.3.1 can be fulfilled as stated in the following lemma whose proof can be found in the appendix.

Lemma 4.3.2 (L. Hörmander [7,9]). *Let V be a bounded open set in $\mathbb{R}^n$ and $\psi \in \mathscr{C}^\infty(\mathbb{R}^n, \mathbb{R})$ be such that $|\psi'| > 0$ in $\overline{V}$. Then $\varphi = e^{\gamma\psi}$ fulfills Assumption 4.3.1 in $\overline{V}$ for the parameter $\gamma > 0$ sufficiently large.*

---

[1]In the sense that only constants are affected. In Theorem 4.3.4 below the constants C and τ_1 change but not the form of the estimate.

The proof of the Carleman estimate will make use of the Gårding inequality. In preparation, we have the following result proven in the appendix that follows from Assumption 4.3.1.

Lemma 4.3.3. *Let $\mu > 0$ and $\rho = \mu(a_2^2 + a_1^2) + \tau\{a_2, a_1\}$. Then, for all $(x, \xi) \in \overline{V} \times \mathbb{R}^n$, we have $\rho(x, \xi) \geq C\lambda^4$, with $C > 0$, for μ sufficiently large.*

We may now prove the following Carleman estimate.

Theorem 4.3.4. *Let V be a bounded open set in $\mathbb{R}^n$ and let φ satisfy Assumption 4.3.1 in $\overline{V}$; then, there exist $\tau_1 \geq \tau_0$ and $C > 0$ such that*

$$\tau^3\|e^{\tau\varphi}u\|_0^2 + \tau\|e^{\tau\varphi}\nabla_x u\|_0^2 \leq C\|e^{\tau\varphi}Au\|_0^2, \tag{4.2}$$

for $u \in \mathscr{C}_c^\infty(\overline{V})$ and $\tau \geq \tau_1$.

Remark 4.3.5. Assumption 4.3.1 appears here as a sufficient condition to obtain such a Carleman estimate. This condition is however necessary (see [7] or [12]).

Proof. We set $v = e^{\tau\varphi}u$. Then, $Au = f$ is equivalent to $A_\varphi v = g = e^{\tau\varphi}f$ or rather $A_2 v + iA_1 v = g$. Observing that $(A_j w_1, w_2) = (w_1, A_j w_2)$ for $w_1, w_2 \in \mathscr{C}_c^\infty(\mathbb{R}^n)$ we then obtain

$$\|g\|_0^2 = \|A_1 v\|_0^2 + \|A_2 v\|_0^2 + 2\Re(A_2 v, iA_1 v) = \left((A_1^2 + A_2^2 + i[A_2, A_1])v, v\right). \tag{4.3}$$

We choose $\mu > 0$ as given in Lemma 4.3.3. Then, for τ such that $\tau^{-1}\mu \leq 1$ we have

$$\tau^{-1}\left(\underbrace{\left(\mu(A_1^2 + A_2^2) + i\tau[A_2, A_1]\right)}_{\text{principal symbol } = \mu(a_1^2 + a_2^2) + \tau\{a_2, a_1\}} v, v\right) \leq \|g\|_0^2.$$

The Gårding inequality and Lemma 4.3.3 then yield, for μ and τ large,

$$\tau^{-1}\|v\|_2^2 \leq C\|g\|_0^2. \tag{4.4}$$

and we obtain $\tau^3\|e^{\tau\varphi}u\|_0^2 + \tau\|\nabla_x(e^{\tau\varphi}u)\|_0^2 \leq C\|e^{\tau\varphi}f\|_0^2$. We write $\nabla_x(e^{\tau\varphi}u) = \tau e^{\tau\varphi}(\nabla_x\varphi)u + e^{\tau\varphi}\nabla_x u$, which yields

$$\tau\|e^{\tau\varphi}\nabla_x u\|_0^2 \leq C\tau^3\|e^{\tau\varphi}u\|_0^2 + C\tau\|\nabla_x(e^{\tau\varphi}u)\|_0^2,$$

since $|\nabla_x\varphi| \leq C$. This concludes the proof. $\square$

Remark 4.3.6. With a density argument the result of Theorem 4.3.4 can be extended to functions $u \in H_0^2(V)$. However, here, we do not treat the case of functions in $H_0^1(V) \cap H^2(V)$. For such a result one needs a local Carleman estimate at the boundary of the open set V as proven in [15, Proposition 2 page 351]. We provide this argument below.

Moreover, a global estimate in V for a function u in $H_0^1(V) \cap H^2(V)$ requires an observation term in the r.h.s. of the Carleman estimate. We shall provide such details below.

4.3.1   The Method of A. Fursikov and O. Yu. Imanuvilov

In the proof of Theorem 4.3.4 we have used Assumption 4.3.1. We give complementary roles to the square terms in (4.3), $\|A_1 u\|_0^2$ and $\|A_2 u\|_0^2$, and to the action of the commutator $i([A_2, A_1]u, u)$. As the square terms approach zero, the commutator term comes into effect and yields positivity. A. Fursikov and O. Yu. Imanuvilov [5] have introduced a modification of the proof that allows one to only consider a term equivalent to the commutator term without using the two square terms.

4.3.1.1   Analysis of the Method

We use the notation of the proof of Theorem 4.3.4, and write

$$\|g + \mu\tau\Delta\varphi v\|_0^2 = \|A_2 v\|_0^2 + \|\underline{A}_1 v\|_0^2 + (i[A_2, A_1]v, v) + 2\Re(A_2 v, \mu\tau\Delta\varphi v),$$

for $0 < \mu < 2$, where $\underline{A}_1 = A_1 - i\mu\tau\Delta\varphi$ and we obtain

$$\|g + \mu\tau\Delta\varphi v\|_0^2 = \|A_2 v\|_0^2 + \|\underline{A}_1 v\|_0^2 + \Re(Rv, v),$$

where $\rho = \sigma(R) = (\{a_2, a_1\} + 2\mu a_2 \tau\Delta\varphi)$. We have the following lemma, which proof can be found in the appendix.

Lemma 4.3.7. *If* $\varphi = e^{\gamma\psi}$, *then for* $\gamma > 0$ *sufficiently large, there exists* $C_\gamma > 0$ *such that*

$$\rho = \{a_2, a_1\} + 2\mu\tau a_2 \Delta\varphi \geq C_\gamma\tau\lambda^2, \quad x \in \overline{V}, \ \xi \in \mathbb{R}^n.$$

With the Gårding inequality we then conclude that $\Re(Rv, v) \geq C'\tau\|v\|_1^2$, for $0 < C' < C_\gamma$ and τ taken sufficiently large. We thus obtain

$$\tau^3\|v\|_0^2 + \tau\|\nabla v\|_0^2 \leq C\|g + \mu\tau\Delta\varphi v\|_0^2$$

The Carleman estimate follows without using the square terms $\|A_2 v\|_0^2$ and $\|\underline{A}_1 v\|_0^2$. In fact we write

$$\|g + \mu\tau\Delta\varphi v\|_0^2 \leq 2\|g\|_0^2 + 2\mu^2\tau^2\|\Delta\varphi v\|_0^2,$$

and the second term in the r.h.s. can be "absorbed" by $\tau^3\|v\|_0^2$ for τ sufficiently large.

Remark 4.3.8. The Fursikov–Imanuvilov method, at the symbol level, is a matter of adding a term of the form $2\mu\tau a_2\Delta\varphi$ to the symbol of the commutator $i[A_2, A_1]$. As the sign of $a_2\Delta\varphi$ is not fixed, a precise choice of the value of μ is crucial.

4.3.1.2 An Alternative Derivation

We shall now compare the microlocal approach we developed above with the "direct"-computational method that A. Fursikov and O. Yu. Imanuvilov follow in their original work [5]. This approach is used very often by authors. Its has some disadvantages and advantages. On the one hand, the computations that are performed hide the positivity argument that we have presented above. They also hide the sufficient sub-ellipticity condition on the weight function. One the other hand, it is based on integrations by parts; this allows to address coefficients with limited regularity, say as low as $W^{1,\infty}$, in the principal part of the operator as done by L. Hörmander in [7].

The direct computation permits also to easily track the dependency on the second large parameter γ which can be useful in some applications. Note however that a modification of the ψDO calculus of Sect. 4.2 also yields the explicit dependency on the second large parameter [11].

Theorem 4.3.9. *Let V be a bounded open set in $\mathbb{R}^n$ and let $\varphi = e^{\gamma\psi}$ with $|\psi'| \geq C > 0$ in $\overline{V}$; then, there exist $\tau_1 \geq \tau_0$, $\gamma_1 > 0$, and $C > 0$ such that*

$$\tau^3\gamma^4 \|e^{\tau\varphi}\varphi^{\frac{3}{2}}u\|_0^2 + \tau\gamma^2 \|e^{\tau\varphi}\varphi^{\frac{1}{2}}\nabla_x u\|_0^2 \leq C \|e^{\tau\varphi} Pu\|_0^2, \qquad (4.5)$$

for $u \in \mathscr{C}_c^\infty(\overline{V})$ and $\tau \geq \tau_1$, $\gamma \geq \gamma_1$.

As mentioned above we preserve the dependency on the second large parameter γ here.

Proof. Above we wrote the conjugated operator A_φ as $A_\varphi = A_2 + iA_1$ with

$$A_2 = D^2 - |\tau\varphi'|^2, \quad A_1 = \tau(\langle\varphi', D\rangle + \langle D, \varphi'\rangle) = 2\tau\langle\varphi', D\rangle - i\tau\Delta\varphi.$$

Following [5] (which corresponds to what is done in Sect. 4.3.1.1) we write

$$A_2v + i\underline{A}_1 v = \underline{g} = g + \mu\tau\Delta\varphi,$$

with $i\underline{A}_1 = 2\tau\langle\varphi', \nabla\rangle + \tau(\mu + 1)\Delta\varphi$. The choice made in [5] is $\mu = 1$. We then write

$$\|\underline{g}\|_0^2 = \|A_2v\|_0^2 + \|\underline{A}_1 v\|_0^2 + 2\Re(A_2v, i\underline{A}_1 v). \qquad (4.6)$$

We focus on the computation of $\Re(A_2v, i\underline{A}_1 v)$ which we write as a sum of 4 terms I_{ij}, $1 \leq i \leq 2$, $1 \leq j \leq 2$, where I_{ij} is the inner product of the ith term in the expression of A_2v and the jth term in the expression of $i\underline{A}_1 v$ above.

Term I_{11}. With two integrations by parts we have

$$I_{11} = 2\tau\Re\big(\nabla v, \nabla(\langle\varphi', \nabla\rangle v)\big) = 2\tau\Re\int \varphi''(\nabla v, \nabla v)\, dx + \tau \int \langle\varphi', \nabla|\nabla v|^2\rangle\, dx$$

$$= 2\tau\Re\int \varphi''(\nabla v, \nabla v)\, dx - \tau \int (\Delta\varphi)|\nabla v|^2\, dx.$$

Term I_{12}. With an integration by parts we have

$$I_{12} = \tau(\mu + 1)\Re\big(-\Delta v, (\Delta\varphi)v\big)$$

$$= \tau(\mu + 1) \int (\Delta\varphi)|\nabla v|^2\, dx + \tau(\mu + 1) \int \langle\nabla\Delta\varphi, \nabla v\rangle v\, dx.$$

Term I_{21}. With an integration by parts we have

$$I_{21} = -2\tau^3\Re\big(|\varphi'|^2 v, \langle\varphi', \nabla v\rangle\big) = -\tau^3 \int |\varphi'|^2\langle\varphi', \nabla|v|^2\rangle\, dx$$

$$= \tau^3 \int \nabla\big(|\varphi'|^2\nabla\varphi\big)|v|^2\, dx.$$

Term I_{22}. We find directly $I_{22} = -\tau^3(\mu + 1) \int |\varphi'|^2(\Delta\varphi)|v|^2\, dx$.
Collecting the different terms together we find

$$\Re(A_2 v, i\underline{A}_1 v) = \int \tau^3\alpha_0|v|^2\, dx, + \int \tau\alpha_1|\nabla v|^2\, dx + X,$$

with $\alpha_0 = \big(\nabla(|\varphi'|^2\nabla\varphi) - (\mu + 1)|\varphi'|^2(\Delta\varphi)\big)$, $\alpha_1 = \mu\Delta\varphi$, and

$$X = 2\tau\Re \int \varphi''(\nabla v, \nabla v)\, dx + \tau(\mu + 1) \int \langle\nabla\Delta\varphi, \nabla v\rangle v\, dx.$$

Using the form of the weight function φ we have the following lemma (see Sect. A.5 for a proof).

Lemma 4.3.10. *For γ sufficiently large we have $\alpha_0 \geq C\gamma^4\varphi^3$ and $\alpha_1 \geq C\gamma^2\varphi$.*

Note that $\varphi''_{jk} = \big(\gamma^2\psi'_j\psi'_k + \gamma\psi''_{jk}\big)\varphi$. For the remainder term it follows that

$$X \geq Y = 2\tau\Re \int \gamma\varphi\psi''(\nabla v, \nabla v)\, dx + \tau(\mu + 1) \int \langle\nabla\Delta\varphi, \nabla v\rangle v\, dx.$$

Noting that $|\varphi^{(k)}| \leq C\gamma^k\varphi$ we then obtain with the Young inequality

$$|Y| \leq \big(C\tau\gamma + \varepsilon\tau\gamma^2\big) \int \varphi|\nabla v|^2\, dx + C_\varepsilon\tau\gamma^4 \int \varphi|v|^2\, dx,$$

for ε as small as needed. For ε small, γ and τ sufficiently large we thus obtain

$$\Re(A_2 v, i\underline{A}_1 v) \geq C\tau^3\gamma^4 \int \varphi^3|v|^2\, dx + C\tau\gamma^2 \int \varphi|\nabla v|^2\, dx.$$

We conclude the proof as in Sect. 4.3.1.1.   □

Remark 4.3.11. In this "computation" oriented proof the key argument lays in Lemma 4.3.10 that assesses the positivity of the coefficients of the leading terms. As mentioned above, the argumentation of this lemma appears disconnected from the necessary and sufficient sub-ellipticity condition of Assumption 4.3.1. In fact the derivation we just performed hides the connection between the weight function φ and the Fourier variable ξ. These two are closely related at the characteristic set of the conjugated operator A_φ, i.e., the set where $a_\varphi = 0$. This precisely where A_φ fails to be elliptic and where the difficulty arises. Some details on this aspect are given in [12].

In the present section we have proven Carleman estimates for smooth compactly supported functions. We avoided complications that arise when taking boundary conditions into account. We postpone such refinements until Sect. 4.5. With the estimates we have at hand we can already prove results in relation with the original motivation for the introduction of Carleman estimates: unique continuation properties.

4.4   Unique Continuation

Let Ω be a bounded open set in $\mathbb{R}^n$. In a neighborhood V of a point $x_0 \in \Omega$, we take a function f such that $\nabla f \neq 0$ in $\overline{V}$. Let $a(x, \xi)$ be a second-order polynomial in ξ that satisfies $a(x, \xi) \geq C|\xi|^2$ with $C > 0$. We define the differential operator $A = a(x, \partial/i)$.

We consider $u \in H^2(V)$ solution of $Au = g(u)$, where g is such that $|g(y)| \leq C|y|$, $y \in \mathbb{R}$. We assume that $u = 0$ in $\{x \in V;\ f(x) \geq f(x_0)\}$. We aim to show that the function u vanishes in a neighborhood of x_0.

We pick a function ψ whose gradient does not vanish near $\overline{V}$ and that satisfies $\langle \nabla f(x_0), \nabla \psi(x_0)\rangle > 0$ and is such that $f - \psi$ reaches a strict local minimum at x_0 as one moves along the level set $\{x \in V;\ \psi(x) = \psi(x_0)\}$. For instance, we may choose $\psi(x) = f(x) - c|x - x_0|^2$. We then set $\varphi = e^{\gamma \psi}$ according to Lemma 4.3.2. In the neighborhood V (or possibly in a smaller neighborhood of x_0) the geometrical situation we have just described is illustrated in Fig. 4.1.

We call W the region $\{x \in V;\ f(x) \geq f(x_0)\}$ (region beneath $\{f(x) = f(x_0)\}$ in Fig. 4.1). We choose V' and V'' neighborhoods of x_0 such that $V'' \Subset V' \Subset V$ and we pick a function $\chi \in \mathscr{C}_c^\infty(V')$ such that $\chi = 1$ in V''. We set $v = \chi u$ and then $v \in H_0^2(V)$. Observe that the Carleman estimate of Theorem 4.3.4 applies to v by Remark 4.3.6. We have

$$Av = A(\chi u) = \chi\, Au + [A, \chi]u,$$

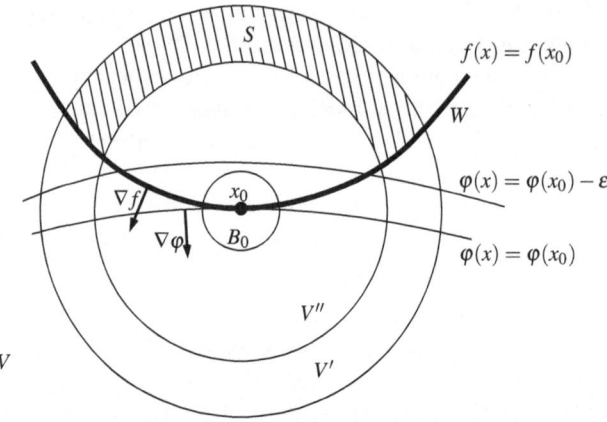

Fig. 4.1 Local geometry for the unique continuation problem. The striped region contains the support of $[A, \chi]u$

where the commutator is a first-order differential operator. We thus obtain

$$\tau^3 \|e^{\tau\varphi}\chi u\|_0^2 + \tau \|e^{\tau\varphi}\nabla_x(\chi u)\|_0^2 \leq C \left(\|e^{\tau\varphi}\chi g(u)\|_0^2 + \|e^{\tau\varphi}[A, \chi]u\|_0^2 \right)$$

$$\leq C' \left(\|e^{\tau\varphi}\chi u\|_0^2 + \|e^{\tau\varphi}[A, \chi]u\|_0^2 \right), \qquad \tau \geq \tau_1.$$

Choosing τ sufficiently large, say $\tau \geq \tau_2$, we may ignore the first term in the r.h.s. of the previous estimate. We then write

$$\tau^3 \|e^{\tau\varphi}u\|_{L^2(V'')}^2 + \tau \|e^{\tau\varphi}\nabla_x u\|_{L^2(V'')}^2 \leq \tau^3 \|e^{\tau\varphi}\chi u\|_0^2 + \tau \|e^{\tau\varphi}\nabla_x(\chi u)\|_0^2$$

$$\leq C \|e^{\tau\varphi}[A, \chi]u\|_{L^2(S)}^2,$$

where $S := V' \setminus (V'' \cup W)$, since the support of $[A, \chi]u$ is confined in the region where χ varies and u does not vanish (see the striped region in Fig. 4.1).

For all $\varepsilon \in \mathbb{R}$, we set $V_\varepsilon = \{x \in V; \; \varphi(x) \leq \varphi(x_0) - \varepsilon\}$. There exists $\varepsilon > 0$ such that $S \Subset V_\varepsilon$. We then choose a ball B_0 with center x_0 such that $B_0 \subset V'' \setminus V_\varepsilon$ and obtain

$$e^{\tau \inf_{B_0} \varphi} \|u\|_{H^1(B_0)} \leq C e^{\tau \sup_S \varphi} \|u\|_{H^1(S)}, \qquad 0 < h < h_2.$$

Since $\inf_{B_0} \varphi > \sup_S \varphi$, letting τ go to $+\infty$, we obtain $u = 0$ in B_0. We have thus proven the following local unique-continuation result.

Proposition 4.4.1. *Let g be such that $|g(y)| \leq C|y|$, $x_0 \in \Omega$ and $u \in H_{loc}^2(\Omega)$ satisfying $Au = g(u)$ and $u = 0$ in $\{x; \; f(x) \geq f(x_0)\}$, in a neighborhood V of x_0. The function f is defined in V and such that $|\nabla f| \neq 0$ in a neighborhood of x_0. Then u vanishes in a neighborhood of x_0.*

With a connectedness argument we then prove the following theorem.

Theorem 4.4.2 (A. Calderón theorem). *Let g be such that $|g(y)| \leq C|y|$. Let Ω be a connected open set in $\mathbb{R}^n$ and let $\omega \Subset \Omega$, with $\omega \neq \emptyset$. If $u \in H^2(\Omega)$ satisfies $Au = g(u)$ in Ω and $u(x) = 0$ in ω, then u vanishes in Ω.*

4.5 Local Carleman Estimates at the Boundary for Elliptic Operators

We are now interested in deriving Carleman estimates for solutions of equations of the type $Au = -\Delta u = f$ on a bounded open subset Ω of $\mathbb{R}^n$ with prescribed boundary conditions. For simplicity we shall treat homogeneous Dirichlet boundary conditions here: $u|_{\partial\Omega} = 0$. Other types of boundary conditions can be more difficult to tackle: inhomogeneous boundary conditions, Neumann type conditions, etc.

Here we are simply interested in proving a local estimate. We thus place ourselves in a neighborhood of a point x_0 of $\partial\Omega$. We denote by n and $d\sigma$ the outward pointing unit normal vector and the Lebesgue measure on $\partial\Omega$ respectively. The approach of A. Fursikov and O. Yu. Imanuvilov is quite direct to extend to the boundary case. We treat this case first (Fig. 4.2).

Theorem 4.5.1. *Let V be an open subset of $\overline{\Omega}$ with $x_0 \in V$. Let $\varphi = e^{\gamma\psi}$ with $|\psi'| \geq C > 0$ in $\overline{V}$ and $\partial_n\psi|_{\partial\Omega\cap V} < 0$; then, there exist $\tau_1 \geq \tau_0$, $\gamma_1 > 0$, and $C > 0$ such that*

$$\tau^3\gamma^4\|e^{\tau\varphi}\varphi^{\frac{3}{2}}u\|_0^2 + \tau\gamma^2\|e^{\tau\varphi}\varphi^{\frac{1}{2}}\nabla_x u\|_0^2 \leq C\|e^{\tau\varphi}Au\|_0^2, \qquad (4.7)$$

for $u \in \mathscr{C}^\infty(\Omega)$ with $\mathrm{supp}(u) \subset \overline{V}$ and $u|_{\partial\Omega\cap V} = 0$, and $\tau \geq \tau_1$, $\gamma \geq \gamma_1$.

Proof. We use the notation of the proof of Theorem 4.3.9. Computing the different terms, as $u|_{\partial\Omega} = 0$, we note that only I_{11} is affected with the integration by parts

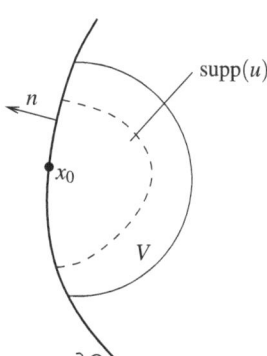

Fig. 4.2 Local geometry for
an estimate at the boundary

yielding boundary terms. In fact we have

$$I_{11} = 2\tau\mathfrak{Re}\big(\nabla v, \nabla(\langle\varphi', \nabla\rangle v)\big) - 2\tau\mathfrak{Re} \int_{\partial\Omega} (\partial_n v)\langle\varphi', \nabla\rangle\bar{v}\,d\sigma$$

$$= 2\tau \int \varphi''(\nabla v, \nabla v)\,dx + \tau \int \langle\varphi', \nabla|\nabla v|^2\rangle\,dx - 2\tau\mathfrak{Re} \int_{\partial\Omega} (\partial_n v)\langle\varphi', \nabla\rangle\bar{v}\,d\sigma$$

$$= 2\tau \int \varphi''(\nabla v, \nabla v)\,dx - \tau \int (\Delta\varphi)|\nabla v|^2\,dx$$

$$- 2\tau\mathfrak{Re} \int_{\partial\Omega} (\partial_n v)\langle\varphi', \nabla\rangle\bar{v}\,d\sigma + \tau \int_{\partial\Omega} (\partial_n\varphi)|\nabla v|^2\,d\sigma.$$

As $u|_{\partial\Omega} = 0$ then $|\nabla u| = |\partial_n u|$ and $\langle\varphi', \nabla\rangle v = (\partial_n\varphi)\partial_n v$ at the boundary. We thus find

$$I_{11} = 2\tau \int \varphi''(\nabla v, \nabla v)\,dx - \tau \int (\Delta\varphi)|\nabla v|^2\,dx - \tau \int_{\partial\Omega} \partial_n\varphi|\partial_n v|^2\,d\sigma.$$

With the condition imposed on the weight function at the boundary we see that the additional term is non-negative as $\partial_n\varphi = \gamma\varphi\partial_n\psi$. The rest of the proof can be carried out as in the proof of Theorem 4.3.9.   □

We shall now use microlocal arguments to prove Carleman estimates at the boundary. As in Sect. 4.3, with such an approach we can truly characterize the required conditions on the weight function. She shall closely follow the original proof given by G. Lebeau and L. Robbiano [15].

In a sufficiently small neighborhood $V \subset \mathbb{R}^n$ of x_0, we place ourselves in normal geodesic coordinates $x = (x', x_n)$. For convenience, we shall take the neighborhood V of the form $V_{x_0'} \times (0, \varepsilon)$, where $V_{x_0'}$ is a sufficiently small neighborhood of x_0'. In such coordinate system, the *principal part* of the operator $-\Delta$ takes the following form [8, Appendix C.5]: $D_{x_n}^2 + r(x, D_{x'})$, with $r(x, \xi')$ a second-order polynomial in ξ' that satisfies

$$r(x, \xi') \in \mathbb{R}, \quad \text{and} \quad C_1|\xi'|^2 \le r(x, \xi') \le C_2|\xi'|^2, \quad x \in V, \xi' \in \mathbb{R}^{n-1}. \quad (4.8)$$

The boundary is now given by $\partial\Omega = \{x_n = 0\}$.

An important fact to note is that the condition in Assumption 4.3.1 is independent of the coordinate system we use [7, Sect. 8.1, page 186].

We need to introduce tangential ψDOs and some of their calculus properties.

Definition 4.5.2. Let $\tau \ge \tau_0 > 0$ and $\lambda_{\mathbb{T}}^2 = |(\tau, \xi')|^2 = \tau^2 + |\xi'|^2$. Let $a(x, \xi', \tau) \in \mathscr{C}^\infty(\mathbb{R}^n \times \mathbb{R}^{n-1} \times \mathbb{R}^+)$ be such that for all multi-indices α, β we have

$$|\partial_x^\alpha \partial_{\xi'}^\beta a(x, \xi', \tau)| \le C_{\alpha,\beta}\lambda_{\mathbb{T}}^{m-|\beta|}, \quad x \in \mathbb{R}^n, \xi' \in \mathbb{R}^{n-1}, \tau \ge \tau_0.$$

We write $a \in S(\lambda_{\mathbb{T}}^m)$.

We then define classes of operators

Definition 4.5.3. If $a \in S(\lambda_{\mathbb{T}}^m)$, we set

$$\mathrm{Op}(a)u(x) := (2\pi)^{1-n} \int e^{i\langle x', \xi' \rangle} a(x, \xi', \tau) \hat{u}(\xi', x_n) \, d\xi',$$

where, here, $\hat{u}$ is the Fourier transform of u in the x' directions. We denote by $\Psi(\lambda_{\mathbb{T}}^m)$ the set of these tangential ψDOs.

The calculus of these operators is similar to those in the whole space.

Proposition 4.5.4 (Symbol calculus). *Let* $a \in S(\lambda_{\mathbb{T}}^m)$ *and* $b \in S(\lambda_{\mathbb{T}}^{m'})$. *Then* $\mathrm{Op}(a) \circ \mathrm{Op}(b) = \mathrm{Op}(c)$ *for a certain* $c \in S(\lambda_{\mathbb{T}}^{m+m'})$ *that admits the following asymptotic expansion*

$$c(x, \xi, \tau) = (a \,\sharp\, b)(x, \xi, \tau) \sim \sum_\alpha \frac{1}{i^{|\alpha|}\alpha!} \, \partial_{\xi'}^\alpha a(x, \xi', \tau) \, \partial_{x'}^\alpha b(x, \xi', \tau),$$

We introduce, for $u = u(x')$,

$$\|u\|_{\mathscr{H}_{\mathbb{T}}^s} = \left\| \mathrm{Op}\big((1 + \tau^2 + |\xi'|^2)^{\frac{s}{2}}\big) u \right\|_{L^2(\mathbb{R}^{n-1})}$$

and set $\mathscr{H}_{\mathbb{T}}^s(\mathbb{R}^{n-1}) := \{u \in \mathscr{S}'(\mathbb{R}^{n-1}); \; \|u\|_{\mathscr{H}_{\mathbb{T}}^s} < \infty\}$. We have the following regularity result.

Proposition 4.5.5. *If* $a(x, \xi', \tau) \in S(\lambda_{\mathbb{T}}^m)$ *and* $s \in \mathbb{R}$, *we then have*

$$\mathrm{Op}(a) : L^2(\mathbb{R}^+, \mathscr{H}_{\mathbb{T}}^s(\mathbb{R}^{n-1})) \to L^2(\mathbb{R}^+, \mathscr{H}_{\mathbb{T}}^{s-m}(\mathbb{R}^{n-1})) \; \text{continuously},$$

uniformly in τ.

Theorem 4.5.6. *Let* φ *satisfy Assumption 4.3.1 in* $\overline{V}$ *and* $\partial_n \varphi|_{\partial\Omega \cap V} < 0$; *then, there exist* $\tau_1 \geq \tau_0$ *and* $C > 0$ *such that*

$$\tau^3 \|e^{\tau\varphi} u\|_0^2 + \tau \|e^{\tau\varphi} \nabla_x u\|_0^2 \leq C \|e^{\tau\varphi} A u\|_0^2, \tag{4.9}$$

for $u \in \mathscr{C}^\infty(\Omega)$ *with* $\mathrm{supp}(u) \subset \overline{V}$, $u|_{\partial\Omega \cap V} = 0$, *and* $\tau \geq \tau_1$.

Proof. Introducing A_2 and A_1 as in Sect. 4.3 we find

$$A_2 = D_{x_n}^2 + \hat{A}_2, \quad A_1 = \tau\big(D_{x_n}\varphi'_{x_n} + \varphi'_{x_n} D_{x_n}\big) + 2\hat{A}_1,$$

with respective principal symbols

$$a_2 = \xi_n^2 + \hat{a}_2, \quad \hat{a}_2 = r(x, \xi') - r(x, \tau\varphi'_{x'}) - \big(\tau\varphi'_{x_n}\big)^2 \in S(\lambda_{\mathbb{T}}^2),$$

$$a_1 = 2\tau\xi_n\varphi'_{x_n} + 2\hat{a}_1, \quad \hat{a}_1 = \tilde{r}(x, \xi', \tau\varphi'_{x'}) \in \tau S(\lambda_{\mathbb{T}})$$

where $\tilde{r}(x, ., .)$ is the symmetric bilinear form associated with the real quadratic form $r(x, \xi')$.

We note that

$$(w_1, A_2 w_2) = (A_2 w_1, w_2)$$
$$- i \left(\left(w_1|_{x_n=0+}, D_{x_n} w_2|_{x_n=0+} \right)_0 + \left(D_{x_n} w_1|_{x_n=0+}, w_2|_{x_n=0+} \right)_0 \right),$$
$$(w_1, A_1 w_2) = (A_1 w_1, w_2) - 2i \left(\tau \varphi'_{x_n} w_1|_{x_n=0+}, w_2|_{x_n=0+} \right)_0,$$

for w_1 and w_2 smooth, where $(., .)_0$ is the L^2 hermitian inner-product in $\{x_n = 0\}$, and we thus obtain

$$\|g\|^2 = \|A_1 v\|^2 + \|A_2 v\|^2 + i([A_2, A_1]v, v) + \mathscr{B}(v),$$

with

$$\mathscr{B}(v) = \left(A_1 v|_{x_n=0+}, D_{x_n} v|_{x_n=0+} \right)_0 + \left((D_{x_n} A_1 - 2\tau\varphi'_{x_n} A_2) v|_{x_n=0+}, v|_{x_n=0+} \right)_0,$$
$$(4.10)$$

which, as $v|_{x_n=0+} = 0$, reduces to

$$\mathscr{B}(v) = 2 \left(\varphi'_{x_n} D_{x_n} v|_{x_n=0+}, D_{x_n} v|_{x_n=0+} \right)_0 \geq 0, \quad \text{as } \varphi'_{x_n}|_{x_n=0+} = -\partial_n\varphi|_{x_n=0+} > 0.$$
$$(4.11)$$

We observe that we have $i[A_2, A_1] = \tilde{H}_0 D_{x_n}^2 + \tilde{H}_1 D_{x_n} + \tilde{H}_2$, where $\tilde{H}_j \in \tau\Psi(\lambda_{\mathbb{T}}^j)$, $j = 0, 1, 2$. We then note that $D_{x_n}^2 - A_2 \in \Psi(\lambda_{\mathbb{T}}^2)$ and $D_{x_n} - \frac{1}{2\tau\varphi'_{x_n}} A_1 \in \Psi(\lambda_{\mathbb{T}})$. We thus find

$$i[A_2, A_1] = \left(H_0 A_2 + \tau^{-1} H_1 A_1 + H_2 \right), \quad H_j \in \tau\Psi(\lambda_{\mathbb{T}}^j), \ j = 0, 1, 2.$$

We have the following lemma, which proof is given below.

Lemma 4.5.7. *For μ and τ sufficiently large, there exists $C > 0$ such that*

$$\tau S(\lambda_{\mathbb{T}}^2) \ni \rho = \mu \frac{\left(\hat{a}_1^2 + (\tau\varphi'_{x_n})^2 \hat{a}_2 \right)^2}{\tau^3 \lambda_{\mathbb{T}}^2} + \sigma(H_2) \geq C\tau\lambda_{\mathbb{T}}^2.$$

Applying the Gårding inequality in the tangential directions we thus obtain, for τ sufficiently large,

$$\|g\|^2 \geq \|A_1 v\|^2 + \|A_2 v\|^2 + \mathfrak{Re}\,(H_0 A_2 v, v) + \mathfrak{Re}\,(\tau^{-1} H_1 A_1 v, v)$$
$$+ C\tau\|v\|_{L^2(\mathbb{R}^+, \mathscr{H}_{\mathbb{T}}^1)}^2 - \mathfrak{Re}\,\tau^{-1} \left((\hat{A}_1^2 + (\tau\varphi'_{x_n})^2 \hat{A}_2) v, Gv \right), \quad (4.12)$$

where $G \in \Psi(\lambda_{\mathbb{T}}^0)$ and $\sigma(G) = \mu \frac{\hat{a}_1^2 + (\tau\varphi'_{x_n})^2 \hat{a}_2}{\tau^2 \lambda_{\mathbb{T}}^2} \in S(\lambda_{\mathbb{T}}^0)$.

We first see that we have

$$\left| (H_0 A_2 v, v) \right| \le C \tau^{-\frac{1}{2}} \|A_2 v\|^2 + C' \tau^{\frac{1}{2}} \|v\|^2_{L^2(\mathbb{R}^+, \mathscr{H}^1_\mathbb{T})}, \tag{4.13}$$

$$\left| (\tau^{-1} H_1 A_1 v, v) \right| \le C \tau^{-\frac{1}{2}} \|A_1 v\|^2 + C' \tau^{\frac{1}{2}} \|v\|^2_{L^2(\mathbb{R}^+, \mathscr{H}^1_\mathbb{T})}. \tag{4.14}$$

From the form of A_1 we deduce the following lemma.

Lemma 4.5.8. *We have* $\tau^{\frac{1}{2}} \|D_{x_n} v\| \le C \tau^{-\frac{1}{2}} \|A_1 v\| + C \tau^{\frac{1}{2}} \|v\|_{L^2(\mathbb{R}^+, \mathscr{H}^1_\mathbb{T})}.$

Next since

$$\hat{A}_1 = \frac{1}{2} \left(A_1 - \tau [D_{x_n}, \varphi'_{x_n}] \right) - \tau \varphi'_{x_n} D_{x_n}, \tag{4.15}$$

we compute

$$\hat{A}_1^2 + (\tau \varphi'_{x_n})^2 \hat{A}_2 = \frac{\hat{A}_1}{2} \left(A_1 - \tau [D_{x_n}, \varphi'_{x_n}] \right) - \hat{A}_1 \tau \varphi'_{x_n} D_{x_n} + (\tau \varphi'_{x_n})^2 (A_2 - D_{x_n}^2).$$

Using (4.15) a second time we have

$$\hat{A}_1^2 + (\tau \varphi'_{x_n})^2 \hat{A}_2 = \frac{\hat{A}_1}{2} \left(A_1 - \tau [D_{x_n}, \varphi'_{x_n}] \right)$$

$$+ \left(\tau \varphi'_{x_n} D_{x_n} - \frac{1}{2} (A_1 - \tau [D_{x_n}, \varphi'_{x_n}]) \right) \tau \varphi'_{x_n} D_{x_n} + (\tau \varphi'_{x_n})^2 (A_2 - D_{x_n}^2),$$

which reads

$$\hat{A}_1^2 + (\tau \varphi'_{x_n})^2 \hat{A}_2 \in (\tau \varphi'_{x_n})^2 A_2 - \frac{1}{2} D_{x_n} \tau \varphi'_{x_n} A_1$$

$$+ \tau \Psi(\lambda^1_\mathbb{T}) A_1 + \tau^2 \Psi(\lambda^0_\mathbb{T}) D_{x_n} + \tau^2 \Psi(\lambda^1_\mathbb{T}). \tag{4.16}$$

We note that

$$\tau^{-1} \left| ((\tau \varphi'_{x_n})^2 A_2 v, Gv) \right| \le \tau^{-\frac{1}{2}} C \|A_2 v\|^2 + \tau^{\frac{1}{2}} C \|v\|^2_{L^2(\mathbb{R}^+, \mathscr{H}^1_\mathbb{T})},$$

and

$$\tau^{-1} \mathfrak{Re} \left(\frac{1}{2} D_{x_n} \tau \varphi'_{x_n} A_1 v, Gv \right) = \frac{1}{2} \mathfrak{Re} \left(\varphi'_{x_n} A_1 v, D_{x_n} Gv \right)$$

$$- \mathfrak{Re} \left(\frac{\varphi'_{x_n}}{2i} A_1 v|_{x_n = 0+}, Gv|_{x_n = 0+} \right)_0,$$

by integration by parts. The last term vanishes as $v|_{x_n=0+} = 0$. We thus obtain

$$\left| \tau^{-1} \mathfrak{Re} \left((\hat{A}_1^2 + (\tau \varphi'_{x_n})^2 \hat{A}_2) v, Gv \right) \right|$$
$$\leq C \left(\tau^{-\frac{1}{2}} \|A_1 v\|^2 + \tau^{-\frac{1}{2}} \|A_2 v\|^2 + \tau^{\frac{1}{2}} \|D_{x_n} v\|^2 + \tau^{\frac{1}{2}} \|v\|^2_{L^2(\mathbb{R}^+, \mathscr{H}_{\mathbb{T}}^1)} \right). \quad (4.17)$$

By choosing τ sufficiently large, from (4.12), (4.13), (4.14) and (4.17), and Lemma 4.5.8, we obtain

$$\|P_\varphi v\|^2 \geq C\tau \left(\|v\|^2_{L^2(\mathbb{R}^+, \mathscr{H}_{\mathbb{T}}^1)} + \|D_{x_n} v\|^2 \right) \geq C\tau \left(\tau^2 \|v\|^2 + \|\nabla v\|^2 \right).$$

This permits to conclude as in the proof of Theorem 4.3.4. $\square$

Proof of Lemma 4.5.7. Observe that a_2 and a_1 are both homogeneous of degree 2 in (τ, ξ'). As a consequence, $\hat{h}_2 = \tau^{-1} \lambda_{\mathbb{T}}^{-2} \sigma(H_2) \in S(\lambda_{\mathbb{T}}^0)$ is homogeneous of degree 0 in (τ, ξ'). The same holds for

$$\hat{R} = \mu \frac{\left(\hat{a}_1^2 + (\tau \varphi'_{x_n})^2 \hat{a}_2 \right)^2}{\tau^4 \lambda_{\mathbb{T}}^4}.$$

We thus work in the compact set $\mathscr{C} = \{\tau^2 + |\xi'|^2 = 1\}$ and prove that $\hat{\rho} = \hat{R} + \hat{h}_2 \geq C > 0$. Note that $\hat{\rho}$ can be continuously extended to $\tau = 0$ on $\mathscr{C}$.

Assume that $\hat{R} = 0$ then choose $\xi_n = -\hat{a}_1/(\tau \varphi'_{x_n})$. This implies $a_1 = 0$ and $a_2 = 0$. Assumption 4.3.1 then yields $\sigma(H_2) > 0$ because of the form of $\{a_2, a_1\}$ we have exhibited. We thus have

$$\hat{R} = 0 \implies \hat{h}_2 > 0.$$

Hence, we are back to the situation described at the end of the proof of Lemma 4.3.3 in Sect. A.3, which yields the conclusion. $\square$

Remark 4.5.9. In Remark 4.3.5 we saw that Assumption 4.3.1 is in fact necessary. One can moreover prove that the condition $\partial_n \varphi|_{\partial\Omega} \leq 0$ is necessary when considering estimates at the boundary.

4.6 From Local to Global Inequalities: Patching Estimates Together

Carleman estimates are local by nature. The previous sections have illustrated how they can be obtained by focusing around a region of interest. They are then applied to functions with support restricted to that region. It can however be useful in some applications to obtain a *global* estimate, that is, an estimate for functions defined in an open set Ω of $\mathbb{R}^n$ where the equation $Au = f$ is satisfied with u

fulfilling some boundary conditions. We present here how such global estimates can be achieved from the previous results. We shall consider homogeneous Dirichlet boundary conditions to exploit the results of Sect. 4.5. In fact one of the important aspects of Carleman estimates is the possibility to stitch them together to handle functions defined on a bigger domain.

As opposed to the local estimates we derived above, global estimates are characterized by an observation region $\omega \subset \Omega$. The necessity of this new feature is quite easy to point out. Let us assume the existence of a smooth weight function φ such that we have

$$\tau^3 \|e^{\tau\varphi} u\|^2_{L^2(\Omega)} + \tau \|e^{\tau\varphi} u\|^2_{L^2(\Omega)} \le C \|e^{\tau\varphi} Au\|^2_{L^2(\Omega)}, \qquad (4.18)$$

for $u \in \mathscr{C}^\infty(\overline{\Omega})$, such that $u|_{\partial\Omega} = 0$, and τ large. By Remark 4.3.5 we necessarily have $|\varphi'| \ge C > 0$ on $\overline{\Omega}$. Hence φ reaches its maximum on $\partial\Omega$, say at $x_0 \in \partial\Omega$. There its tangential derivative vanishes and thus $|\varphi'(x_0)| = |\partial_n\varphi(x_0)|$. By Remark 4.5.9 we necessarily have $\partial_n\varphi|_{\partial\Omega} \le 0$. Hence, $\partial_n\varphi(x_0) < 0$. This contradicts the fact that $\varphi(x_0) = \max_{\overline{\Omega}} \varphi$. We thus conclude that (4.18) cannot hold. As we shall see in Theorem 4.6.3 below, this can be repaired by adding an observation term

$$\tau^3 \|e^{\tau\varphi} u\|^2_{L^2(\omega)}$$

in the r.h.s. of the estimate, where ω is an open subset of Ω.

To patch local estimates together to form a global estimate, we choose a global weight function that can be used to derive each of the local estimates of Sects. 4.3 and 4.5 by satisfying the following requirements.

Assumption 4.6.1 *Let* $\omega_0 \Subset \omega \Subset \Omega$. *The weight function* φ *satisfies*

$$\varphi|_{\partial\Omega} = \text{Cst}, \qquad \partial_n\varphi|_{\partial\Omega} < 0, \qquad |\varphi'(x)| \ne 0, \ x \in \Omega \setminus \omega_0,$$
$$a_2 = 0 \ \text{and} \ a_1 = 0 \ \Rightarrow \{a_2, a_1\} > 0, \qquad x \in \Omega \setminus \omega_0.$$

Such conditions can be fulfilled by taking φ of the form

$$\varphi(x) = e^{\gamma\psi(x)}, \quad \text{with } |\psi'(x)| \ne 0, \ x \in \Omega \setminus \omega_0, \ \text{and}$$
$$\psi|_{\partial\Omega} = 0, \qquad \partial_n\psi|_{\partial\Omega} < 0, \qquad \psi(x) > 0, \ x \in \Omega,$$

and by taking the positive parameter γ sufficiently large. For the construction of such a function ψ we refer to [5, Lemma 1.1]. The construction makes use of Morse functions and the associated approximation theorem [2].

Remark 4.6.2. Here, the global weight function we choose is used to prove the local estimates. This is standard. In some situations, such global weight function may not exist. This is for instance the case of Carleman estimates for operators with principal part $\nabla(c(x)\nabla)$ where c has jumps across several interfaces. Local estimates may be proved and can still be stitched together. Yet, the resulting global weight function may not be locally equal to the weight function used for the proof of the local estimates. We refer to [13] for more details.

Theorem 4.6.3 (Global Carleman estimate). *Let φ be a function that satisfies Assumption 4.6.1. Then there exist $\tau_1 > 0$ and $C \geq 0$ such that*

$$\tau^3 \|e^{\tau\varphi} u\|^2_{L^2(\Omega)} + \tau \|e^{\tau\varphi} u\|^2_{L^2(\Omega)} \leq C \left(\|e^{\tau\varphi} Au\|^2_{L^2(\Omega)} + \tau^3 \|e^{\tau\varphi} u\|^2_{L^2(\omega)} \right),$$

for $u \in \mathscr{C}^\infty(\overline{\Omega})$, such that $u|_{\partial\Omega} = 0$, and $\tau \geq \tau_1$.

Proof. Let ω_1 be such that $\omega_0 \Subset \omega_1 \Subset \omega$. For all $x \in \overline{\Omega} \setminus \omega_1$, there exist an open subset V_x of $\overline{\Omega}$, with $x \in V_x \subset \overline{\Omega} \setminus \omega_0$ for which the local Carleman estimate, in the interior or at the boundary, holds with the weight function φ, for smooth functions with support in the compact $K_x = \overline{V_x}$.

From the covering of $\overline{\Omega} \setminus \omega_1$ by the open sets V_x, $x \in \overline{\Omega} \setminus \omega_1$ we can extract a finite covering $(V_i)_{i \in \mathscr{I}}$, such that for all $i \in \mathscr{I}$ the Carleman estimate in V_i holds for $\tau \geq \tau_i > 0$, $C = C_i > 0$ and $\text{supp}(u) \subset K_i = \overline{V_i}$.

Let $(\chi_i)_{i \in \mathscr{I}}$ be a partition of unity subordinated to the covering V_i, $i \in \mathscr{I}$, [10, 21], i.e., $\chi_i \in \mathscr{C}^\infty(\overline{\Omega})$,

$$\text{supp}(\chi_i) \subset K_i = \overline{V_i}, \qquad 0 \leq \chi_i \leq 1, \ i \in \mathscr{I}, \qquad \text{and} \qquad \sum_{i \in \mathscr{I}} \chi_i = 1 \text{ in } \Omega \setminus \omega_1.$$

Note that we have $\text{supp}(\chi_i) \cap \omega_0 = \emptyset$. For all $i \in \mathscr{I}$, we set $u_i = \chi_i u$. Then for each u_i we have a local Carleman estimate. We now observe that we have

$$Au_i = A(\chi_i u) = \chi_i Au + [A, \chi_i]u,$$

where the commutator is a first-order differential operator. For all $i \in \mathscr{I}$, we thus obtain

$$\|e^{\tau\varphi} Au_i\|^2_{L^2(\Omega)} \leq C \|e^{\tau\varphi} Au\|^2_{L^2(\Omega)} + C \|e^{\tau\varphi} u\|^2_{L^2(\Omega)} + C \|e^{\tau\varphi} \nabla u\|^2_{L^2(\Omega)}. \quad (4.19)$$

We note that we have

$$\tau^3 \|e^{\tau\varphi} u\|^2_{L^2(\Omega)} + \tau \|e^{\tau\varphi} \nabla u\|^2_{L^2(\Omega)} \leq C \sum_{i \in \mathscr{I}} \left(\tau^3 \|e^{\tau\varphi} u_i\|^2_{L^2(\Omega)} + \tau \|e^{\tau\varphi} \nabla u_i\|^2_{L^2(\Omega)} \right)$$
$$+ C\tau^3 \|e^{\tau\varphi} u\|^2_{L^2(\omega_1)} + C\tau \|e^{\tau\varphi} \nabla u\|^2_{L^2(\omega_1)},$$

From the local Carleman estimates and (4.19) we then obtain

$$\tau^3 \|e^{\tau\varphi} u\|^2_{L^2(\Omega)} + \tau \|e^{\tau\varphi} \nabla u\|^2_{L^2(\Omega)} \leq C \Big(\|e^{\tau\varphi} Au\|^2_{L^2(\Omega)} + \|e^{\tau\varphi} u\|^2_{L^2(\Omega)}$$
$$+ \|e^{\tau\varphi} \nabla u\|^2_{L^2(\Omega)} + \tau^3 \|e^{\tau\varphi} u\|^2_{L^2(\omega_1)}$$
$$+ \tau \|e^{\tau\varphi} \nabla u\|^2_{L^2(\omega_1)} \Big).$$

For τ sufficiently large we have

$$\tau^3 \|e^{\tau\varphi} u\|_{L^2(\Omega)}^2 + \tau \|e^{\tau\varphi} \nabla u\|_{L^2(\Omega)}^2 \leq C \left(\|e^{\tau\varphi} Au\|_{L^2(\Omega)}^2 + \tau^3 \|e^{\tau\varphi} u\|_{L^2(\omega_1)}^2 \right.$$
$$\left. + \tau \|e^{\tau\varphi} \nabla u\|_{L^2(\omega_1)}^2 \right).$$

We now aim to remove the last term in the r.h.s. of the previous estimation. Let $\chi \in \mathscr{C}_c^\infty(\omega)$, $0 \leq \chi \leq 1$, be such that $\chi = 1$ in a neighborhood of $\overline{\omega}_1$. If $Au = f$, after multiplication by $e^{2\tau\varphi} \chi \overline{u}$, and integration over Ω, we obtain

$$-\mathfrak{Re} \int_\Omega e^{2\tau\varphi} \chi \overline{u} \Delta u \, dx = \mathfrak{Re} \int_\Omega e^{2\tau\varphi} \chi \overline{u} f \, dx \tag{4.20}$$

The r.h.s. can be estimated as

$$\left| \mathfrak{Re} \int_\Omega e^{2\tau\varphi} \chi \overline{u} f \, dx \right| \leq C \|e^{\tau\varphi} f\|_{L^2(\Omega)}^2 + C \|e^{\tau\varphi} u\|_{L^2(\omega)}^2.$$

For the l.h.s. of (4.20), with integration by parts in x, we have

$$-\mathfrak{Re} \int_\Omega e^{2\tau\varphi} \chi \overline{u} \Delta u \, dx = \int_\Omega e^{2\tau\varphi} \chi |\nabla u|^2 \, dx + \mathfrak{Re} \int_\Omega \nabla(e^{2\tau\varphi} \chi) \overline{u} \nabla u \, dx$$

$$\geq \|e^{\tau\varphi} \nabla u\|_{L^2(\omega_1)}^2 - \frac{1}{2} \int_\Omega \Delta(e^{2\tau\varphi} \chi) |u|^2 \, dx,$$

$$\geq \|e^{\tau\varphi} \nabla u\|_{L^2(\omega_1)}^2 - C\tau^2 \|e^{\tau\varphi} u\|_{L^2(\omega)}^2.$$

The previous estimates and (4.20) then yield

$$\tau \|e^{\tau\varphi} \nabla u\|_{L^2(\omega_1)}^2 \leq C \|e^{\tau\varphi} Au\|_{L^2(\Omega)}^2 + C\tau^3 \|e^{\tau\varphi} u\|_{L^2(\omega)}^2.$$

The proof is complete. □

If we patch together the estimates with two large parameters of Theorems 4.3.9 and 4.5.1 we obtain the following result (note that with the approach used in the proofs of those theorems such an estimate can be obtained globally in the first place as is done in [5]).

Theorem 4.6.4 (Global Carleman estimate). *Let* $\omega_0 \Subset \omega \subset \Omega$. *Let* ψ *be a function that satisfies* $|\psi'| \geq C > 0$ *in* $\overline{V} \setminus \omega_0$, $\psi|_{\partial\Omega} = Cst$, *and* $\partial_n \psi|_{\partial\Omega \cap V} < 0$. *Let* $\varphi = e^{\gamma\psi}$. *Then there exist* $\tau_1 > 0$, $\gamma_1 > 0$, *and* $C \geq 0$ *such that*

$$\tau^3 \gamma^4 \|\varphi^{\frac{3}{2}} e^{\tau\varphi} u\|_{L^2(\Omega)}^2 + \tau\gamma^2 \|\varphi^{\frac{1}{2}} e^{\tau\varphi} u\|_{L^2(\Omega)}^2$$

$$\leq C \left(\|e^{\tau\varphi} Au\|_{L^2(\Omega)}^2 + \tau^3 \gamma^4 \|\varphi^{\frac{3}{2}} e^{\tau\varphi} u\|_{L^2(\omega)}^2 \right),$$

for $u \in \mathscr{C}^\infty(\overline{\Omega})$, *such that* $u|_{\partial\Omega} = 0$, $\tau \geq \tau_1$ *and* $\gamma \geq \gamma_1$.

4.7 Estimates for Parabolic Operators

To prove Carleman estimates for parabolic operators, say $P = \partial_t - \Delta$, we can proceed with similar approaches as for elliptic operators.

As in the previous sections Ω is a bounded open set in $\mathbb{R}^n$. We set $Q = (0, T) \times \Omega$. We start by proving local (in space) estimates, away from the boundary $\partial\Omega$.

4.7.1 Local Estimate Away from the Boundary

We set $\theta(t) = (t(T - t))^{-1}$. For a *negative* weight function $\varphi(x)$ we define $P_\varphi = e^{\tau\theta\varphi} P e^{-\tau\theta\varphi}$. This type of conjugation with a exponential weight that exhibits a singular behavior at $t = 0$ and $t = T$ was first introduced in [5]. Note in particular the $e^{\tau\theta\varphi}$ vanishes very rapidly at these times.

We set here $\lambda^2 = (\tau\theta)^2 + |\xi|^2$. The function $\tau\theta$ will play the role of the large parameter here. We shall obtain $\tau\theta$ large below by imposing τ/T^2 large.

From the definition of θ, and observing that $\theta'(t) = -(T - 2t)\theta^2(t)$ and that $\theta'' = 2\theta^2 + 2(T - 2t)^2\theta^3$, we have

$$\frac{4}{T^2} \leq \theta, \qquad |\theta'| \leq CT\theta^2, \qquad |\theta''| \leq CT^2\theta^3. \tag{4.21}$$

We have

$$P_\varphi = \partial_t - \iota\varphi(x)\theta'(t) - \Delta - |\tau\theta\varphi'|^2 + \langle\tau\theta\varphi', \nabla_x\rangle + \langle\nabla_x, \tau\theta\varphi'\rangle.$$

We define the following symmetric operators

$$P_2 = (P_\varphi + P_\varphi^*)/2, \quad P_1 = (P_\varphi - P_\varphi^*)/(2i),$$

which gives

$$P_2 = -\tau\varphi(x)\theta'(t) + \underbrace{D^2 - |\tau\theta\varphi'|^2}_{=A_2}, \quad P_1 = D_t + \underbrace{\tau\theta(\langle\varphi', D\rangle + \langle D, \varphi'\rangle)}_{=A_1},$$

with respective symbols

$$\tilde{p}_2 = -\tau\varphi(x)\theta'(t) + |\xi|^2 - |\tau\theta\varphi'|^2, \quad \tilde{p}_1 = \sigma + 2\tau\theta\langle\varphi', \xi\rangle - i\tau\theta\Delta\varphi.$$

For the principal symbols we have

$$p_2 = -\tau\varphi(x)\theta'(t) + a_2, \quad a_2 = |\xi|^2 - |\tau\theta\varphi'|^2, \quad p_1 = \sigma + a_1, \quad a_1 = 2\tau\theta\langle\varphi', \xi\rangle.$$

Note that $p_2, a_2 \in S(\lambda^2)$ and $p_2 - a_2 \in T\tau^{-1}S(\lambda^2)$.

Assumption 4.7.1 *The weight function* $\varphi \in \mathscr{C}^{\infty}(\mathbb{R}^n, \mathbb{R})$ *satisfies* $\varphi < 0$, $|\varphi'| > 0$ *in* $\overline{V}$ *and*

$$\forall (x, \xi) \in \overline{V} \times \mathbb{R}^n, \quad a_2(x, \xi) = 0 \implies \{a_2, a_1\}(x, \xi) \geq C > 0.$$

Such a weight function can be obtained in the form $\varphi = e^{\gamma \psi} - e^{\gamma K}$ with $|\psi'| \geq C > 0$, $\|\psi\|_{\infty} < K$ and γ sufficiently large (see e.g. [12]). The proof is in fact very close to that of Lemma 4.3.2.

Remark 4.7.2. Note that the condition we invoke is stronger than its counterpart in the elliptic case, Assumption 4.3.1. However, this condition is necessary and sufficient for the Carleman estimate to hold. We refer to [12] for the details.

Lemma 4.7.3. *There exists* $C > 0$ *and* τ_1 *such that*

$$\rho = \mu p_2^2 - \tau \theta \partial_t p_2 + \tau \theta \{p_2, a_1\} \geq C \lambda^4,$$

for $\tau \geq \tau_1 T$, *uniformly in* $t \in [0, T]$.

Proof. The proof of Lemma 4.3.3 gives $\rho_a = \mu a_2^2 + \tau \theta \{a_2, a_1\} \geq C \lambda^4$ for μ sufficiently large, uniformly in $\tau \theta \gtrsim \tau / T^2$, hence uniformly in $t \in [0, T]$.

Now observe that $p_2^2 - a_2^2 \in (T \tau^{-1} + T^2 \tau^{-2}) S(\lambda^4)$. We also have

$$\partial_t p_2 = -\tau \varphi(x) \theta'' - 2\theta \theta' |\tau \varphi'|^2,$$

which gives $\tau \theta \partial_t p_2 \in (T \tau^{-1} + T^2 \tau^{-2}) S(\lambda^4)$ by (4.21). Finally we write

$$\{p_2, a_1\} = \{a_2, a_1\} + \underbrace{\{p_2 - a_2, a_1\}}_{\in T\tau^{-1} S(\lambda^3)}.$$

In fine, we find $\rho - \rho_a \in (T \tau^{-1} + T^2 \tau^{-2}) S(\lambda^4)$. We thus conclude by choosing τ / T sufficiently large. □

Theorem 4.7.4. *Let K be a compact set of Ω and V be an open subset of Ω that is a neighborhood of K. Let φ be a weight function that satisfies Assumption 4.7.1 in $\overline{V}$. Then there exist $C > 0$ and $\tau_1 > 0$ such that*

$$\tau^3 \|\theta^{\frac{3}{2}} e^{\tau \theta \varphi} u\|_{L^2(Q)}^2 + \tau \|\theta^{\frac{1}{2}} e^{\tau \theta \varphi} \nabla_x u\|_{L^2(Q)}^2 + \tau^{-1} \|\theta^{-\frac{1}{2}} e^{\tau \theta \varphi} \partial_t u\|_{L^2(Q)}^2$$

$$+ \tau^{-1} \sum_{1 \leq j,k \leq n} \|\theta^{-\frac{1}{2}} e^{\tau \theta \varphi} \partial_{jk}^2 u\|_{L^2(Q)}^2 \leq C \|e^{\tau \theta \varphi} P u\|_{L^2(Q)}^2,$$

for $u \in \mathscr{C}^{\infty}([0, T] \times \Omega)$, with $u(t) \in \mathscr{C}_c^{\infty}(K)$ for all $t \in [0, T]$, and $\tau \geq \tau_1(T + T^2)$.

The elegant and short proof we provide here, with a positivity argument uniform in time, is due to G. Lebeau [14]. It can be found in [12].

Proof. We introduce $v = e^{\tau \theta \varphi} u$. We observe that v, along with all its time derivatives, vanishes at time $t = 0$ and $t = T$, since $\varphi \leq -C < 0$ in K. We have $P_\varphi v = e^{\tau \theta \varphi} P u = g$ and we write, similarly to (4.3),

$$\|g\|_{L^2(Q)}^2 = \|P_1 v\|_{L^2(Q)}^2 + \|P_2 v\|_{L^2(Q)}^2 + i([P_2, P_1]v, v)_{L^2(Q)},$$

which yields,

$$\|g\|_{L^2(Q)}^2 \geq \tau^{-1} \big(\theta^{-1} (\mu P_2^2 + i\tau \theta [P_2, P_1]) v, v \big)_{L^2(Q)},$$

for $\mu > 0$ and $0 < \mu(\tau\theta)^{-1} < 1$. We note that $\mu P_2^2 + i\tau\theta[P_2, P_1]$ has for principal symbol $\mu p_2^2 - \tau\theta \partial_t p_2 + \tau\theta \{p_2, a_1\}$. For μ large, that is $\tau\theta$ large, Lemma 4.7.3 with τT^{-1} large, and the Gårding inequality (uniform in t) yield a constant $C > 0$ such that

$$C\|g\|_{L^2(Q)}^2 \geq \tau^{-1} \int_0^T \theta^{-1} \|v\|_2^2 dt$$

$$\sim \tau^3 \|\theta^{\frac{3}{2}} v\|_0^2 + \tau \|\theta^{\frac{1}{2}} \nabla_x v\|_0^2 + \tau^{-1} \sum_{1 \leq j,k \leq n} \|\theta^{-\frac{1}{2}} \partial_{jk}^2 v\|_0^2.$$

We conclude as in the proof of Theorem 4.3.4. $\square$

4.7.2 Alternative Derivation

Similarly to what we did in Sect. 4.3 in the elliptic case we can provide an alternative proof of the Carleman estimate, simply based on integration by parts.

Let $\tilde{\psi}$ be such that $|\tilde{\psi}'| \geq C > 0$. We set $\psi = \tilde{\psi} + K_0$ with $K_0 = m\|\tilde{\psi}\|_\infty$ and $m > 1$. We then introduce

$$\phi = e^{\gamma\psi}, \quad \varphi = e^{\gamma\psi} - e^{\gamma\overline{\psi}} < 0,$$

with $\overline{\psi} = 2m\|\tilde{\psi}\|_\infty$ (see [4]). We then have

$$\varphi' = \gamma\psi'\phi, \quad |\varphi| \leq C\phi^2. \tag{4.22}$$

We recall that we have

$$1 \leq CT^2\theta, \quad |\theta'| \leq CT\theta^2, \quad |\theta''| \leq CT^2\theta^3. \tag{4.23}$$

Theorem 4.7.5. *Let K be a compact set of Ω and V an open subset of Ω that is a neighborhood of K. There exist $C > 0$, $\gamma_1 > 0$, $\tau_1 > 0$ such that*

$$\tau^3\gamma^4\|(\theta\phi)^{\frac{3}{2}}e^{\tau\theta\varphi}u\|^2_{L^2(Q)}+\tau\gamma^2\|(\theta\phi)^{\frac{1}{2}}e^{\tau\theta\varphi}\nabla_x u\|^2_{L^2(Q)}+\tau^{-1}\|(\theta\phi)^{-\frac{1}{2}}e^{\tau\theta\varphi}\partial_t u\|^2_{L^2(Q)}$$

$$+\tau^{-1}\sum_{1\le j,k\le n}\|(\theta\phi)^{-\frac{1}{2}}e^{\tau\theta\varphi}\partial^2_{jk}u\|^2_{L^2(Q)}\le C\|e^{\tau\theta\varphi}Pu\|^2_{L^2(Q)},$$

for $u \in \mathscr{C}^\infty([0,T]\times\Omega)$, with $u(t)\in\mathscr{C}^\infty_c(K)$ for all $t\in[0,T]$, and $\gamma\ge\gamma_1$ and $\tau\ge\tau_1(T+T^2)$.

Proof. Above we wrote the conjugated operator P_φ as $P_\varphi = P_2 + iP_1$ with

$$P_2 = D_x^2 - |\tau\theta\varphi'|^2 - \tau\varphi(x)\theta'(t),$$
$$P_1 = \tau\theta(\langle\varphi',D_x\rangle+\langle D_x,\varphi'\rangle)+D_t = 2\tau\theta\langle\varphi',D_x\rangle-i\tau\theta\Delta\varphi+D_t.$$

Following [5] (which corresponds to what is done in Sect. 4.3.1.1) we write

$$P_2 v + i\underline{P}_1 v = \underline{g} = g + \mu\tau\theta\Delta\varphi,$$

with $i\underline{P}_1 = 2\tau\theta\langle\varphi',\nabla_x\rangle + \tau(\mu+1)\theta\Delta\varphi+\partial_t$. The choice made in [5] is $\mu=1$. We then write

$$\|\underline{g}\|^2_{L^2(Q)} = \|P_2 v\|^2_{L^2(Q)} + \|\underline{P}_1 v\|^2_{L^2(Q)} + 2\Re(P_2 v, i\underline{P}_1 v)_{L^2(Q)}. \qquad (4.24)$$

We focus on the computation of $\Re(P_2 v, i\underline{P}_1 v)$ which we write as a sum of 4 terms I_{ij}, $1\le i\le 3$, $1\le j\le 3$, where I_{ij} is the inner product of the ith term in the expression of $P_2 v$ and the jth term in the expression of $i\underline{P}_1 v$ above. The term $I_{i,j}$, $1\le i,j\le 2$, can be recovered from the proof of Theorem 4.3.9:

$$I_{11} = 2\tau\Re\iint_Q \theta\varphi''(\nabla_x v,\nabla_x v)\,dt dx - \tau\iint_Q \theta(\Delta_x\varphi)|\nabla_x v|^2\,dt dx,$$

$$I_{12} = \tau(\mu+1)\iint_Q \theta(\Delta_x\varphi)|\nabla_x v|^2\,dt dx + \tau(\mu+1)\iint_Q \theta\langle\nabla_x\Delta_x\varphi,\nabla_x v\rangle v\,dt dx,$$

$$I_{21} = \tau^3\iint_Q \theta^3\nabla_x(|\varphi'|^2\nabla_x\varphi)|v|^2\,dt dx,$$

$$I_{22} = -\tau^3(\mu+1)\iint_Q \theta^3|\varphi'|^2(\Delta_x\varphi)|v|^2\,dt dx.$$

and moreover

$$I_{11} + I_{12} + I_{21} + I_{22} \geq C\tau^3\gamma^4 \iint_Q (\theta\phi)^3 |v|^2 \, dt\,dx + C\tau\gamma^2 \iint_Q \theta\phi |\nabla_x v|^2 \, dt\,dx + Y,$$

$$(4.25)$$

for γ sufficiently large, and

$$Y = 2\tau\gamma\Re \iint_Q \theta\phi\psi''(\nabla_x v, \nabla_x v) \, dt\,dx + \tau(\mu + 1) \iint_Q \theta\langle\nabla_x\Delta\varphi, \nabla_x v\rangle v \, dt\,dx.$$

We have, for $\varepsilon > 0$,

$$|Y| \leq C \iint_Q (\tau\gamma\theta + \varepsilon\tau\theta\gamma^2)\phi|\nabla_x v|^2 \, dt\,dx + C_\varepsilon\tau T^4\gamma^3 \iint_Q \theta^3\phi|v|^2 \, dt\,dx \quad (4.26)$$

We now compute the additional terms.

Term I_{13}. By integration by parts in x and t we simply find (using the Dirichlet boundary condition):

$$I_{13} = \Re \iint_Q D_x^2 v\partial_t v \, dt\,dx = \frac{1}{2} \iint_Q \partial_t |D_x v|^2 \, dt\,dx = 0.$$

Term I_{23}. We have

$$I_{23} = -\frac{1}{2} \iint_Q |\tau\theta\varphi'|^2 \partial_t |v|^2 \, dt\,dx = \iint_Q \theta\theta'|\tau\varphi'|^2 |v|^2 \, dt\,dx,$$

which gives

$$|I_{23}| \leq \tau^2\gamma^2 T \iint_Q \theta^3\phi|v|^2 \, dt\,dx. \qquad (4.27)$$

Term I_{33}. Similarly we have

$$I_{33} = -\frac{1}{2} \iint_Q \tau\varphi\theta'\partial_t|v|^2 \, dt\,dx = \frac{1}{2} \iint_Q \tau\varphi\theta''|v|^2 \, dt\,dx,$$

and we have

$$|I_{33}| \leq C\tau T^2 \iint_Q \phi^2\theta^3|v|^2 \, dt\,dx. \qquad (4.28)$$

Term I_{31}.

$$I_{31} = -\tau^2 \iint_Q \varphi \theta' \theta \langle \varphi', \nabla_x |v|^2 \rangle \, dtdx = \tau^2 \iint_Q \theta' \theta \nabla_x (\varphi \nabla_x \varphi) |v|^2 \, dtdx,$$

and we have

$$|I_{31}| \leq C \tau^2 T \gamma^2 \iint_Q \theta^3 \phi^3 |v|^2 \, dtdx. \tag{4.29}$$

Term I_{32}. Finally $I_{32} = -\tau^2 (\mu + 1) \iint_Q \varphi \theta' \theta \Delta \varphi |v|^2 \, dtdx$, which gives

$$|I_{32}| \leq C \tau^2 T \gamma^2 \iint_Q \phi^3 \theta^3 |v|^2 \, dtdx. \tag{4.30}$$

With inequality (4.25) and estimates (4.26)–(4.30), for γ, τ/T and τ/T^2 sufficiently large we obtain

$$C \mathfrak{Re}(P_2 v, i \underline{P}_1 v) \geq \tau^3 \gamma^4 \iint_Q (\theta \phi)^3 |v|^2 \, dtdx + \tau \gamma^2 \iint_Q \theta \phi |\nabla_x v|^2 \, dtdx.$$

To incorporate the additional terms involving higher order derivatives we observe that

$$\tau^{-1} \| (\theta \phi)^{-\frac{1}{2}} \partial_t v \|_{L^2(Q)}^2 \leq C \tau^{-1} T^2 \| \underline{P}_1 v \|_{L^2(Q)}^2 + C \tau \gamma^2 \| (\theta \phi)^{\frac{1}{2}} \nabla_x v \|_{L^2(Q)}^2$$

$$+ C \tau \gamma^4 \iint_Q \underbrace{\theta}_{\leq T^4 \theta^3} \phi |v|^2 \, dtdx.$$

As $\tau \geq C T^2$ we obtain

$$C \| \underline{g} \|^2 \geq \tau^3 \gamma^4 \| (\theta \phi)^{\frac{3}{2}} v \|_{L^2(Q)}^2 + \tau \gamma^2 \| (\theta \phi)^{\frac{1}{2}} \nabla_x v \|_{L^2(Q)}^2 + \tau^{-1} \| (\theta \phi)^{-\frac{1}{2}} \partial_t v \|_{L^2(Q)}^2.$$

An additional term with second-order derivatives in space can be integrated similarly with the term $P_2 v$ (using elliptic estimates). We conclude as in Sect. 4.3.1.1. □

4.7.3   Local Carleman Estimates at the Boundary

Here, we work in the vicinity of the boundary $\partial \Omega$ with homogeneous Dirichlet boundary condition. Microlocal arguments can be used to obtain a local estimate at the boundary for a parabolic operator. Such a proof can be found in [12, 13]; the ideas of the proof are similar to the proof of Theorem 4.5.6. However, the positivity argument used in the proof (with the Gårding inequality) implies the time direction. This requires much more elaborate ψDO calculus techniques than those we have presented here. We thus refer to the references given above and simply state the result. A nice result would be the derivation of the parabolic estimate at the boundary following the G. Lebeau's idea of the proof of Theorem 4.7.4, that is, a positivity argument uniform in time.

Theorem 4.7.6. *Let* $x_0 \in \partial\Omega$ *and* K *be a compact set of* $\overline{\Omega}$, $x_0 \in K$, *and* V *an open subset of* $\overline{\Omega}$ *that is a neighborhood of* K *in* $\overline{\Omega}$, *with* K *and* V *chosen sufficiently small. Let* φ *satisfy Assumption 4.7.1 in* $\overline{V}$ *and* $\partial_n\varphi|_{\partial\Omega\cap V} < 0$; *then, there exist* $\tau_1 \geq \tau_0$ *and* $C > 0$ *such that*

$$\tau^3\|\theta^{\frac{3}{2}}e^{\tau\theta\varphi}u\|^2_{L^2(Q)} + \tau\|\theta^{\frac{1}{2}}e^{\tau\theta\varphi}\nabla_x u\|^2_{L^2(Q)} + \tau^{-1}\|\theta^{-\frac{1}{2}}e^{\tau\theta\varphi}\partial_t u\|^2_{L^2(Q)}$$

$$+ \tau^{-1}\sum_{1\leq j,k\leq n}\|\theta^{-\frac{1}{2}}e^{\tau\theta\varphi}\partial^2_{jk}u\|^2_{L^2(Q)} \leq C\|e^{\tau\theta\varphi}Pu\|^2_{L^2(Q)},$$

for $u \in \mathscr{C}^\infty([0,T]\times\overline{\Omega})$, *with* $\text{supp}(u(t)) \subset K$ *for all* $t \in [0,T]$, *and* $u|_{(0,T)\times(\partial\Omega\cap V)} = 0$, *and* $\tau \geq \tau_1(T + T^2)$.

For the second more explicit method we have presented here, an inspection of the proof of Theorem 4.7.5 shows that we can obtain a similar form of estimate at the boundary.

We choose $x_0 \in \partial\Omega$, K a compact set of $\overline{\Omega}$, $x_0 \in K$, and an neighborhood of K, V, open subset of $\overline{\Omega}$. Let $\tilde{\psi} \in \mathscr{C}^\infty(\overline{V})$ be such that $|\tilde{\psi}'| \geq C > 0$. As in Sect. 4.7 we set $\psi = \tilde{\psi} + K_0$ with $K_0 = m\|\tilde{\psi}\|_\infty$ and $m > 1$. We then introduce

$$\phi = e^{\gamma\psi}, \quad \varphi = e^{\gamma\psi} - e^{\gamma\overline{\psi}} < 0,$$

with $\overline{\psi} = 2m\|\tilde{\psi}\|_\infty$.

Theorem 4.7.7. *If* $\partial_n\psi|_{\partial\Omega\cap V} < 0$, *there exist* $\tau_1 \geq \tau_0$, $\gamma_1 > 0$, *and* $C > 0$ *such that*

$$\tau^3\gamma^4\|(\theta\phi)^{\frac{3}{2}}e^{\tau\theta\varphi}u\|^2_{L^2(Q)} + \tau\gamma^2\|(\theta\phi)^{\frac{1}{2}}e^{\tau\theta\varphi}\nabla_x u\|^2_{L^2(Q)} + \tau^{-1}\|(\theta\phi)^{-\frac{1}{2}}e^{\tau\theta\varphi}\partial_t u\|^2_{L^2(Q)}$$

$$+ \tau^{-1}\sum_{1\leq j,k\leq n}\|(\theta\phi)^{-\frac{1}{2}}e^{\tau\theta\varphi}\partial^2_{jk}u\|^2_{L^2(Q)} \leq C\|e^{\tau\theta\varphi}Pu\|^2_{L^2(Q)},$$

for $u \in \mathscr{C}^\infty([0,T]\times\Omega)$, *with* $u(t) \in \mathscr{C}^\infty_c(K)$ *for all* $t \in [0,T]$, $u|_{(0,T)\times(\partial\Omega\cap V)} = 0$, *and* $\gamma \geq \gamma_1$ *and* $\tau \geq \tau_1(T + T^2)$.

4.7.4 Global Estimates for Parabolic Operators

As in Sect. 4.6 the local estimates we prove above can be patched together. We choose a global weight function that satisfies the following requirements.

Assumption 4.7.8 *Let* $\omega_0 \Subset \omega \Subset \Omega$. *The weight function* $\varphi(x)$ *satisfies*

$$\varphi|_{\partial\Omega} = \text{Cst}, \qquad \partial_n\varphi|_{\partial\Omega} < 0, \qquad |\varphi'(x)| \neq 0, \; x \in \Omega \setminus \omega_0,$$
$$a_2 = 0 \;\Rightarrow\; \{a_2, a_1\} > 0, \qquad x \in \Omega \setminus \omega_0.$$

Theorem 4.7.9 (Global Carleman estimate—parabolic case). *Let* φ *be a function that satisfies Assumption 4.7.8. Then there exist* $\tau_1 > 0$ *and* $C \geq 0$ *such that*

$$\tau^3 \|\theta^{\frac{3}{2}} e^{\tau\theta\varphi} u\|^2_{L^2(Q)} + \tau \|\theta^{\frac{1}{2}} e^{\tau\theta\varphi} \nabla_x u\|^2_{L^2(Q)}$$

$$+ \tau^{-1} \|\theta^{-\frac{1}{2}} e^{\tau\theta\varphi} \partial_t u\|^2_{L^2(Q)} + \tau^{-1} \sum_{1 \leq j,k \leq n} \|\theta^{-\frac{1}{2}} e^{\tau\theta\varphi} \partial^2_{jk} u\|^2_{L^2(Q)}$$

$$\leq C \left(\|e^{\tau\theta\varphi} Pu\|^2_{L^2(Q)} + \tau^3 \|\theta^{\frac{3}{2}} e^{\tau\theta\varphi} u\|^2_{L^2((0,T)\times\omega)} \right),$$

for $u \in \mathscr{C}^\infty([0,T] \times \overline{\Omega})$, *such that* $u|_{(0,T)\times\partial\Omega} = 0$, *and* $\tau \geq \tau_1(T + T^2)$.

Similarly we have the following result.

Theorem 4.7.10 (Global Carleman estimate—parabolic case). *Let* $\omega_0 \Subset \omega \subset \Omega$. *Let* ψ *be a function that satisfies* $|\tilde{\psi}'| \geq C > 0$ *in* $\overline{V} \setminus \omega_0$, $\tilde{\psi}|_{\partial\Omega} = \text{Cst}$, *and* $\partial_n\tilde{\psi}|_{\partial\Omega \cap V} < 0$. *Set* $\psi = \tilde{\psi} + K_0$ *with* $K_0 = m\|\tilde{\psi}\|_\infty$ *and* $m > 1$ *and*

$$\phi = e^{\gamma\psi}, \qquad \varphi = e^{\gamma\psi} - e^{\gamma\overline{\psi}} < 0,$$

with $\overline{\psi} = 2m\|\tilde{\psi}\|_\infty$
Then there exist $\tau_1 > 0$, $\gamma_1 > 0$, *and* $C \geq 0$ *such that*

$$\tau^3\gamma^4 \|(\theta\phi)^{\frac{3}{2}} e^{\tau\theta\varphi} u\|^2_{L^2(Q)} + \tau\gamma^2 \|(\theta\phi)^{\frac{1}{2}} e^{\tau\theta\varphi} \nabla_x u\|^2_{L^2(Q)}$$

$$+ \tau^{-1} \|(\theta\phi)^{-\frac{1}{2}} e^{\tau\theta\varphi} \partial_t u\|^2_{L^2(Q)} + \tau^{-1} \sum_{1 \leq j,k \leq n} \|(\theta\phi)^{-\frac{1}{2}} e^{\tau\theta\varphi} \partial^2_{jk} u\|^2_{L^2(Q)}$$

$$\leq C \left(\|e^{\tau\theta\varphi} Pu\|^2_{L^2(Q)} + \tau^3\gamma^4 \|(\theta\phi)^{\frac{3}{2}} e^{\tau\theta\varphi} u\|^2_{L^2((0,T)\times\omega)} \right),$$

for $u \in \mathscr{C}^\infty([0,T] \times \overline{\Omega})$, *such that* $u|_{(0,T)\times\partial\Omega} = 0$, $\gamma \geq \gamma_1$, *and* $\tau \geq \tau_1(T + T^2)$.

As already explain in Sect. 4.6, the construction of the global weight function we invoke here makes use of Morse functions and the associated approximation theorem.

4.8 Controllability Results for Parabolic Equations

We denote by $S(t)$ the heat semigroup for homogeneous Dirichlet boundary conditions that is $S(t)y_0 = y(t)$ solution to

$$(\partial_t - \Delta)y = 0, \text{ in } Q = (0,T) \times \Omega, \qquad y|_{(0,T) \times \partial \Omega} = 0, \qquad y|_{t=0} = y_0.$$

By the Hille–Yoshida theorem, $-\Delta$ generates a C_0-semigroup. This semigroup is in fact analytic. We refer to [17] for all these notions.

Let $\omega \subset \Omega$. We introduce the following operator:

$$L_t : L^2((0,t) \times \Omega) \to L^2(\Omega)$$

$$f \mapsto \int_0^t S(t-s) 1_\omega f(s) \, ds.$$

The solution to

$$(\partial_t - \Delta)y = 1_\omega f, \text{ in } Q = (0,T) \times \Omega, \qquad y|_{(0,T) \times \partial \Omega} = 0, \qquad y|_{t=0} = y_0, \tag{4.31}$$

is then of the form

$$y(t) = S(t)y_0 + L_t f.$$

Definition 4.8.1. We say that system (4.31) is approximately controllable at time T if for any $y_0, y_T \in L^2(\Omega)$, and $\varepsilon > 0$, there exists $f \in L^2(Q)$ such that

$$\|y_T - y(T)\|_{L^2(\Omega)} \leq \varepsilon.$$

Note that this is equivalent to having the range of L_T, $R(L_T)$, dense in $L^2(\Omega)$.

Definition 4.8.2. We say that system (4.31) is null-controllable at time T if for any $y_0 \in L^2(\Omega)$, there exists $f \in L^2(Q)$ such that

$$y(T) = 0.$$

In other words the range of L_T, $R(L_T)$, contains all the natural trajectories at time T:

$$\forall y_0 \in L^2, \quad S(T)y_0 \in R(L_T).$$

Hence this is often referred to the controllability to the trajectories.

4.8.1 Unique Continuation and Applications to Approximate Controllability

Here we shall prove the approximate controllability of the heat equation in the case of a control function only acting in ω.

Theorem 4.8.3. *The range of L_T, $R(L_T)$, is dense in $L^2(\Omega)$.*

Proof. The density of $R(L_T)$ means

$$\left(\forall f \in L^2((0,T) \times \Omega), \quad (L_T(f), v)_{L^2(\Omega)} = 0 \right) \quad \Rightarrow \quad v = 0.$$

Observe that

$$(L_T(f), v)_{L^2(\Omega)} = \int_0^T (S(T-s)1_\omega f(s), v)ds = \int_0^T (f(s), 1_\omega S(T-s)v)ds,$$

as $S(t)$ is selfadjoint. The density of $R(L_T)$ can now be written as

$$S(t)v = 0 \text{ in } L^2((0,T) \times \omega) \quad \Rightarrow \quad v = 0.$$

As $q(t) = S(t)v$ is solution to

$$(\partial_t - \Delta)q = 0, \text{ in } Q = (0,T) \times \Omega, \qquad q|_{(0,T) \times \partial\Omega} = 0, \qquad q|_{t=0} = v,$$

This follows from the unique continuation result of Proposition 4.8.4 below. □

Proposition 4.8.4. *Let g be such that $|g(y)| \leq C|y|$. Let Ω be an connected open set in $\mathbb{R}^n$ and let $\omega \subset \Omega$, with $\omega \neq \emptyset$. If $q \in L^2(0,T, H_0^1(\Omega))$ satisfies $(\partial_t - \Delta)q = g(q)$ in $(0,T) \times \Omega$ and $q(t,x) = 0$ in $(0,T) \times \omega$, then q vanishes in $(0,T) \times \Omega$.*

This result follows from the local Carleman estimates we have obtained in Sect. 4.7. The proof is similar to what is done in Sect. 4.4 in the elliptic case.

4.8.2 Null Controllability for the Heat Equation

Here we shall prove the null controllability of the heat equation.

Theorem 4.8.5. *Let $\omega \subset \Omega$. The heat equation is null controllable for a control function acting in $(0,T) \times \omega$.*

This result was proven by G. Lebeau and L. Robbiano in [15] and A. Fursikov and O. Yu. Imanuvilov in [5]. The two approaches they developed are different and complementary. For a synthetic presentation of the Lebeau–Robbiano method we refer to [12]. Here we follow the second approach, using that a global parabolic

Carleman estimate directly implies an observability estimate, which, in turn, yields the null controllability.

Remark 4.8.6. There are still not well understood discrepancies between these two approaches: The Fursikov–Imanuvilov methods allows one to treat time-dependent coefficients. The Lebeau–Robbiano method yields a fine knowledge of the cost of the control for the lower eigenmodes of the elliptic operator and an asymptotic estimation of the control cost at short time T (see e.g. [16, 20]).

The following result is central in the proof the null controllability of the heat equation.

Proposition 4.8.7. *The heat equation is null controllable at time T if and only if there exists $C_0 = C_0(T)$ such that*

$$\|S(T)q\|^2_{L^2(\Omega)} \le C_0 \|1_\omega S(t)q\|^2_{L^2((0,T)\times\Omega)}. \tag{4.32}$$

Inequality (4.32) is referred to as an observability inequality. The constant C_0 is observability constant.

Proof. The proof relies on the following property for bounded operators in Hilbert spaces. Let X, Y, Z be three Hilbert spaces, $A \in \mathscr{L}(X, Z)$, and $B \in \mathscr{L}(Y, Z)$. Then

$$R(A) \subset R(B) \quad \Leftrightarrow \quad \exists C > 0, \ \forall q \in Z, \ \|A^*q\|_X \le C\|B^*q\|_Y.$$

Here $X = L^2(\Omega)$, $Y = L^2((0, T) \times \Omega)$ and $Z = L^2(\Omega)$. We take $A = S(T)$ and $B = L_T$. Recall that $S(T)^* = S(T)$. We find

$$L_T^* : L^2(\Omega) \to L^2((0, T) \times \Omega),$$
$$q \mapsto 1_\omega S(T - t)q.$$

The null controllability is thus equivalent to having: there exists $C > 0$ such that

$$\|S(T)q\|^2_{L^2(\Omega)} \le C\|1_\omega S(T - t)q\|^2_{L^2((0,T)\times\Omega)}$$
$$= C\|1_\omega S(t)q\|^2_{L^2((0,T)\times\Omega)}. \qquad \square$$

We shall now prove such an observability inequality to conclude to the null controllability of the heat equation. Let $u(t) = S(t)q$, that is u is solution to

$$\partial_t u - \Delta u = 0, \text{ in } Q = (0, T) \times \Omega, \quad u|_{(0,T)\times\partial\Omega} = 0, \quad u|_{t=0} = q.$$

From the global Carleman inequality of Theorems 4.7.9 or 4.7.10 we have

$$\tau^3 \|e^{\tau\theta\varphi}\theta^{\frac{3}{2}}u\|^2_{L^2(Q)} \lesssim \tau^3 \|e^{\tau\theta\varphi}\theta^{\frac{3}{2}}u\|^2_{L^2((0,T)\times\omega)},$$

for $\tau = \tau_1(T + T^2)$.

In particular we have

$$\int_{T/4}^{3T/4} e^{2\tau\theta \inf_{\Omega} \varphi} \theta^3 \|u(t)\|_{L^2(\Omega)}^2 \, dt \lesssim \int_0^T e^{2\tau\theta \sup_{\Omega} \varphi} \theta^3 \|u(t)\|_{L^2(\omega)}^2 \, dt.$$

As $\tau = \tau_1(T + T^2)$ and $\theta(t) = (t(T - t))^{-1}$, and $\sup_{\Omega} \varphi \leq -C < 0$, we check that we have the following inequalities

Lemma 4.8.8. *We have*

$$e^{2\tau\theta \inf_{\Omega} \varphi} \theta^3 \gtrsim T^{-6} e^{-C\left(1+1/T\right)}, \qquad t \in [T/4, 3T/4],$$

$$e^{2\tau\theta \sup_{\Omega} \varphi} \theta^3 \lesssim T^{-6}, \qquad t \in (0, T).$$

It thus follows that we have

$$e^{-C\left(1+1/T\right)} \int_{T/4}^{3T/4} \|u(t)\|_{L^2(\Omega)}^2 \, dt \lesssim \int_0^T \|u(t)\|_{L^2(\omega)}^2 \, dt.$$

The natural parabolic energy decays gives

$$\frac{1}{2} T \|u(T)\|_{L^2(\Omega)}^2 \leq \int_{T/4}^{3T/4} \|u(t)\|_{L^2(\Omega)}^2 \, dt.$$

Hence we obtain

$$\|u(T)\|_{L^2(\Omega)} \lesssim e^{C\left(1+1/T\right)} \|u\|_{L^2((0,T)\times\omega)}.$$

From Proposition 4.8.7 we obtain the null controllability result of Theorem 4.8.5.

Appendix: Proofs of Intermediate Results

A.1   Proof of the Gårding Inequality

The symbol $a(x, \xi, \tau)$ is of the form $a(x, \xi, \tau) = a_m(x, \xi, \tau) + a_{m-1}(x, \xi, \tau)$, with $a_{m-1} \in S(\lambda^{m-1})$. For τ sufficiently large, say $\tau > \tau_1$, the full symbol $a(x, \xi, \tau)$ satisfies

$$\mathfrak{Re}\, a(x, \xi, \tau) \geq C'' \lambda^m, \qquad x \in K, \, \xi \in \mathbb{R}^n, \, \tau \geq \tau_1,$$

with $C' < C'' < C$. Let U be a neighborhood of K such that the previous inequality holds for $(x, \xi) \in U \times \mathbb{R}^n$ with the constant C'' replaced by C''' that satisfies

$C' < C''' < C'' < C$. Let $\chi(x) \in \mathscr{C}_c^\infty(U)$ be such that $0 \leq \chi \leq 1$ and $\chi = 1$ in a neighborhood of K. We then set $\tilde{a}(x, \xi, \tau) = \chi(x)a(x, \xi, \tau) + C'''(1 - \chi)(x)\lambda^m$ that satisfies

$$\tilde{a} \in S(\lambda^m) \quad \text{and} \quad \mathfrak{Re}\tilde{a}(x, \xi, h) \geq C'''\lambda^m, \qquad x \in \mathbb{R}^n, \ \xi \in \mathbb{R}^n, \ \tau \geq \tau_1. \quad (4.33)$$

We note that $(\text{Op}(\tilde{a})u, u) = (\text{Op}(a)u, u)$ if $\text{supp}(u) \subset K$. Without any loss of generality we may thus consider that the symbol a satisfies (4.33) in the remaining of the proof.

We then choose $L > 0$ such that $C' < L < C'''$ and we set

$$b(x, \xi, \tau) := \left(\mathfrak{Re}a(x, \xi, \tau) - L\lambda^m\right)^{\frac{1}{2}} \in S(\lambda^{m/2}), \qquad \text{and } B = \text{Op}(b).$$

The ψDO symbolic calculus gives $B^* \circ B = \mathfrak{Re}\text{Op}(a) - L\Lambda^m + R$, with $R \in \Psi(\lambda^{m-1})$, where $\mathfrak{Re}\text{Op}(a)$ actually means $(\text{Op}(a) + \text{Op}(a)^*)/2$. We then have

$$\mathfrak{Re}(\text{Op}(a)u, u) = (\mathfrak{Re}\text{Op}(a)u, u) \geq L(\Lambda^m u, u) - (Ru, u)$$

$$\geq L\|\Lambda^{m/2}u\|_0^2 - L'\|u\|_{(m-1)/2}^2$$

$$\geq \int_{\mathbb{R}^n} \lambda^m(L - L'/\lambda)|\hat{u}|^2 d\xi.$$

We conclude the proof by taking τ sufficiently large.

Alternative proof using the Sharp Gårding inequality (Theorem 4.2.8)

We have $\alpha(x, \xi, \tau) = a(x, \xi, \tau) - C'''\lambda^m$ such that $\mathfrak{Re}(\alpha) \geq 0$. Then $\mathfrak{Re}(\text{Op}(\alpha)u, u) \geq -C_0\|u\|_{\frac{m-1}{2}}^2$, and hence

$$\mathfrak{Re}(\text{Op}(a)u, u) \geq C'''\|u\|_{\frac{m}{2}}^2 - C_0\|u\|_{\frac{m-1}{2}}^2.$$

The conclusion thus follows as above. $\square$

A.2 Example of Functions Fulfilling the Sub-ellipticity Condition: Proof of Lemma 4.3.2

We shall actually prove the following stronger lemma here.

Lemma 4.8.9. *Let V be a bounded open set in $\mathbb{R}^n$ and $\psi \in \mathscr{C}^\infty(\mathbb{R}^n, \mathbb{R})$ such that $|\psi'| > 0$ in $\overline{V}$. Then for $\gamma > 0$ sufficiently large, $\varphi = e^{\gamma\psi}$ satisfies $|\varphi'| \geq C > 0$ in $\overline{V}$ and*

$$\forall (x,\xi) \in \overline{V} \times \mathbb{R}^n, \quad a_2(x,\xi) = 0 \quad \Rightarrow \quad \{a_2, a_1\}(x,\xi) \geq C > 0. \qquad (4.34)$$

Proof. The computation of the Poisson bracket

$$\{a_2, a_1\} = \sum_j \partial_{\xi_j} a_2 \partial_{x_j} a_1 - \partial_{x_j} a_2 \partial_{\xi_j} a_1$$

gives

$$\{a_2, a_1\} = 4 \sum_{1 \leq j,k \leq n} \tau \varphi''_{j,k} (\xi_j \xi_k + \tau^2 \varphi'_j \varphi'_k) = 4(\tau \varphi''(\xi,\xi) + \tau^3 \varphi''(\varphi',\varphi')).$$

Here we have $\varphi = e^{\gamma \psi}$, and thus $\varphi' = \gamma \varphi \psi'$ and $\varphi''_{jk} = \gamma \varphi \psi''_{jk} + \gamma^2 \varphi \psi'_j \psi'_k$, $j,k = 1,\ldots,n$, which yields

$$\{a_2, a_1\} = 4(\gamma \tau \varphi)^3 \Big(\gamma |\psi'|^4 + \psi''(\psi',\psi') + \psi''((\tau \gamma)^{-1}\xi, (\tau \gamma)^{-1}\xi)$$

$$+ \underbrace{\gamma^{-1}(\tau \varphi)^{-2} \langle \psi', \xi \rangle^2}_{\geq 0} \Big).$$

When $a_2 = 0$ we have $|\xi| = \gamma \tau \varphi |\psi'|$. We then note that

$$|\psi''((\gamma \tau \varphi)^{-1}\xi, (\gamma \tau \varphi)^{-1}\xi)| \leq C|\psi'|^2, \qquad |\psi''(\psi',\psi')| \leq C|\psi'|^2.$$

We deduce

$$\{a_2, a_1\} \geq 4(\gamma \tau \varphi)^3 \left(\gamma |\psi'|^4 - C|\psi'|^2 \right).$$

We then see that for γ sufficiently large we have $\{a_2, a_1\} \geq C_\gamma > 0$, since $|\psi'| \geq C > 0$. □

A.3 Proof of Lemma 4.3.3

We note that ρ is homogeneous of degree 4 in (τ, ξ). As ρ can be continuously extended to $\tau = 0$, we shall thus prove that $\rho \geq C$ on the compact set $\mathscr{C} = \{(x,\tau,\xi); \ x \in \overline{V}, \ \lambda = |(\tau,\xi)| = 1\}$.

In a more general framework, consider two continuous functions, f and g, defined in a compact set $\mathscr{K}$, and assume that $f \geq 0$ and $f(y) = 0 \Rightarrow g(y) \geq L > 0$. We set $h_\mu = \mu f + g$.

For all $y \in \mathscr{K}$, either $f(y) = 0$ and thus $h_\mu(y) > L$, or $f(y) > 0$ and thus there exists $\mu_y > 0$ such that $h_{\mu_y}(y) > 0$. This inequality holds locally in an open neighborhood V_y of y. From the covering of $\mathscr{K}$ by the open sets V_y, we select a finite covering $V_{y_1}, \ldots, V_{y_n}$ and set $\mu = \max_{1 \leq j \leq n} \mu_j$. We then obtain $h_\mu \geq C > 0$. We simply apply this result to ρ on $\mathscr{C}$. □

A.4 Proof of Lemma 4.3.7

We saw in Sect. A.2 that

$$\{a_2, a_1\} = 4(\gamma\tau\varphi)^3\big(\gamma|\psi'|^4 + \psi''(\psi', \psi') + \psi''((\gamma\tau\varphi)^{-1}\xi, (\gamma\tau\varphi)^{-1}\xi)$$
$$+ \gamma^{-1}(\tau\varphi)^{-2}\langle\psi', \xi\rangle^2\big).$$

We observe that $a_2\Delta\varphi = \big(|\xi|^2 - (\gamma|\psi'|\tau\varphi)^2\big)\big(\gamma^2|\psi'|^2\varphi + \gamma(\Delta\psi)\varphi\big)$, which yields

$$\rho = (\gamma\tau\varphi)^3\bigg(4\psi''((\gamma\tau\varphi)^{-1}\xi, (\gamma\tau\varphi)^{-1}\xi) + 2\mu(\gamma|\psi'|^2 + \Delta\psi)\left|\frac{\xi}{\gamma\tau\varphi}\right|^2$$
$$+ 4\gamma^{-1}(\tau\varphi)^{-2}\langle\psi', \xi\rangle^2 + (4 - 2\mu)\gamma|\psi'|^4 + 4\psi''(\psi', \psi')$$
$$- 2\mu|\psi'|^2\Delta\psi\bigg),$$

which, as $0 < \mu < 2$, we can make larger than $C_\gamma\tau\lambda^2$, with $C_\gamma > 0$ by taking γ sufficiently large. ◻

A.5 Proof of Lemma 4.3.10

With $\varphi = e^{\gamma\psi}$ we find

$$\alpha_0 = \nabla\big(\gamma^3\varphi^3|\psi'|^2\nabla\psi\big) - (\mu + 1)\gamma^2\varphi^2|\psi'|^2\big(\gamma^2|\psi'|^2\varphi + \gamma(\Delta\psi)\varphi\big)$$
$$= (2 - \mu)\gamma^4|\psi'|^4\varphi^3 + \gamma^3\varphi^3\big(\nabla(|\psi'|^2\nabla\psi) - (\mu + 1)|\psi'|^2\Delta\psi\big).$$

as $|\psi'| \geq C > 0$ and $2 - \mu > 0$ we obtain $\alpha_0 \geq C\gamma^4\varphi^3$ for γ sufficiently large. We also find $\alpha_1 = \mu\big(\gamma^2|\psi'|^2\varphi + \gamma(\Delta\psi)\varphi\big)$ and we obtain $\alpha_1 \geq C\gamma^2\varphi$ for γ sufficiently large. ◻

Acknowledgements The author was partially supported by l'Agence Nationale de la Recherche under grant ANR-07-JCJC-0139-01. The author is grateful to Assia Benabdallah, Gilles Lebeau, Nicolas Lerner and Luc Robbiano for manifold fruitful discussions on Carleman estimates. The authors wishes to deeply thank the organizers of the CIME school in July 2010, Piermaco Cannarsa and Jean-Michel Coron, for giving him the opportunity to present the material of these course notes. The authors also expresses his appreciation to the great work of the CIME staff that made his stay in Cetraro most enjoyable.

References

1. S. Alinhac, P. Gérard, *Opérateurs Pseudo-Différentiels et Théorème de Nash-Moser*, Editions du CNRS, (Meudon, France, 1991)
2. J.P. Aubin, I. Ekeland, *Applied Non Linear Analysis*, (Wiley, New York, 1984)
3. T. Carleman, Sur une problème d'unicité pour les systèmes d'équations aux dérivées partielles à deux variables indépendantes. Ark. Mat. Astr. Fys. **26B**(17), 1–9 (1939)
4. E. Fernández-Cara, S. Guerrero, Global Carleman inequalities for parabolic systems and application to controllability. SIAM J. Control Optim. **45**(4), 1395–1446 (2006)
5. A. Fursikov, O.Y. Imanuvilov, Controllability of evolution equations, vol. 34. (Seoul National University, Korea, 1996); Lecture notes
6. A. Grigis, J. Sjöstrand, *Microlocal Analysis for Differential Operators*, (Cambridge University Press, Cambridge, 1994)
7. L. Hörmander, *Linear Partial Differential Operators*, (Springer, Berlin 1963)
8. L. Hörmander, *The Analysis of Linear Partial Differential Operators*, vol. 3. (Springer, Berlin, 1985); Second printing 1994
9. L. Hörmander, *The Analysis of Linear Partial Differential Operators*, vol. 4. (Springer, Berlin, 1985)
10. L. Hörmander, *The Analysis of Linear Partial Differential Operators*, vol. 1, 2nd edn. (Springer, Berlin, 1990)
11. J. Le Rousseau, On carleman estimates with a second large parameter. in prep.
12. J. Le Rousseau, G. Lebeau, On Carleman estimates for elliptic and parabolic operators. Applications to unique continuation and control of parabolic equations. ESAIM Control Optim. Calc. Var. (2011). doi:10.1051/cocv/2011168
13. J. Le Rousseau, L. Robbiano, Local and global Carleman estimates for parabolic operators with coefficients with jumps at interfaces. Invent. Math. **183**, 245–336 (2011)
14. G. Lebeau, Cours sur les inégalités de Carleman (2005). Mastère Equations aux Dérivées Partielles et Applications, Faculté des Sciences de Tunis, Tunisie
15. G. Lebeau, L. Robbiano, Contrôle exact de l'équation de la chaleur. Comm. Part. Differ. Equat. **20**, 335–356 (1995)
16. L. Miller, A direct lebeau–robbiano strategy for the observability of heat-like semigroups. Discrete Continuous Dyn. Syst. Series B **14**, 1465–1485 (2010)
17. A. Pazy, *Semigroups of Linear Operators and Applications to Partial Differential Equations*, (Springer, New York, 1983)
18. M.A. Shubin, *Pseudodifferential Operators and Spectral Theory*, 2nd edn. (Springer, Berlin, Heidelberg, 2001)
19. M.E. Taylor, *Pseudodifferential Operators*, (Princeton University Press, Princeton, New Jersey, 1981)
20. G. Tenenbaum, M. Tucsnak, On the null-controllability of diffusion equations. ESAIM Control Optim. Calc. Var., **17**(4), (2010)
21. F. Treves, *Topological Vector Spaces, Distributions and Kernels*, (Academic Press, New York, 1967)

Chapter 5
The Wave Equation: Control and Numerics

Sylvain Ervedoza and Enrique Zuazua

Abstract In these Notes we make a self-contained presentation of the theory that has been developed recently for the numerical analysis of the controllability properties of wave propagation phenomena and, in particular, for the constant coefficient wave equation. We develop the so-called discrete approach. In other words, we analyze to which extent the semidiscrete or fully discrete dynamics arising when discretizing the wave equation by means of the most classical scheme of numerical analysis, shear the property of being controllable, uniformly with respect to the mesh-size parameters and if the corresponding controls converge to the continuous ones as the mesh-size tends to zero. We focus mainly on finite-difference approximation schemes for the one-dimensional constant coefficient wave equation. Using the well known equivalence of the control problem with the observation one, we analyze carefully the second one, which consists in determining the total energy of solutions out of partial measurements. We show how spectral analysis and the theory of non-harmonic Fourier series allows, first, to show that high frequency wave packets may behave in a pathological manner and, second, to design efficient filtering mechanisms. We also develop the multiplier approach that allows to provide energy identities relating the total energy of solutions and the energy concentrated on the boundary. These observability properties obtained after filtering, by duality, allow to build controls that, normally, do not control the

S. Ervedoza (✉)
CNRS; Institut de Mathématiques, de Toulouse UMR 5219 31062 Toulouse, France

Université de Toulouse; UPS, INSA, INP, ISAE, UT1, UTM; IMT 31062 Toulouse, France
e-mail: sylvain.ervedoza@math.univ-toulouse.fr

E. Zuazua
BCAM - Basque Center for Applied Mathematics, Alameda de Mazarredo, 14 E-48009 Bilbao, Basque Country, Spain

Ikerbasque Research Professor, Ikerbasque - Basque Foundation for Science, E48011 Bilbao, Basque Country, Spain
e-mail: zuazua@bcamath.org

F. Alabau-Boussouira et al., *Control of Partial Differential Equations*,
Lecture Notes in Mathematics 2048, DOI 10.1007/978-3-642-27893-8_5,
© Springer-Verlag Berlin Heidelberg 2012

full dynamics of the system but rather guarantee a relaxed controllability property. Despite of this they converge to the continuous ones. We also present a minor variant of the classical Hilbert Uniqueness Method allowing to build smooth controls for smooth data. This result plays a key role in the proof of the convergence rates of the discrete controls towards the continuous ones. These results are illustrated by means of several numerical experiments.

5.1 Introduction

In these notes, we make a survey presentation of the work done in the last years on the problems of controllability and observability of waves from a numerical analysis viewpoint. In particular, we explain that, even for numerical schemes that converge in the classical sense of numerical analysis, one cannot expect them to automatically be well behaved for observation and control purposes. This paper is essentially an updated version of [114], in which we collect most of the more recent developments.

Problems of control and observation of waves arise in many different contexts and for various models but, to be more precise and better present the milestones of the theory that has been developed so far, we will focus our analysis on the wave equation, and mainly in the 1-dimensional setting where several methods can be used to get rather explicit and complete results. We shall mainly focus on the finite difference method on a regular grid. Some of these results can be extended to several space dimensions but, still, a lot remains to be done to deal with general variable coefficient wave equations and with schemes on non-uniform grids in one and several space dimensions.

Controllability refers to the possibility of driving the system under consideration (here, the wave equation) to a prescribed final state at a given final time using a control function. Of course, this question is interesting when the control function does not act everywhere but is rather located in some part of the domain or on its boundary through suitable actuators.

On the other hand, observability refers to the possibility of measuring the whole energy of the solutions of the free trajectories (i.e., without control) through partial measurements. Again, one easily understands that such a property is interesting and non-trivial only when the measurements are not complete and done on the whole domain where waves propagate, but they are rather localized in part of the domain or on its boundary through suitable sensors.

It turns out that these two properties are equivalent and dual one from another. This is the basis of the so-called Hilbert Uniqueness Method [68, 69] introduced by J.-L. Lions, that we shall recall more precisely in Sect. 5.2, first in a finite dimensional setting and then for abstract conservative systems such as the wave equation.

In particular, on the basis of the Hilbert Uniqueness Method (HUM) one can also build algorithms to compute the optimal control, the one of minimal norm, in a sense to be made precise (see details in Sect. 5.2). We shall in particular explain how, using the observability property, one can slightly modify HUM with a weight function in time, vanishing for the initial and final time, so that the control obtained minimizing this functional preserves the regularity properties of the data to be controlled, see Sect. 5.2.3 and [35]. Curiously enough, these results, inspired in [24] where the regularity of the control for the wave equation is analyzed through microlocal analysis, are very recent. The abstract version of these results was proved in [35] using a simplified proof without requiring microlocal analysis tools. Note however that the results in [24], which are specific to the wave equation, are stronger than the ones in [35] since they yield also a very precise dyadic decomposition of the controls.

In the context of wave propagation phenomena, observability and controllability properties are very much related to the propagation of rays, that, for the constant coefficient wave equation, are straight space-time lines traveling at velocity one and bouncing on the boundary according to the Descartes–Snell law of Geometric Optics (see Sect. 5.3.4 and [6, 14]). In view of the finite velocity of propagation of rays, as we shall explain, one needs the observation/control time to be large enough to allow all the characteristics to meet the observation/control region and ensure the observability/controllability properties to hold.

In 1-d, these properties of propagation and reflection are the essence of the method of characteristics leading to D'Alembert's formula. That is why our presentation of the observability/controllability of waves focuses mainly in the 1-d setting, see Sect. 5.3.

As we said above, HUM characterizes the optimal control through a minimization process of a quadratic coercive functional for the solutions of the adjoint wave equation. This allows characterizing the controls through the corresponding Euler–Lagrange equations or Optimality System and building efficient algorithms for computing them. This is the so called *continuous approach* in which one first derives a complete characterization of the controls for the continuous wave model to later use numerical analysis tools to approximate them. There are of course different ways of implementing this continuous approach for the construction of numerical approximations of the controls. The first article devoted to this issue is, probably [5]. Recently, the continuous approach to numerical control was developed differently by Cindae et al. [20]. They adapted at the numerical level the well-known iterative algorithm by D. Russell [96] in which the property of controllability is obtained from the stabilization one by an iterated back and forth application of the dissipative semigroup. We also refer to the recent works of D. Auroux et J. Blum [3] that have developed a similar approach in the context of the control of nonlinear viscous conservation laws. As we shall further explain in [32], the methods in [20] lead to very similar algorithms to those one would get by applying numerical approximations in the conjugate gradient algorithm associated to the continuous minimization problem that HUM leads to. While in [20] the back and forth iteration is done always on the dissipative system, when following HUM one alternates between the state equation and the adjoint one in dual functional settings. Both approaches lead to similar convergence rates but the HUM one is more flexible since it can be adapted to a large class of problems, including those in which the control operator is unbounded as it happens often in practice and in particular for the boundary control of the wave equation.

But, very often in practice, one frequently applies the more direct, so called, *discrete approach* which consists on, first, discretizing the equation using a convergent numerical approximation scheme, to later compute a control for this numerical approximation. The model obtained after numerical discretization being a finite dimensional time continuous or discrete system, the computation of its control can be performed using standard existing finite-dimensional methods and software. But *this natural approach often fails.* In particular, in the context of the wave equation under consideration, as we shall see below, for some initial data, this approach yields discrete controls that are not even bounded as the mesh-size goes to zero, see Theorem 5.4.7.

Note that this point of view was systematically developed by R. Glowinski, J. L. Lions and coworkers (see [43, 44]) to build numerical approximation algorithms. In their works they developed and implemented conjugate gradient descent algorithms combined with Finite Element Methods for approximating the wave equations. They observed the bad conditioning of the corresponding discrete problems and indicated the need of filtering the high frequencies. This was done in particular using two-grid filtering techniques (see R. Glowinski [40]) and motivated a substantial part of the work that we present in this article.

Part of this paper is devoted to develop a thorough study of this divergence or blow up phenomenon for the space finite-difference semidiscrete 1-d wave equation as a model example since other classical schemes, such as the ones given by finite element methods, exhibit the same behavior.

Our approach is based on the analysis of the observability property of these finite-dimensional systems that approximate the 1-d wave equation. In particular, as we shall see, even when the convergence of the numerical method in the classical sense of numerical analysis is guaranteed, the discrete systems are not uniformly observable with respect to the space discretization parameter, see Theorem 5.4.2. As a consequence, by duality, there are initial data for which the sequence of corresponding discrete controls diverge (Theorem 5.4.7). In other words, the stability in what concerns the solvability of the initial-boundary value problem is not sufficient to guarantee the stability with respect to the observability property.

The lack of uniform observability can be explained and understood by looking to the propagation properties of the solutions of the numerical approximation schemes. In Sect. 5.4.3, we will explain that the numerical schemes generate spurious solutions traveling with the so called group velocity which, for high frequency numerical solutions, is of the order of the mesh-size parameter [72, 101, 104]. To be more precise, the high-frequencies involved in these wave packets are of the order of $1/h$, h being the space mesh-size. Asymptotically, as h tends to zero, they weakly converge to zero, thus being compatible with the convergence of the numerical scheme in the classical sense of numerical analysis, while being an obstruction for the observability property to hold uniformly with respect to the mesh size parameter h. This is so since the time that these wave packets need to get into the observation region is of the order of $1/h$. Actually, for $T > 0$ fixed, the observability constant for the semidiscrete problems is of the order of $\exp(C/h)$, see [77].

Our analysis of the lack of uniform observability for the discrete waves also indicates the path to avoid these divergence and blow up phenomena to occur. A careful analysis of the velocity of propagation of numerical waves shows that low frequency components propagate with a uniform velocity, a fact that is compatible with uniform observability properties. Here, by low-frequencies we refer to those covering a fixed percentage of the spectrum of the corresponding discrete dynamics, independent of the mesh-size. These "low" frequencies end up filling all the frequency range as the discretization parameter goes to zero. This shows how, through filtering, i.e. focusing on the low-frequency components, one can prove uniform observability results and still, by letting h tend to zero, recover the full dynamics of the continuous model. The need of filtering the high-frequency components to focus on the low-frequency ones was already observed in the papers by R. Glowinski, J.-L. Lions & all [40, 43, 44]. Among the different ways of doing that, in this paper we shall present Fourier filtering techniques [53], Tychonoff regularization methods and bi-grid techniques [2, 82, 83].

In Sect. 5.5 we show how these ideas yield observability properties that hold uniformly with respect to the discretization parameters within the subspace of filtered solutions. We will also briefly present the results in [111] and in [27] in the multi-dimensional setting.

These uniform observability properties lead to controllability results with uniformly bounded controls. However, the controls one obtains in this manner do not control the full state but only suitable low-frequency projections of the numerical solutions. We shall then show how to prove the convergence of the discrete controls towards the continuous ones and to derive convergence rates. The procedure we describe is general and can be adapted to various situations, i.e. different models and numerical approximation schemes.

Note that, to the best of our knowledge, this is the first time that convergence rates are proved for numerical approximation methods of the HUM control. This requires, in particular, a systematic method to build controls preserving the regularity of the data to be controlled. This is done by a suitable weighted version of the HUM-method, see Theorem 5.2.12 and [35].

The paper is organized as follows. Section 5.2 recalls the basic facts on the Hilbert Uniqueness Method. It also includes the main results of [35] on the regularity of controls for smooth data. In Sect. 5.3 we present the main observability/controllability results for the constant coefficient one-dimensional (1-d) wave equation, and briefly comment on the works [6, 14] in the multi-dimensional setting, using microlocal analysis techniques. The main results on the lack of observability/controllability of finite difference semidiscretizations are then presented in Sect. 5.4. In Sect. 5.5, we discuss several methods for curing high-frequency pathologies and getting weak observability estimates. In Sect. 5.6 we describe how these observability results can be used to develop numerical methods to obtain discrete controls that converge towards the continuous ones, with explicit convergence rates whenever the data to be controlled are smoother. In Sect. 5.7 we discuss several other related issues and present a list of interesting open problems.

Notations

In all these notes, we shall use several different notations:

- In an abstract setting, x is the state, solution of the controlled equation, A is the operator that prescribes the dynamics, B is the control operator, v is a generic control and φ is the solution of the adjoint equation.
- When considering the wave equation, the state is denoted by y and the adjoint state by u.
- Several controls shall appear. The notation v is used to denote a generic control function. V_{hum} refers to the control given by the Hilbert Uniqueness Method and V to the control given by the method in [35], which will be explained hereafter in Sect. 5.2.1.
- Indexes h will refer to the space mesh size and all the above notations will be denoted with indexes h when denoting quantities related to the semidiscrete system. Furthermore, all vectorial quantities depending on h are noted in bold characters.

5.2 Control and Observation of Finite-Dimensional and Abstract Systems

5.2.1 Control of Finite-Dimensional Systems

Numerical approximation schemes and, more precisely, those that are semidiscrete (discrete in space and continuous in time) yield finite-dimensional systems of Ordinary Differential Equations (ODEs).

There is by now an extensive literature on the control of finite-dimensional systems, and the problem is well understood for linear ones (see [66, 98]). In this section we recall the basics ingredients of the theory and we present it in a manner well suited to be extended to the PDE setting and to the limit process from finite to infinite dimensions that numerical analysis requires (see [36, 78] for more details). Indeed, the problem of convergence of controls as the mesh size in the numerical approximation tends to zero is very closely related to passing to the limit as the dimension of finite-dimensional systems tends to infinity. The latter topic is widely open and this article aims at describing some of its key aspects.

Consider the finite-dimensional system of dimension N:

$$x' = Ax + Bv, \quad 0 \le t \le T; \quad x(0) = x^0, \tag{5.1}$$

where $x = x(t) \in \mathbb{R}^N$ is the N-dimensional state and $v = v(t) \in \mathbb{R}^M$ is the M-dimensional control, with $M \le N$.

Here A is an $N \times N$ matrix with constant real coefficients and B is an $N \times M$ matrix. The matrix A determines the dynamics of the system and B models the way M controls act on it.

In practice, it is desirable to control the N components of the system with a low number of controls, and the best would be to do it by a single one, in which case $M = 1$.

System (5.1) is said to be *controllable* in time T when every initial datum $x^0 \in \mathbb{R}^N$ can be driven to any final datum x^T in $\mathbb{R}^N$ in time T. In other words, we ask if for any $(x^0, x^T) \in (\mathbb{R}^N)^2$, there exists a control function $v : [0, T] \to \mathbb{R}^M$ so that the solution x of (5.1) satisfies

$$x(T) = x^T. \tag{5.2}$$

Since we are in a linear finite dimensional setting, it is easy to check that system (5.1) is controllable in time $T > 0$ if and only if it is null-controllable in time $T > 0$, i.e. if for any $x^0 \in \mathbb{R}^N$, there exists a control function $v : [0, T] \to \mathbb{R}^M$ so that the solution x of (5.1) satisfies

$$x(T) = 0. \tag{5.3}$$

In the following we shall focus on the null-controllability and we shall refer to it simply as controllability.

There is a necessary and sufficient condition for controllability which is purely algebraic in nature. It is the so-called *Kalman condition*: System (5.1) is controllable in some time $T > 0$ if and only if

$$\operatorname{rank}[B, AB, \dots, A^{N-1}B] = N. \tag{5.4}$$

There is a direct proof of this result which uses the representation of solutions of (5.1) by means of the variations of constants formula. However, the methods we shall develop along this article rely more on the dual (but completely equivalent!) problem of observability of the adjoint system that we discuss now.

Consider the *adjoint system*

$$-\varphi' = A^*\varphi, \quad 0 \le t \le T; \quad \varphi(T) = \varphi^T. \tag{5.5}$$

Multiplying (5.1) by φ and integrating it on $(0, T)$, one immediately gets that for all $\varphi^T \in \mathbb{R}^N$,

$$\langle x(T), \varphi^T \rangle_{\mathbb{R}^N} = \int_0^T \langle v, B^*\varphi \rangle_{\mathbb{R}^M} \, dt + \langle x^0, \varphi(0) \rangle_{\mathbb{R}^N}. \tag{5.6}$$

Hence v is a control function for (5.1) if and only if for all $\varphi^T \in \mathbb{R}^N$,

$$0 = \int_0^T \langle v, B^*\varphi \rangle_{\mathbb{R}^M} \, dt + \langle x^0, \varphi(0) \rangle_{\mathbb{R}^N}. \tag{5.7}$$

This characterization of the controls for (5.1) is the heart of the duality methods we shall use in all these notes, the so-called Hilbert Uniqueness Method (HUM), introduced by J. L. Lions in [68, 69] and that has tremendously influenced the recent development of the field of PDE control and related topics.

Theorem 5.2.1. *System* (5.1) *is controllable in time T if and only if the adjoint system* (5.5) *is observable in time T, i.e., if there exists a constant $C_{obs} = C_{obs}(T) > 0$ such that, for every solution φ of* (5.5) *with initial data φ^T it holds:*

$$|\varphi(0)|^2_{\mathbb{R}^N} \le C_{obs}^2 \int_0^T |B^*\varphi|^2_{\mathbb{R}^M} dt. \tag{5.8}$$

Both properties hold in all time T if and only if the Kalman rank condition (5.4) *is satisfied.*

Remark 5.2.2. The equivalence between the controllability of the state equation and the observability of the adjoint one is one of the most classical ingredients of the controllability theory of finite-dimensional systems (see, for instance, Theorem 1.10.2 in [57]). In general, observability refers to the possibility of recovering the full solution by means of some partial measurements or observations. In the present context, i.e. in (5.8), one is allowed to measure the output $B^*\varphi$ during the time interval $(0, T)$ and wishes to recover complete information on the initial datum $\varphi(0)$. Since in finite-dimensions all norms are equivalent, and the ODEs under consideration are well-posed in the forward and backward sense of time, observing the value of the solution of the adjoint state equation $\varphi(0)$ at $t = 0$ as in (5.8) is equivalent to observing its datum φ^T at time $t = T$ or both of them.

Proof. We proceed in several steps.

Step 1. Construction of controls as minimizers of a quadratic functional. The proof we present here provides a constructive method for building controls from the observability inequality (5.8). Indeed, assume (5.8) holds and consider the quadratic functional $J : \mathbb{R}^N \to \mathbb{R}$:

$$J(\varphi^T) = \frac{1}{2} \int_0^T |B^*\varphi(t)|^2_{\mathbb{R}^M} dt + \langle x^0, \varphi(0) \rangle_{\mathbb{R}^N}. \tag{5.9}$$

If Φ^T is a minimizer for J, since $DJ(\Phi^T) = 0$, then the control

$$V_{hum} = B^*\Phi, \tag{5.10}$$

where Φ is the solution of (5.5) with initial datum Φ^T at time $t = T$ satisfies (5.7). Hence the corresponding solution x of (5.1) satisfies the control requirement $x(T) = 0$.

Thus, to build the control it is sufficient to minimize the functional J. For, we apply the direct method of the calculus of variations. The functional J being continuous, quadratic, and nonnegative, since we are in finite space dimensions, it is sufficient to prove its coercivity, which holds if and only if (5.8) holds.

Step 2. Equivalence between the observability inequality (5.8) and the Kalman condition.

Since we are in finite-dimensions and all norms are equivalent, (5.8) is equivalent to the following uniqueness or unique continuation property: *Does the fact that $B^*\varphi$ vanish for all $0 \leq t \leq T$ imply that $\varphi \equiv 0$?*

Taking into account that solutions φ are analytic in time, $B^*\varphi$ vanishes for all $t \in (0, T)$ if and only if all the derivatives of $B^*\varphi$ of any order at time $t = T$ vanish. Since $\varphi = e^{-A^*(t-T)}\varphi^T$ this is equivalent to $B^*[A^*]^k\varphi^T \equiv 0$ for all $k \geq 0$. But, according to the Cayley–Hamilton theorem, this holds if and only if it is satisfied for all $k = 0, \ldots, N-1$. Therefore $B^*\varphi \equiv 0$ is equivalent to $\varphi^T \in \cap_{k\in\{0,N-1\}} Ker(B^*[A^*]^k)$. Hence (5.8) holds if and only if $\cap_{k\in\{0,\ldots,N-1\}} Ker(B^*[A^*]^k) = \{0\}$, which is obviously equivalent to (5.4).

Step 3. Lack of controllability when unique continuation fails. If the observability estimate (5.8) does not hold, there exists a non-trivial $\hat{\varphi}^T \neq 0$ so that $B^*\hat{\varphi}(t) = 0$ for all $t \in (0, T)$. We claim that the initial data $x^0 = \hat{\varphi}(0)$ cannot be steered to $x^T = 0$. Otherwise, for some control function v, one would have from (5.7) that

$$0 = \int_0^T \langle v, B^*\hat{\varphi}\rangle_{\mathbb{R}^M} + \langle x^0, \hat{\varphi}(0)\rangle_{\mathbb{R}^N},$$

which would imply $|\hat{\varphi}(0)|^2 = 0$ and then contradict the fact that $\hat{\varphi}^T \neq 0$. □

Remark 5.2.3. The problem of observability can be formulated as that of determining uniquely the adjoint state everywhere in terms of partial measurements. The property of observability of the adjoint system (5.5) is equivalent to the inequality (5.8) because of the linear character of the system. In the context of infinite-dimensional systems or PDE this issue is sensitive to the norms under consideration.

Remark 5.2.4. It is important to note that in this finite-dimensional context, the value of time T plays no role in what concerns the property of controllability. In particular, whether a system is controllable (or its adjoint observable) is independent of the time T of control. Note that the situation is totally different for the wave equation. There, due to the finite velocity of propagation, the time needed to control/observe waves from the boundary needs to be large enough, of the order of the size of the ratio between the size of the domain and the velocity of propagation.

In fact, the main task to be undertaken to pass to the limit in numerical approximations of control problems for wave equations as the mesh size tends to zero is to explain why, even though at the finite-dimensional level the control time T is irrelevant, it may play a key role for PDEs.

Note however that, even at the level of finite-dimensional systems, the problem of how the size of controls depends on the control time T and in particular how they behave as $T \to 0$ is an interesting issue, see [97].

Remark 5.2.5. Using (5.7) with $\varphi = \Phi$ given by the minimization of the functional J in (5.9), one easily checks that any control for (5.1)–(5.3) satisfies

$$\int_0^T \langle v, V_{hum} \rangle_{\mathbb{R}^M} \, dt = \int_0^T \langle v, B^*\Phi \rangle_{\mathbb{R}^M} \, dt = -\langle x^0, \Phi(0) \rangle_{\mathbb{R}^N}$$

$$= \int_0^T |B^*\Phi|_{\mathbb{R}^M}^2 \, dt = \int_0^T |V_{hum}|_{\mathbb{R}^M}^2 \, dt.$$

This immediately yields that the HUM control V_{hum} is the one of minimal $L^2(0, T; \mathbb{R}^M)$-norm.

The proof of Theorem 5.2.1 also yields the following important result:

Corollary 5.2.6. *Given $T > 0$, we assume that (5.8) holds.*
Then for any $x^0 \in \mathbb{R}^N$, there is only one control function satisfying (5.3) that can be written as $B^*\varphi$ for φ solution of (5.5). This is the so-called HUM control V_{hum} constructed in (5.9)–(5.10).

Proof. Such a control should satisfy (5.7), hence $\varphi(T)$ should be a critical point of J defined in (5.9). But J is strictly convex because of (5.8) and therefore has only one critical point. □

Again, this has an important consequence:

Corollary 5.2.7. *Given $T > 0$, we assume that (5.8) holds.*
Then the map constructed in Theorem 5.2.1

$$\mathbb{V}_{hum} : x^0 \in \mathbb{R}^N \mapsto V_{hum} \in L^2(0, T; \mathbb{R}^M), \tag{5.11}$$

where V_{hum} is the control computed in (5.10), is linear.

Proof. Given any pairs x_1^0, x_2^0, obviously, by linearity, the solution x of (5.1) with initial data $(x_1^0 + \lambda x_2^0)$ and control function $\mathbb{V}_{hum}(x_1^0) + \lambda \mathbb{V}_{hum}(x_2^0)$ satisfies $x(T) = 0$. Moreover, using Corollary (5.2.6), one easily deduces that $\mathbb{V}_{hum}(x_1^0 + \lambda x_2^0)$ coincides with $\mathbb{V}_{hum}(x_1^0) + \lambda \mathbb{V}_{hum}(x_2^0)$. □

The norm of this map can even be characterized:

Theorem 5.2.8. *Given $T > 0$, we assume that (5.8) holds, the norm of the control map $\mathbb{V}_{hum} : \mathbb{R}^N \to L^2(0,T;\mathbb{R}^M)$ coincides with C_{obs}, the observability constant in (5.8).*

Proof. The proof of the controllability in Theorem 5.2.1 yields explicit bounds on the controls V_{hum} in (5.10) in terms of the observability constant in (5.8). Indeed, plugging $\varphi = \Phi$ in (5.7), the control V_{Hum} given by (5.10) can be seen to satisfy

$$\|V_{hum}\|_{L^2(0,T;\mathbb{R}^M)} \le C_{obs}|x^0|_{\mathbb{R}^N}, \tag{5.12}$$

C_{obs} being the same constant as in (5.8). Therefore, $\|\mathbb{V}_{hum}\|_{\mathfrak{L}((\mathbb{R}^N)^2;L^2(0,T;\mathbb{R}^M))} \le C_{obs}$.

We shall now prove the reverse inequality. Take $\hat{\varphi}$ non-trivial such that it saturates (5.8), and set $x^0 = -\hat{\varphi}(0)$. Then, using (5.7) with $\varphi = \hat{\varphi}$, any control v for x^0 should satisfy

$$|\hat{\varphi}(0)|^2_{\mathbb{R}^N} = |x^0|^2_{\mathbb{R}^N} = \int_0^T \langle v, B^*\hat{\varphi}\rangle_{\mathbb{R}^M}\, dt \le \|v\|_{L^2(0,T;\mathbb{R}^M)} \|B^*\hat{\varphi}\|_{L^2(0,T;\mathbb{R}^M)}.$$

Using that $\hat{\varphi}$ is non-trivial and saturates (5.8), we find out that the control function $\mathbb{V}_{hum}(-\hat{\varphi}(0))$ should be of norm at least $C_{obs}|\hat{\varphi}(0)|$, hence the result. $\qquad\square$

Remark 5.2.9. Step 3 of the proof of Theorem 5.2.1 and the proof of Theorem 5.2.8 rely on the same idea, that data that are difficult to observe correspond to the ones that are the most difficult ones to control.

5.2.2 Controllability and Observability for Abstract Conservative Systems

In this section, let X be a Hilbert space endowed with the norm $\|\cdot\|_X$ and let $\mathbb{T} = (\mathbb{T}_t)_{t\in\mathbb{R}}$ be a strongly continuous group on X, with generator $A : \mathscr{D}(A) \subset X \to X$.

We further assume that A is a skew-adjoint operator $A^* = -A$.

For convenience, we also assume that A is invertible with continuous inverse in X. This can be done without loss of generality by translating the semigroup if necessary using $\beta \in \mathbb{R}$ and replacing A by $A - i\beta I$.

Define then the Hilbert space $X_1 = \mathscr{D}(A)$ of elements of X such that $\|Ax\|_X < \infty$, endowed with the norm $\|\cdot\|_1 = \|A\cdot\|_X$. Also define X_{-1} as the completion of X with respect to the norm $\|\cdot\|_{-1} = \|A^{-1}\cdot\|_X$.

Let us then consider the control system

$$x' = Ax + Bv, \quad t \geq 0, \qquad x(0) = x^0 \in X, \tag{5.13}$$

where $B \in \mathcal{L}(\mathcal{U}, X_{-1})$, $\mathcal{U}$ is an Hilbert space which describes the possible actions of the control, and $v \in L^2_{loc}([0, \infty); \mathcal{U})$ is a control function.

We assume that the operator B is admissible in the sense of [102, Def. 4.2.1]:

Definition 5.2.10. The operator $B \in \mathcal{L}(\mathcal{U}, X_{-1})$ is said to be an admissible control operator for $\mathbb{T}$ if for some $\tau > 0$, the operator $\mathcal{R}_\tau$ defined on $L^2(0, T; \mathcal{U})$ by

$$\mathcal{R}_\tau v = \int_0^\tau \mathbb{T}_{\tau-s} B v(s) \, ds$$

satisfies $\operatorname{Ran} \mathcal{R}_\tau \subset X$, where $\operatorname{Ran} \Phi_\tau$ denotes the range of the map Φ_τ.

When B is an admissible control operator for $\mathbb{T}$, system (5.13) is called admissible.

Note that, obviously, if B is a bounded operator, that is if $B \in \mathcal{L}(\mathcal{U}, X)$, then B is admissible for $\mathbb{T}$. But there are non-trivial examples as, for instance, the boundary control of the wave equation with Dirichlet boundary conditions, in which B is unbounded but admissible, see [68]. In such cases, the admissibility property is then a consequence of a suitable hidden regularity result for the solutions of the adjoint system.

To be more precise, the admissibility of B for $\mathbb{T}$ is equivalent to the existence of a time $T > 0$ and a constant $K_T > 0$ such that any solution of

$$\varphi' = A\varphi, \quad t \in (0, T), \qquad \varphi(T) = \varphi^T \tag{5.14}$$

satisfies

$$\int_0^T \|B^* \varphi(t)\|_{\mathcal{U}}^2 \, dt \leq K_T \|\varphi^T\|_X^2. \tag{5.15}$$

In this section we will always assume that B is an admissible control operator for $\mathbb{T}$. Then, for every $x^0 \in X$ and $v \in L^2_{loc}([0, \infty); \mathcal{U})$, (5.13) has a unique mild solution x which belongs to $C([0, \infty); X)$ (see [102, Proposition 4.2.5]).

Our purpose is to study the controllability of system (5.13).

System (5.13) is said to be null controllable in time T if for any $x^0 \in X$, there exists a control function $v \in L^2(0, T; \mathcal{U})$ such that the solution of (5.13) satisfies

$$x(T) = 0. \tag{5.16}$$

System (5.13) is said to be null controllable if it is null controllable in some time $T > 0$.

Note that since system (5.13) is linear and time-reversible, an easy argument left to the reader shows that system (5.13) is null controllable in time T if and only if it is exactly controllable, i.e. for all x^0, x^T in X, there exists a control function $v \in L^2(0, T; \mathcal{U})$ such that the solution x of (5.13) with initial data x^0 and control function v satisfies $x(T) = x^T$. Hence we will focus on the null-controllability property in the sequel, and we shall refer to it simply as controllability.

Here again, we claim that system (5.13) is controllable in time T if and only if there exists a constant C_{obs} such that all solutions φ of the adjoint (5.14) satisfy

$$\|\varphi(0)\|_X^2 \le C_{obs}^2 \int_0^T \|B^*\varphi(t)\|_{\mathcal{U}}^2 \, dt. \tag{5.17}$$

We shall refer the interested reader to [68] for the proof of the fact that the exact controllability in time T implies the observability (5.17) for the adjoint system (5.14). This is based on a closed graph theorem.

The other implication is actually proved in Step 1 of the proof of Theorem 5.2.1, which describes the Hilbert Uniqueness Method. The idea is to find a minimizer of the functional

$$J(\varphi^T) = \frac{1}{2} \int_0^T \|B^*\varphi(t)\|_{\mathcal{U}}^2 \, dt + \langle x^0, \varphi(0)\rangle_X, \quad \varphi^T \in X. \tag{5.18}$$

Note that such a minimizer exists and is unique due to the observability property (5.17). Then, if Φ^T denotes the minimizer of J, since $DJ(\Phi^T) = 0$, the function

$$V_{hum} - B^*\Phi, \tag{5.19}$$

where Φ is the corresponding solution of (5.14) is a control function. Indeed, it satisfies for all $\varphi^T \in X$,

$$\int_0^T \langle V_{hum}, B^*\varphi(t)\rangle_{\mathcal{U}} \, dt + \langle x^0, \varphi(0)\rangle_X = 0, \tag{5.20}$$

which, as in (5.7), characterizes the controls of (5.13).

5.2.3 Smoothness Results for HUM Controls

In this section, we assume that the adjoint system (5.14) satisfies the observability assumption (5.17) in some time T^*. We also assume that the admissibility property holds.

We now address the issue of the regularity of the control function V_{hum} obtained by minimizing the functional J in (5.18). To be more precise, we analyze whether this control preserves the smoothness of the initial data to be controlled.

According to a counterexample that we will present later on in Sect. 5.3.3, we will see that, under the very general assumptions under consideration, no smoothness of the control computed by the minimization of the functional J in (5.9) can be expected.

We thus propose an alternate method, based on HUM, which yields a control of minimal norm in some weighted (in time) L^2 space, and for which we prove that, with no further assumptions, if $x^0 \in X_1$, then this control function belongs to $H_0^1(0, T; \mathcal{U})$. Thus, this result can be readily applied to the most relevant examples, as it is for instance the case of the wave equation with Dirichlet boundary control. In particular, this implies that the controlled solution x of (5.13) belongs to $C^1([0, T], X)$ and also, in various situations (see Sect. 5.3.5), to a strict subspace of X for all time $t \in [0, T]$, which will reflect the extra regularity of the initial data to be controlled. In particular, if BB^* maps X_1 into X_1, then the controlled solution x will belong to $C^0([0, T]; X_1)$.

Fix $T > T^*$ and choose $\delta > 0$ such that $T - 2\delta \geq T^*$. Let $\eta = \eta(t) \in L^\infty(\mathbb{R})$ be such that

$$\eta : \mathbb{R} \to [0, 1], \qquad \eta(t) = \begin{cases} 0 & \text{if } t \notin (0, T), \\ 1 & \text{if } t \in [\delta, T - \delta]. \end{cases} \tag{5.21}$$

In particular, there exists a positive constant C_{obs} such that any solution φ of (5.14) satisfies

$$\|\varphi(0)\|_X^2 \leq C_{obs}^2 \int_0^T \eta(t) \|B^*\varphi(t)\|_{\mathcal{U}}^2 \, dt. \tag{5.22}$$

Then define the functional J by

$$J(\varphi^T) = \frac{1}{2} \int_0^T \eta(t) \|B^*\varphi(t)\|_{\mathcal{U}}^2 \, dt + \langle x^0, \varphi(0) \rangle_X, \tag{5.23}$$

where φ denotes the solution of the adjoint system (5.14) with initial data φ^T.

Inequality (5.22) then implies the strict convexity of the functional J and its coercivity, but with respect to the norm

$$\|\varphi^T\|_{obs}^2 = \int_0^T \eta(t) \|B^*\varphi(t)\|_{\mathcal{U}}^2 \, dt. \tag{5.24}$$

Let us now remark that, since we assumed that $\mathbb{T}$ is a strongly continuous unitary group, the three norms $\left\|\varphi^T\right\|_X$, $\|\varphi(0)\|_X$ and $\left\|\varphi^T\right\|_{obs}$ (in view of (5.15)–(5.22)) are equivalent.

We are now in position to state our first result:

Proposition 5.2.11. *Let* $x^0 \in X$. *Assume that system* (5.13) *is admissible and exactly observable in some time* T^*. *Let* $T > T^*$ *and* $\eta \in L^\infty(\mathbb{R})$ *as in* (5.21).

Then the functional J *in* (5.23) *has a unique minimizer* $\Phi^T \in X$ *on* X. *Besides, the function* V *given by*

$$V(t) = \eta(t)B^*\Phi(t), \tag{5.25}$$

where $\Phi(t)$ *is the solution of* (5.14) *with initial datum* Φ^T, *is a control function for system* (5.13). *This control can also be characterized as the one of minimal* $L^2(0, T; dt/\eta; \mathscr{U})$*-norm among all possible controls for which the solution of* (5.13) *satisfies the control requirement* (5.16). *Besides,*

$$\int_0^T \|V(t)\|_{\mathscr{U}}^2 \frac{dt}{\eta(t)} = \left\|\Phi^T\right\|_{obs}^2 \leq C_{obs}^2 \|x_0\|_X^2, \tag{5.26}$$

where C_{obs} *is the constant in the observability inequality* (5.22).

Moreover, this process defines linear maps

$$\mathbb{V}_a : \begin{cases} X \longrightarrow X^* = X \\ x^0 \mapsto \Phi^T \end{cases} \quad and \quad \mathbb{V} : \begin{cases} X \longrightarrow L^2\left(0, T, \dfrac{dt}{\eta(t)}; \mathscr{U}\right) \\ x^0 \mapsto V. \end{cases} \tag{5.27}$$

Besides, V *is the unique admissible control function that can be written* $v(t) = \eta(t)B^*\varphi(t)$ *for some* φ *solution of the adjoint* (5.14).

This result is similar to those obtained in the context of HUM (see [68] and previous paragraphs) and follows the same lines as Step 1 of the proof of Theorem 5.2.1. Normally the weight η is simply taken to be $\eta \equiv 1$ on $[0, T]$ while in the present formulation, the fact that it vanishes at $t = 0, T$ plays a key role.

The main novelty and advantage of using the weight function η is that, with no further assumption on the control operator B, the control inherits the regularity of the data to be controlled.

To state our results, it is convenient to introduce, for $s \in \mathbb{R}_+$, some notations: $\lceil s \rceil$ denotes the smallest integer satisfying $\lceil s \rceil \geq s$, $\lfloor s \rfloor$ is the largest integer satisfying $\lfloor s \rfloor \leq s$ and $\{s\} = s - \lfloor s \rfloor$. Finally, the space C^s denotes the classical Hölder space.

Theorem 5.2.12 ([35]). *Assume that the hypotheses of Proposition 5.2.11 are satisfied.*

Let $s \in \mathbb{R}_+$ *be a nonnegative real number and further assume that* $\eta \in C^{\lceil s \rceil}(\mathbb{R})$.

If the initial datum x^0 to be controlled belongs to $\mathscr{D}(A^s)$, then the minimizer Φ^T given by Proposition 5.2.11 and the control function V given by (5.25), respectively, belong to $\mathscr{D}(A^s)$ and $H_0^s(0, T; \mathscr{U})$.

Besides, there exists a positive constant $C_s = C_s(\eta, C_{obs}, K_T)$ independent of $x^0 \in \mathscr{D}(A^s)$ such that

$$\left\| \Phi^T \right\|_{\mathscr{D}(A^s)}^2 + \|V\|_{H_0^s(0,T;\mathscr{U})}^2 \leq C_s \left\| x^0 \right\|_{\mathscr{D}(A^s)}^2. \qquad (5.28)$$

In other words, the maps $\mathbb{V}_a$ and $\mathbb{V}$ defined in (5.27) satisfy:

$$\mathbb{V}_a : \mathscr{D}(A^s) \longrightarrow \mathscr{D}(A^s), \qquad \mathbb{V} : \mathscr{D}(A^s) \longrightarrow H_0^s(0, T; \mathscr{U}). \qquad (5.29)$$

In other words, the constructive method we have proposed, strongly inspired by HUM, naturally reads the regularity of the initial data to be controlled, and provides smoother controls for smoother initial data. Note however that if one is interested to the regularity in space of the controlled trajectory, one needs to work slightly more.

Indeed, one of the main consequences of Theorem 5.2.13 is the following regularity result for the controlled trajectory:

Corollary 5.2.13 ([35]). *Under the assumptions of Theorem 5.2.12, if the initial datum x^0 to be controlled belongs to $\mathscr{D}(A^s)$, then the controlled solution x of (5.13) with the control function V given by Proposition 5.2.11 belongs to*

$$C^s([0, T]; X) \overset{\lfloor s \rfloor}{\underset{k=0}{\cap}} C^k([0, T]; \mathscr{X}_{s-k}), \qquad (5.30)$$

where the spaces $(\mathscr{X}_j)_{j \in \mathbb{N}}$ are defined by induction by

$$\mathscr{X}_0 = X, \qquad \mathscr{X}_j = A^{-1}(\mathscr{X}_{j-1} + BB^* \mathscr{D}(A^j)), \qquad (5.31)$$

and the spaces $\mathscr{X}_s$ for $s \geq 0$ are defined by interpolation by

$$\mathscr{X}_s = [\mathscr{X}_{\lfloor s \rfloor}, \mathscr{X}_{\lceil s \rceil}]_{\{s\}}.$$

The spaces $\mathscr{X}_j$ are not explicit in general. However, there are several cases in which they can be shown to be included in Hilbert spaces of the form $\mathscr{D}(A^j)$, which in practical applications to PDE are constituted by functions that are smoother than X with respect to the space variable.

In particular, if BB^* maps $\mathscr{D}(A^j)$ to itself for all $j \in \mathbb{N}$, then the spaces $\mathscr{X}_j$ can be shown to coincide with $\mathscr{D}(A^j)$ for all $j > 0$. Of course, this is sharp, since one cannot expect the controlled solution to be better than $C^0([0, T]; \mathscr{D}(A^s))$ for initial data $x^0 \in \mathscr{D}(A^s)$.

Proof (Sketch of the proof of Theorem 5.2.12). We focus on the case $s = 1$, the others being completely similar.

Since $V(t) = \eta(t)B^*\Phi(t)$ is a control function, for any $\varphi^T \in X$, identity (5.20) holds. Then, assuming that $\varphi^T = A^2\Phi^T \in X$, we get

$$\int_0^T \eta(t)\langle B^*\Phi(t), B^*A^2\Phi(t)\rangle_{\mathscr{U}} \, dt + \langle x^0, A^2\Phi(0)\rangle_X = 0.$$

But

$$\int_0^T \eta(t)\langle B^*\Phi(t), B^*A^2\Phi(t)\rangle_{\mathscr{U}} \, dt$$

$$= \int_0^T \eta(t)\langle B^*\Phi(t), B^*\Phi''(t)\rangle_{\mathscr{U}} \, dt$$

$$= -\int_0^T \eta'(t)\langle B^*\Phi(t), B^*\Phi'(t)\rangle_{\mathscr{U}}^2 \, dt - \int_0^T \eta(t)\|B^*\Phi'(t)\|_{\mathscr{U}}^2 \, dt \quad (5.32)$$

and

$$\langle x^0, A^2\Phi(0)\rangle_X = -\langle Ax^0, A\Phi(0)\rangle_X.$$

Therefore, assuming some regularity on Φ_T, namely $\Phi_T \in \mathscr{D}(A^2)$, one can prove

$$\int_0^T \eta(t)\,\|B^*\Phi'(t)\|_U^2 \, dt + \int_0^T \eta'(t)\langle B^*\Phi(t), B^*\Phi'(t)\rangle_U \, dt + \langle Ax^0, A\Phi(0)\rangle_X = 0.$$

$$(5.33)$$

But, since $\eta \in C^1(\mathbb{R})$, for any $\varepsilon > 0$, (the constants C below denote various positive constants which do not depend on ε and that may change from line to line)

$$\left|\int_0^T \eta'(t)\langle B^*\Phi(t), B^*\Phi'(t)\rangle_{\mathscr{U}} \, dt\right| \leq \frac{C}{\varepsilon} \int_0^T \|B^*\Phi(t)\|_{\mathscr{U}}^2 \, dt + \varepsilon \int_0^T \|B^*\Phi'(t)\|_{\mathscr{U}}^2 \, dt$$

$$\leq \frac{C}{\varepsilon}\,\|\Phi^T\|_X^2 + C\varepsilon\,\|\Phi'(T)\|_X^2$$

$$\leq \frac{C}{\varepsilon}\,\|\Phi^T\|_{obs}^2 + C\varepsilon\,\|\Phi'(0)\|_X^2$$

$$\leq \frac{C}{\varepsilon}\,\|\Phi^T\|_X^2 + C\varepsilon \int_0^T \eta(t)\,\|B^*\Phi'(t)\|_{\mathscr{U}}^2 \, dt,$$

where we used the equivalence of the norms $\|\varphi^T\|_X$, $\|\varphi(0)\|_X$, $\|B^*\varphi\|_{L^2(0,T;\mathscr{U})}$ and $\|\varphi^T\|_{obs}$, the admissibility and observability inequalities (5.15) and (5.22) and estimate (5.26).

In particular, taking $\varepsilon > 0$ small enough,

$$\left| \int_0^T \eta'(t) \langle B^* \Phi(t), B^* \Phi'(t) \rangle_{\mathcal{U}} \, dt \right| \leq C \left\| x^0 \right\|_X^2 + \frac{1}{2} \int_0^T \eta(t) \left\| B^* \Phi(t) \right\|_{\mathcal{U}}^2 \, dt. \tag{5.34}$$

It then follows from (5.33) that

$$\frac{1}{2} \int_0^T \eta(t) \left\| B^* \Phi'(t) \right\|_{\mathcal{U}}^2 \, dt \leq C \left\| x^0 \right\|_X^2 + \left\| A x^0 \right\|_X \left\| A \Phi^T \right\|_X. \tag{5.35}$$

But $\left\| x^0 \right\|_X \leq C \left\| A x^0 \right\|_X$ and, applying the observability inequality (5.22) to $A\Phi(0)$, which reads

$$\left\| A\Phi(0) \right\|_X^2 \leq C_{obs}^2 \int_0^T \eta(t) \left\| B^* \Phi'(t) \right\|_{\mathcal{U}}^2 \, dt,$$

we obtain

$$\left\| A \Phi^T \right\|_X^2 = \left\| A \Phi(0) \right\|_X^2 \leq C_{obs}^2 \int_0^T \eta(t) \left\| B^* \Phi'(t) \right\|_{\mathcal{U}}^2 \, dt \leq C \left\| A x^0 \right\|_X^2. \tag{5.36}$$

Since $V' = \eta' B^* \Phi + \eta(t) B^* \Phi'$,

$$\int_0^T \left\| V'(t) \right\|_{\mathcal{U}}^2 \, dt \leq 2 \int_0^T \eta'(t)^2 \left\| B^* \Phi(t) \right\|_{\mathcal{U}}^2 \, dt + 2 \int_0^T \eta(t)^2 \left\| B^* \Phi'(t) \right\|_{\mathcal{U}}^2 \, dt. \tag{5.37}$$

But

$$\int_0^T \eta'(t)^2 \left\| B^* \Phi(t) \right\|_{\mathcal{U}}^2 \, dt \leq C \int_0^T \left\| B^* \Phi(t) \right\|_{\mathcal{U}}^2 \, dt$$

$$\leq C \left\| \Phi^T \right\|_X^2 \leq C \left\| \Phi^T \right\|_{obs}^2 \leq C \left\| x^0 \right\|_X^2, \tag{5.38}$$

where we used (5.15), the equivalence of the norms $\left\| \Phi^T \right\|_X$ and $\left\| \Phi(0) \right\|_X$, (5.22) and (5.26) in the last estimate.

Then estimate (5.28) follows from estimates (5.36), (5.37) and (5.38).

To make the arguments in the formal proof above rigorous, one should take, instead of $\varphi^T = A^2 \Phi^T$, which is a priori not allowed in (5.7),

$$\varphi_\tau^T = \frac{1}{\tau^2}\left(\Phi(T+\tau) + \Phi(T-\tau) - 2\Phi^T\right),$$

and then pass to the limit in $\tau \to 0$.

As we said, the proof for integers $s \in \mathbb{N}$ can be made following the same lines. And the general case $s \geq 0$ can then be deduced using interpolation results. Details can be found in [35]. □

Remark 5.2.14. When the operator B is bounded from X to $\mathcal{U}$, the HUM functional J in (5.18), without the time cut-off function η, satisfies the same regularity results as the one in Theorem 5.2.12 for $s = 1$. For larger s, and if one furthermore assumes that $BB^* \in \mathcal{L}(\mathcal{D}(A^k))$ for all $k \leq s$, then Theorem 5.2.12 holds. One immediately deduces Corollary 5.2.13 as well. Of course, in this case, an easy induction argument shows that $\mathcal{X}_k = \mathcal{D}(A^k)$ for all $k \leq s$.

The main difference appearing in the proof when $\eta \equiv 1$ is that, when integrating by parts, boundary terms appear at $t = 0, T$. But they can be suitably bounded when B is bounded. Note that when the cut-off function η is introduced, these boundary terms vanish and are transformed into time-integrated terms that are bounded by the weaker admissibility condition.

Remark 5.2.15. Note that such regularity results can be found in [24] for the wave equation with internal control and a control operator satisfying $BB^* \in \cap_{k\geq 0} \mathcal{L}(\mathcal{D}(A^k))$. There, the authors propose a thorough study of the operator $\mathbb{V}_a$ in (5.29) and give precise estimates on how it acts on each range of frequencies. This is of course much more precise than the results presented in Theorem 5.2.12.

But the proof of the results in [24] requires the use of very deep technical tools such as microlocal analysis and Littlewood–Paley decomposition.

Let us also point out the article [64] which illustrates numerically the estimates obtained in [24] on the operator $\mathbb{V}_a$ in (5.27).

Also remark however that our approach, though it yields less precise results in the context of the distributed control wave equation, is much more robust and applies also for boundary control problems and any linear conservative equations.

5.3 The Constant Coefficient Wave Equation

5.3.1 Problem Formulation: The 1-d Case

Let us first consider the constant coefficient 1-d wave equation:

$$\begin{cases} u_{tt} - u_{xx} = 0, & 0 < x < 1, \, 0 < t < T, \\ u(0,t) = u(1,t) = 0, & 0 < t < T, \\ u(x,0) = u^0(x), \, u_t(x,0) = u^1(x), & 0 < x < 1. \end{cases} \tag{5.39}$$

In (5.39) $u = u(x,t)$ describes the displacement of a vibrating string occupying $(0, 1)$.

The energy of solutions of (5.39) is conserved in time, i.e.,

$$E(t) = \frac{1}{2} \int_0^1 \left[|u_x(x,t)|^2 + |u_t(x,t)|^2 \right] dx = E(0) \quad \forall 0 \le t \le T. \tag{5.40}$$

The problem of boundary observability of (5.39) can be formulated as follows: *To give sufficient conditions on T such that there exists $C_{obs}(T) > 0$ for which the following inequality holds for all solutions of (5.39):*

$$E(0) \le C_{obs}(T)^2 \int_0^T |u_x(1,t)|^2 \, dt. \tag{5.41}$$

Inequality (5.41), when it holds, guarantees that the total energy of solutions can be "observed" from the boundary measurement on the extreme $x = 1$. The best constant $C_{obs}(T)$ in (5.41) is the so-called *observability constant*.[1]

Similarly as in the previous section, the observability problem above is equivalent to the following boundary controllability property: For any $(y^0, y^1) \in L^2(0,1) \times H^{-1}(0,1)$ there exists $v \in L^2(0,T)$ such that the solution of the controlled wave equation

$$\begin{cases} y_{tt} - y_{xx} = 0, & 0 < x < 1, \, 0 < t < T, \\ y(0,t) = 0; \; y(1,t) = v(t), & 0 < t < T, \\ y(x,0) = y^0(x), \; y_t(x,0) = y^1(x), & 0 < x < 1, \end{cases} \tag{5.42}$$

satisfies

$$y(x,T) = y_t(x,T) = 0, \quad 0 < x < 1. \tag{5.43}$$

Note that system (5.39) fits in the abstract setting given in (5.14) with

$$\varphi = \begin{pmatrix} u \\ u_t \end{pmatrix}, \quad A = \begin{pmatrix} 0 & Id \\ \partial_{xx} & 0 \end{pmatrix},$$

with $\quad X = H_0^1(0,1) \times L^2(0,1), \quad \mathscr{D}(A) = H^2 \cap H_0^1(0,1) \times H_0^1(0,1).$
$$\tag{5.44}$$

Hence the corresponding control system should be given by duality as in Sect. 5.2. However, in the PDE context, it is classical to identify $L^2(0,1)$ with its dual.

---

[1] Inequality (5.41) is just an example of a variety of similar observability problems: (a) one could observe the energy concentrated on the extreme $x = 0$ or in the two extremes $x = 0$ and 1 simultaneously; (b) the $L^2(0,T)$-norm of $u_x(1,t)$ could be replaced by some other norm; (c) one could also observe the energy concentrated in a subinterval (α, β) of $(0,1)$, etc.

Of course, once this identification is done, though X is an Hilbert space, its dual X^* cannot be identified anymore with itself. That explains why the control system (5.42) is considered with initial data $L^2(0, 1) \times H^{-1}(0, 1)$, which is a natural candidate for X^*. But our presentation in the abstract setting in Sect. 5.2 can still be done in that case, but that would require the introduction of further notations that may be confusing.

Thus, we directly address this example showing why controllability of (5.42) is a consequence of (5.41) by a minimization method which yields the control of minimal $L^2(0, T)$-norm, similarly as the one developed in the previous section.

Given $(y^0, y^1) \in L^2(0, 1) \times H^{-1}(0, 1)$, a control $V_{hum} \in L^2(0, T)$ can be computed as

$$V_{hum}(t) = U_x(1, t), \tag{5.45}$$

where U is the solution of (5.39) corresponding to initial data $(U^0, U^1) \in H_0^1(0, 1) \times L^2(0, 1)$ minimizing the functional

$$J((u^0, u^1)) = \frac{1}{2} \int_0^T |u_x(1, t)|^2 dt + \int_0^1 y^0 u^1 dx - \langle y^1, u^0 \rangle_{H^{-1} \times H_0^1} \tag{5.46}$$

in the space $H_0^1(0, 1) \times L^2(0, 1)$.

Note that J is convex. The continuity of J in $H_0^1(0, 1) \times L^2(0, 1)$ is guaranteed by the fact that the solutions of (5.39) satisfy $u_x(1, t) \in L^2(0, T)$ (the so-called hidden regularity property, that holds also for the Dirichlet problem for the wave equation in several space dimensions; see [60, 68, 69]). More, precisely, for all $T > 0$ there exists a constant $K(T) > 0$ such that, for all solution of (5.39),

$$\int_0^T \left[|u_x(0, t)|^2 + |u_x(1, t)|^2 \right] dt \leq K(T) E(0). \tag{5.47}$$

Thus, to prove the existence of a minimizer for J, it is sufficient to prove that it is coercive. This is guaranteed by the observability inequality (5.41). Also note that the observability inequality (5.41) also guarantees the strict convexity of J and then the uniqueness of a minimizer for J.

Let us see that the minimum of J provides the control. The functional J is of class C^1. Consequently, the gradient of J at the minimizer vanishes:

$$0 = \langle DJ((U^0, U^1)), (w^0, w^1) \rangle = \int_0^T U_x(1, t) w_x(1, t) dt$$

$$+ \int_0^1 y^0 w^1 dx - \langle y^1, w^0 \rangle_{H^{-1} \times H_0^1}$$

for all $(w^0, w^1) \in H_0^1(0, 1) \times L^2(0, 1)$, where w stands for the solution of (5.39) with initial data (w^0, w^1). By choosing the control as in (5.45) this identity yields

$$\int_0^T V_{hum}(t) w_x(1, t) dt + \int_0^1 y^0 w^1 dx - \langle y^1, w^0 \rangle_{H^{-1} \times H_0^1} = 0. \qquad (5.48)$$

On the other hand, multiplying in (5.42) by w and integrating by parts, we get

$$\int_0^T v(t) w_x(1, t) dt + \int_0^1 y^0 w^1 dx - \langle y^1, w^0 \rangle_{H^{-1} \times H_0^1}$$

$$- \int_0^1 y(T) w_t(T) dx + \langle y_t(T), w(T) \rangle_{H^{-1} \times H_0^1} = 0. \qquad (5.49)$$

Combining these two identities we get $\int_0^1 y(T) w_t(T) dx - \langle y_t(T), w(T) \rangle_{H^{-1} \times H_0^1}$
$= 0$ for all $(w^0, w^1) \in H_0^1(0, 1) \times L^2(0, 1)$, which is equivalent to the exact controllability condition (5.43).

This argument shows that *observability implies controllability*. The reverse is also true. If controllability holds, then, using Banach closed graph Theorem, the linear map that to all initial data $(y^0, y^1) \in L^2(0, 1) \times H^{-1}(0, 1)$ of the state (5.42) associates the control of the minimal $L^2(0, T)$-norm, which can be still denoted by V_{hum} in view of Remark 5.2.4, is bounded. Multiplying the state (5.42) with that control by u, solution of (5.39), and using (5.43), we obtain

$$\int_0^T V_{hum}(t) u_x(1, t) dt + \int_0^1 y^0 u^1 dx - \langle y^1, u^0 \rangle_{H^{-1} \times H_0^1} = 0. \qquad (5.50)$$

Consequently,

$$\left| \int_0^1 [y^0 u^1 - y^1 u^0] dx \right| = \left| \int_0^T V_{hum}(t) u_x(1, t) dt \right| \leq \|v\|_{L^2(0,T)} \|u_x(1, t)\|_{L^2(0,T)}$$

$$\leq C \|(y^0, y^1)\|_{L^2(0,1) \times H^{-1}(0,1)} \|u_x(1, t)\|_{L^2(0,T)} \qquad (5.51)$$

for all $(y^0, y^1) \in L^2(0, 1) \times H^{-1}(0, 1)$, which implies the observability inequality (5.41).

Throughout this paper we shall mainly focus on the problem of observability. However, in view of the equivalence above, all the results we present have immediate consequences for controllability. The most important ones will also be stated. Note, however, that controllability is not the only application of the observability inequalities, which are also of systematic use in the context of inverse problems [12, 56, 58, 59]. We shall discuss this issue briefly in open problem # 6 in Sect. 5.7.2.

Remark 5.3.1. Note that here, we consider the adjoint (5.39) with initial data at time $t = 0$, whereas in the previous section, we have considered the adjoint (5.14) with initial data at time $t = T$. This can be done because of the time-reversibility of the wave equation under consideration.

5.3.2   Observability for the 1-d Wave Equation

The following holds.

Proposition 5.3.2. *For any* $T \geq 2$, *system* (5.39) *is observable. In other words, for any* $T \geq 2$ *there exists* $C_{obs}(T) > 0$ *such that* (5.41) *holds for any solution of* (5.39). *Conversely, if* $T < 2$, (5.39) *is not observable, or, equivalently,*

$$\inf_{u \text{ solution of } (5.39)} \left[\frac{1}{E(0)} \int_0^T | u_x(1,t) |^2 \, dt \right] = 0. \tag{5.52}$$

The proof of observability for $T \geq 2$ can be carried out in several ways, including Fourier series (and generalizations to non-harmonic Fourier series, see [105]), multipliers (Komornik [60]; Lions [68, 69]), sidewise energy estimates [110], Carleman inequalities (Zhang [106]), and microlocal[2] tools (Bardos, Lebeau, and Rauch [6]; Burq and Gérard [14]).

Let us explain how it can be proved using Fourier series. Solutions of (5.39) can be written in the form

$$u(x,t) = \sum_{k \geq 1} \left(a_k \cos(k\pi t) + \frac{b_k}{k\pi} \sin(k\pi t) \right) \sin(k\pi x), \tag{5.53}$$

$$u^0(x) = \sum_{k \geq 1} a_k \sin(k\pi x), \quad u^1(x) = \sum_{k \geq 1} b_k \sin(k\pi x).$$

---

[2]Microlocal analysis deals, roughly speaking, with the possibility of localizing functions and its singularities not only in the physical space but also in the frequency domain. Localization in the frequency domain may be done according to the size of frequencies but also to sectors in the euclidean space in which they belong to. This allows introducing the notion of microlocal regularity; see, for instance, [48].

It follows that

$$E(0) = \frac{1}{4} \sum_{k \geq 1} \left[a_k^2 k^2 \pi^2 + b_k^2 \right].$$

On the other hand,

$$u_x(1,t) = \sum_{k \geq 1} (-1)^k \left[k \pi a_k \sin(k\pi t) + b_k \cos(k\pi t) \right].$$

Using the orthogonality properties of $\sin(k\pi t)$ and $\cos(k\pi t)$ in $L^2(0,2)$, it follows that

$$\int_0^2 |u_x(1,t)|^2 \, dt = \sum_{k \geq 1} \left(\pi^2 k^2 a_k^2 + b_k^2 \right).$$

The two identities above show that the observability inequality holds when $T = 2$ and therefore for any $T > 2$ as well. In fact, in this particular case, we even have the identity

$$E(0) = \frac{1}{4} \int_0^2 |u_x(1,t)|^2 \, dt. \tag{5.54}$$

On the other hand, for $T < 2$ the observability inequality does not hold. Indeed, suppose that $T = 2 - 2\delta$ with $\delta \in (0,2)$. Solve

$$u_{tt} - u_{xx} = 0, \quad (x,t) \in (0,1) \times (0,T), \quad u(0,t) = u(1,t) = 0, \quad 0 < t < T, \tag{5.55}$$

with data at time $t = T/2 = 1 - \delta$ with support in the subinterval $(0, \delta)$. This solution is such that $u_x(1,t) = 0$ for $0 < t < T = 2 - 2\delta$ since the segment $x = 1, t \in (0,T)$ remains outside the domain of influence of the space segment $t = T/2, x \in (0, \delta)$ (see Fig. 5.1).

Note that the observability time ($T = 2$) is twice the length of the string. This is due to the fact that an initial disturbance concentrated near $x = 1$ may propagate to the left (in the space variable) as t increases and only reach the extreme $x = 1$ of the interval after bouncing at the left extreme $x = 0$ (as described in Fig. 5.1).

As we have seen, in one dimension and with constant coefficients, the observability inequality is easy to understand. The same results are true for sufficiently smooth coefficients (BV-regularity suffices). However, when the coefficients are simply Hölder continuous, these properties may fail, thereby contradicting an initial intuition (see [19]).

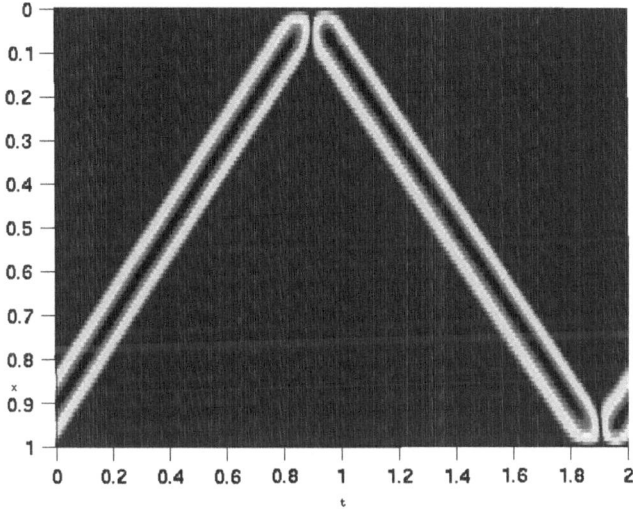

Fig. 5.1 Wave localized at $t = 0$ near the endpoint $x = 1$ that propagates with velocity 1 to the left, bounces at $x = 0$, and reaches $x = 1$ again in a time of the order of 2

5.3.3   Computing the Boundary Control

In this section, we compute explicitly the control function given by HUM. As a consequence, we will explain in particular that one cannot hope similar regularity results as the ones in Corollary 5.2.13 when no cut-off function in time is introduced within the functional J as in (5.23).

Let us consider the 1-d wave (5.42) controlled by the boundary in time $T = 4$. The time $T = 4$ is larger than the critical time of controllability, corresponding to $T^* = 2$, which is the time needed by the waves to go from $x = 1$ to $x = 0$ and bounce back at $x = 0$.

The application of the classical *Hilbert Uniqueness Method* in this case consists in minimizing the functional J given by (5.46) to obtain a control V_{hum} from (5.45) in terms of the minimizer of J.

We now use the fact that, when the control time horizon is $T = 4$ (actually it is true for any even integer), the functional J acts diagonally on the Fourier coefficients of the solutions u of (5.39) and then the minimizer of J can be computed explicitly.

Using (5.53) for the solutions of (5.39), one easily checks that

$$\frac{1}{2} \int_0^T |u_x(1,t)|^2 \, dt = \frac{1}{2} \int_0^4 \left| \sum_{|k|=1}^{\infty} \left(a_k \cos(k\pi t) + \frac{b_k}{k\pi} \sin(k\pi t) \right) k\pi (-1)^k \right|^2 dt$$

$$= \sum_{k=1}^{\infty} \left(|a_k|^2 k^2 \pi^2 + |b_k|^2 \right).$$

The initial datum to be controlled $(y^0, y^1) \in L^2(0, 1) \times H^{-1}(0, 1)$ can be written in Fourier series as

$$(y^0, y^1) = \sum_{k=1}^{\infty} (\hat{y}_k^0, \hat{y}_k^1) \sin(k\pi x), \tag{5.56}$$

with

$$\sum_{k=1}^{\infty} \left(|\hat{y}_k^0|^2 + \frac{|\hat{y}_k^1|^2}{k^2\pi^2} \right) < \infty.$$

Thus, for (u^0, u^1) as in (5.53),

$$J((u^0, u^1)) = \sum_{k=1}^{\infty} \left(|a_k|^2 k^2 \pi^2 + |b_k|^2 \right) + \frac{1}{2} \sum_{k=1}^{\infty} \left(\hat{y}_k^0 b_k - \hat{y}_k^1 a_k \right). \tag{5.57}$$

Therefore the minimizer (U^0, U^1) of J can be given as

$$(U^0, U^1) = \sum_{k=1}^{\infty} (A_k, B_k) \sin(k\pi x), \quad \text{with} \quad \begin{cases} A_k = \dfrac{\hat{y}_k^1}{4k^2\pi^2}, \\ B_k = -\dfrac{\hat{y}_k^0}{4}, \end{cases} \tag{5.58}$$

and the control function V_{hum} is simply

$$V_{hum}(t) = \partial_x U(1, t) = \frac{1}{4} \sum_{k=1}^{\infty} (-1)^k \left(\frac{\hat{y}_k^1}{k\pi} \cos(k\pi t) - \hat{y}_0^k \sin(k\pi t) \right).$$

In particular, it is obvious that, for $(y^0, y^1) \in H_0^1(0, 1) \times L^2(0, 1)$, this method yields $(U^0, U^1) \in H^2 \cap H_0^1(0, 1) \times H_0^1(0, 1)$ and $V \in H^1(0, T)$.

However,

$$V_{hum}(0) = \frac{\sqrt{2}}{4} \sum_{k=1}^{\infty} (-1)^k \frac{\hat{y}_k^1}{k\pi}.$$

Therefore, if $y^0 \in H_0^1(0, 1)$, the controlled solution y of (5.41) with that control function cannot be a strong solution if

$$\sum_{k=1}^{\infty} (-1)^k \frac{\hat{y}_k^1}{k\pi} \neq 0,$$

whatever the regularity of the initial datum to be controlled is, because the compatibility condition $y(1, 0) = V(0)$ does not hold.

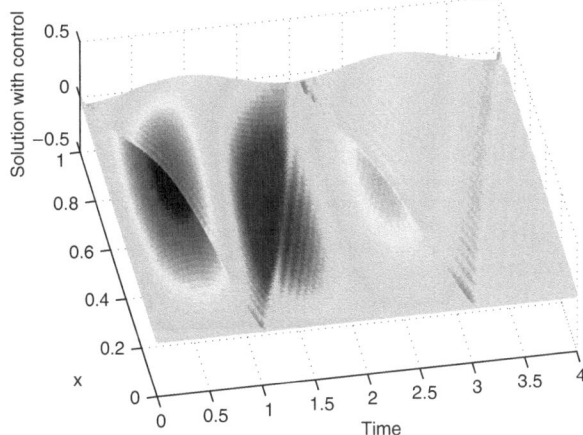

Fig. 5.2 The controlled trajectory for the wave equation with initial data $(y^0(x), y^1(x)) = (0, \sin(\pi x))$ for the HUM control in time $T = 4$. A kick is introduced by the control function at $(t, x) = (0, 1)$ and travels in the domain, hence making the solution non-smooth

Of course, such case happens, for instance when the initial datum to be controlled simply is $(y^0(x), y^1(x)) = (0, \sin(\pi x))$. This is illustrated in Fig. 5.2. There, with the control given by HUM, we see that the controlled solution is singular along the characteristic line emanating from $(t, x) = (0, 1)$.

As this example shows, the regularity of the initial datum does not yield additional regularity for the controlled wave equation when using the HUM control.

5.3.4   The Multidimensional Wave Equation

In several space dimensions the observability problem for the wave equation is much more complex and cannot be solved using Fourier series except in some particular geometries. The velocity of propagation is still one for all solutions but energy propagates along bicharacteristic rays.

Before going further, let us give the precise definition of *bicharacteristic ray*. Consider the wave equation with a scalar, positive, and smooth variable coefficient $a = a(x)$:

$$u_{tt} - \operatorname{div}(a(x)\nabla u) = 0. \tag{5.59}$$

Bicharacteristic rays $s \mapsto (x(s), t(s), \xi(s), \tau(s))$ solve the Hamiltonian system

$$\begin{cases} x'(s) = -a(x)\xi, & t'(s) = \tau, \\ \xi'(s) = \nabla a(x)|\xi|^2, & \tau'(s) = 0, \end{cases} \tag{5.60}$$

on the characteristic set $\tau^2 - a(x)|\xi|^2 = 0$.

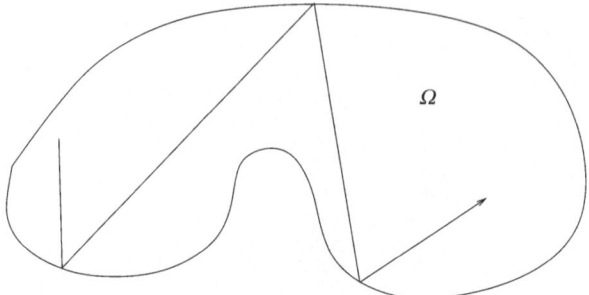

Fig. 5.3 Ray that propagates inside the domain Ω following straight lines that are reflected on the boundary according to the laws of Geometric Optics

Rays describe the microlocal propagation of energy. The projections of the bicharacteristic rays in the (x, t) variables are the rays of Geometric Optics that play a fundamental role in the analysis of the observation and control properties through the Geometric Control Condition (GCC), that will be introduced below. As time evolves, the rays move in the physical space according to the solutions of (5.60). Moreover, the direction in the Fourier space (ξ, τ) in which the energy of solutions is concentrated as they propagate is given precisely by the projection of the bicharacteristic ray in the (ξ, τ) variables. When the coefficient $a = a(x)$ is constant, the ray is a straight line and carries the energy outward, which is always concentrated in the same direction in the Fourier space, as expected. But for variable coefficients the dynamics is more complex. This Hamiltonian system describes the dynamics of rays in the interior of the domain where the equation is satisfied. When rays reach the boundary they are reflected according to the Snell–Descartes laws of Geometric Optics (Fig. 5.3).[3]

When the coefficient $a = a(x)$ varies in space, the dynamics of this system may be quite complex and can lead to some unexpected behavior [74].

Let us now address the control problem for smooth domains[4] in the constant coefficient case.

Let Ω be a bounded domain of $\mathbb{R}^n$, $n \geq 1$, with boundary $\partial\Omega$ of class C^2, let ω be an open and nonempty subset of Ω, and let $T > 0$. Consider the linear controlled wave equation in the cylinder $Q = \Omega \times (0, T)$:

$$\begin{cases} y_{tt} - \Delta y = f 1_\omega & \text{in} \quad Q, \\ y = 0 & \text{on} \quad \partial\Omega \times (0, T) \\ y(x, 0) = y^0(x), \, y_t(x, 0) = y^1(x) & \text{in} \quad \Omega. \end{cases} \quad (5.61)$$

---

[3]Note, however, that tangent rays may be diffractive or even enter the boundary. We refer to [6] for a deeper discussion of these issues.

[4]We refer to Grisvard [45] for a discussion of these problems in the context of non-smooth domains.

In (5.61), 1_ω is the characteristic function of the set ω, $y = y(x, t)$ is the state, and $f = f(x, t)$ is the control variable. Since f is multiplied by 1_ω the action of the control is localized in ω.

When $(y^0, y^1) \in H_0^1(\Omega) \times L^2(\Omega)$ and $f \in L^2(Q)$, the system (5.59) has a unique solution $y \in C\left([0, T]; H_0^1(\Omega)\right) \cap C^1\left([0, T]; L^2(\Omega)\right)$.

The problem of *controllability*, generally speaking, is as follows: *Given* $(y^0, y^1) \in H_0^1(\Omega) \times L^2(\Omega)$, *to find* $f \in L^2(Q)$ *such that the solution of system (5.61) satisfies*

$$y(T) \equiv y_t(T) \equiv 0. \qquad (5.62)$$

The method of Sect. 5.3, the so-called HUM, shows that the exact controllability property is equivalent to the following *observability inequality*:

$$\left\|\left(u^0, u^1\right)\right\|_{H_0^1(\Omega) \times L^2(\Omega)}^2 \leq C \int_0^T \int_\omega u_t^2 \, dx \, dt \qquad (5.63)$$

for every solution of the adjoint uncontrolled system

$$\begin{cases} u_{tt} - \Delta u = 0 & \text{in} \quad \Omega \times (0, T), \\ u = 0 & \text{on} \quad \partial\Omega \times (0, T), \\ u(x, 0) = u^0(x), u_t(x, 0) = u^1(x) & \text{in} \quad \Omega. \end{cases} \qquad (5.64)$$

The main result concerning (5.63) is that the observability inequality holds if and only if the GCC is satisfied (see, for instance, Bardos, Lebeau, and Rauch [6] and Burq and Gérard [14]): Roughly speaking, the GCC for (Ω, ω, T) states that all rays of Geometric Optics should enter in the domain ω in a time smaller than T.

For instance, when the domain is a ball, the subset of the boundary where the control is being applied needs to contain a point of each diameter. Otherwise, if a diameter skips the control region, it may support solutions that are not observed (see Ralston [87]). In the case of the square domain Ω, observability/controllability fails if the control is supported on a set which is strictly smaller than two adjacent sides, as shown in Fig. 5.4.

Several remarks are in order.

Remark 5.3.3. Since we are dealing with solutions of the wave equation, for the GCC to hold, the control time T has to be sufficiently large due to the finite speed of propagation, the trivial case $\omega = \Omega$ being the exception. However, the time being large enough does not suffice, since the control subdomain ω needs to satisfy the GCC in a finite time. Figure 5.4 provides an example of this fact.

Remark 5.3.4. Most of the literature on the controllability of the wave equation has been written in the framework of the *boundary control* problem discussed in the previous section in the 1-dimensional setting. The control problems formulated

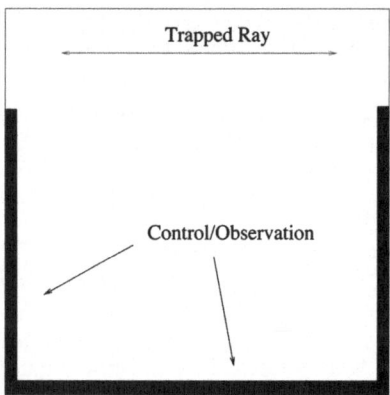

Fig. 5.4 A geometric configuration in which the GCC is not satisfied, whatever $T > 0$ is. The domain where waves evolve is a square. The control is located on a subset of three adjacent sides of the boundary, leaving a small horizontal subsegment uncontrolled. There is a horizontal line that constitutes a ray that bounces back and forth for all time perpendicularly on two points of the vertical boundaries where the control does not act

above for (5.59) are usually referred to as *internal controllability* problems since the control acts on the subset ω of Ω. The latter is easier to deal with since it avoids considering non homogeneous boundary conditions, in which case solutions have to be defined in the sense of transposition [68, 69] and lie in $C^0([0, T]; L^2(\Omega)) \cap C^1([0, T]; H^{-1}(\Omega))$ for boundary controls in $L^2((0, T) \times \partial\Omega)$.

Note that, if Γ denotes an open subset of the boundary $\partial\Omega$, the HUM then expresses the link between controllability of data in $L^2(\Omega) \times H^{-1}(\Omega)$ with controls in $L^2((0, T) \times \Gamma)$ with the following observability inequality, of course similar to (5.41): There exists a constant C_{obs} such that every solution of the adjoint control system (5.64) satisfies

$$\left\| \left(u^0, u^1\right) \right\|^2_{H_0^1(\Omega) \times L^2(\Omega)} \leq C_{obs}^2 \int_0^T \int_\Gamma |\partial_n u|^2 d\sigma dt \qquad (5.65)$$

for every solution of the adjoint uncontrolled system (5.64).

Let us now discuss what is known about (5.65):

(a) Using multiplier techniques, Ho [47] proved that if one considers subsets of $\partial\Omega$ of the form $\Gamma(x^0) = \left\{ x \in \partial\Omega : (x - x^0) \cdot n(x) > 0 \right\}$ for some $x^0 \in \mathbb{R}^n$ (we denote by $n(x)$ the outward unit normal to $\partial\Omega$ in $x \in \partial\Omega$ and by $\cdot$ the scalar product in $\mathbb{R}^n$) and if $T > 0$ is large enough, the following boundary observability inequality holds:

$$\left\| \left(u^0, u^1\right) \right\|^2_{H_0^1(\Omega) \times L^2(\Omega)} \leq C_{obs}^2 \int_0^T \int_{\Gamma(x^0)} |\partial_n u|^2 d\sigma dt \qquad (5.66)$$

for all $(u^0, u^1) \in H_0^1(\Omega) \times L^2(\Omega)$, which is the observability inequality that is required to solve the boundary controllability problem.

Later, (5.66) was proved in [68, 69] for any $T > T(x^0) = 2 \| x - x^0 \|_{L^\infty(\Omega)}$. This is the optimal observability time that one may derive by means of this multiplier (see Osses [84] for other variants).

Proceeding as in [68], one can easily prove that (5.66) implies (5.63) when ω is a neighborhood of $\Gamma(x^0)$ in Ω, i.e., $\omega = \Omega \cap \Theta$, where Θ is a neighborhood of $\Gamma(x^0)$ in $\mathbb{R}^n$, with $T > 2\|x - x^0\|_{L^\infty(\Omega\setminus\omega)}$. In particular, exact controllability holds when ω is a neighborhood of the boundary of Ω.

(b) Bardos, Lebeau, and Rauch [6] proved that, in the class of C^∞ domains, the observability inequality (5.65) holds if and only if (Ω, Γ, T) satisfies the GCC: Every ray of Geometric Optics that propagates in Ω and is reflected on its boundary $\partial\Omega$ intersects Γ at a non-diffractive point in time less than T.

This result was proved by means of microlocal analysis. Later the microlocal approach was simplified by Burq [7] by using the microlocal defect measures introduced by Gérard [38] in the context of homogenization and kinetic equations. In [7] the GCC was shown to be sufficient for exact controllability for domains Ω of class C^3 and equations with C^2 coefficients. The result for variable coefficients is the same: The observability inequality and, thus, the exact controllability property hold if and only if all rays of Geometric Optics intersect the control region before the control time. However, it is important to note that, although in the constant coefficient equation all rays are straight lines, in the variable coefficient case this is no longer the case, which makes it harder to gain intuition about the GCC.

5.3.5 Smoothness Properties

Note that the results in Sect. 5.2.3 also apply once observability (5.63) holds. In particular, adding a cut-off function in time $\eta(t)$ as in (5.21) within the functional J in (5.46) implies gentle regularity results for the corresponding minimizers of J and the corresponding control functions.

5.3.5.1 Internal Control Operators

Assume that, for some time T^*,

$$\left\|(u^0, u^1)\right\|^2_{H_0^1(\Omega)\times L^2(\Omega)} \leq C_{obs}^2 \int_0^{T^*}\int_\Omega \chi_\omega^2 u_t^2 \, dx \, dt, \tag{5.67}$$

for all solutions u of (5.64), where $\chi_\omega = \chi_\omega(x)$ is a non-negative function on Ω which is localized in ω.

Let $T > T^*$, choose $\delta > 0$ such that $T - 2\delta \geq T^*$ and fix a function η satisfying (5.21).

Then the functional J introduced in (5.23) is defined for $(u^0, u^1) \in L^2(\Omega) \times H^{-1}(\Omega)$ by

$$
J((u^0, u^1)) = \frac{1}{2} \int_0^T \int_\Omega \eta(t) \chi_\omega^2(x) |u(x,t)|^2 \, dx dt + \int_\Omega y^1 u^0 dx
$$
$$
- \langle y^0, u^1 \rangle_{H_0^1(\Omega) \times H^{-1}(\Omega)}, \tag{5.68}
$$

where u is the solution of (5.64) with initial data $(u^0, u^1) \in L^2(\Omega) \times H^{-1}(\Omega)$.

This functional is not exactly the one corresponding to the abstract presentation above since we did not identify the energy space $H_0^1(\Omega) \times L^2(\Omega)$ with its dual. We have rather shifted by one derivative the regularity of the adjoint solutions under consideration so that their initial data lie in $L^2(\Omega) \times H^{-1}(\Omega)$. Note that this functional is more natural when doing PDE because of the classical identification of $L^2(\Omega)$ with its dual.

But now, the relevant estimate is, instead of (5.67),

$$
\| (u^0, u^1) \|_{L^2(\Omega) \times H^{-1}(\Omega)}^2 \leq C_{obs}^2 \int_0^{T^*} \int_\Omega \chi_\omega^2 |u|^2 dx dt, \tag{5.69}
$$

Let us also emphasize that the two estimates (5.67) and (5.69) are completely equivalent and can be deduced one from another by differentiating or integrating the solutions of (5.64) with respect to the time t.

To state our results precisely, we define the operator A as in (5.44). In particular, $\mathscr{D}(A^s)$ is the space $H^{s+1}(\Omega) \times H^s(\Omega)$ for $s \geq 0$ with compatibility boundary conditions depending on $s \geq 0$. To be more precise, $(y^0, y^1) \in \mathscr{D}(A^s)$ if and only if $(y^0, y^1) \in H^{s+1}(\Omega) \times H^s(\Omega)$ and satisfies

$$
y_{|\partial\Omega}^0 = -\Delta y_{|\partial\Omega}^0 = (-\Delta)^j y_{|\partial\Omega}^0 = 0 \quad j \in \{0, \ldots, \lfloor s/2 + 1/4 \rfloor\}
$$

and

$$
y_{|\partial\Omega}^1 = -\Delta y_{|\partial\Omega}^1 = (-\Delta)^j y_{|\partial\Omega}^1 = 0 \quad j \in \{0, \ldots, \lfloor s/2 - -1/4 \rfloor\}.
$$

To simplify the notations in a consistent way, we also introduce $\mathscr{D}(A^s)$ for $s \in [-1, 0]$, which is, for $s = -1$, $\mathscr{D}(A^{-1}) = L^2(\Omega) \times H^{-1}(\Omega)$, for $s = 0$, $\mathscr{D}(A^0) = X = H_0^1(\Omega) \times L^2(\Omega)$ and for $s \in (-1, 0)$, $\mathscr{D}(A^s)$ is the corresponding interpolation between $\mathscr{D}(A^{-1})$ and $X = \mathscr{D}(A^0)$.

Actually, for explaining these notations, we emphasize that we did not identify X with its dual. Therefore, we shall introduce the space $X^* = L^2(\Omega) \times H^{-1}(\Omega)$, the operator

$$A^* = \begin{pmatrix} 0 & I \\ \Delta & 0 \end{pmatrix}, \quad \text{with } \mathscr{D}(A^*) = H_0^1(\Omega) \times L^2(\Omega).$$

Of course, with the above notations, for all $s \geq 0$, $\mathscr{D}((A^*)^s) = \mathscr{D}(A^{s-1})$.

Theorem 5.2.12 and its corollaries then imply:

Theorem 5.3.5. *Let η be a smooth weight function satisfying (5.21). Let χ_ω be a cut-off function as above localizing the support of the control. Then, under the controllability conditions above, given any $(y^0, y^1) \in H_0^1(\Omega) \times L^2(\Omega)$, there exists a unique minimizer (U^0, U^1) of J over $L^2(\Omega) \times H^{-1}(\Omega)$. The function*

$$V(x, t) = \eta(t)\chi_\omega(x)U(x, t) \tag{5.70}$$

is a control for

$$\begin{cases} y_{tt} - \Delta y = V\chi_\omega, & \text{in } \Omega \times (0, \infty), \\ y = 0, & \text{on } \partial\Omega \times (0, \infty), \\ (y(0), y_t(0)) = (y^0, y^1) & \in H_0^1(\Omega) \times L^2(\Omega), \end{cases} \tag{5.71}$$

which is characterized as the control function of minimal $L^2(0, T; dt/\eta; L^2(\omega))$-norm, defined by

$$\|v\|_{L^2(0,T;dt/\eta;L^2(\omega))}^2 = \int_0^T \int_\omega |v(x, t)|^2 dx \frac{dt}{\eta(t)}.$$

Furthermore, if the weight function η satisfies $\eta \in C^\infty(\mathbb{R})$, then if (y^0, y^1) belongs to $\mathscr{D}(A^s)$ for some $s \in \mathbb{R}_+$, $(U_0, U_1) \in \mathscr{D}((A^*)^s) = \mathscr{D}(A^{s-1})$.

In particular, when χ_ω is smooth and all its normal derivatives vanish at the boundary, the control function V given by (5.70) belongs to

$$V \in H^s(0, T; L^2(\omega)) \cap \bigcap_{k=0}^{\lfloor s-1 \rfloor} C^k([0, T]; H^{s-k}(\omega)), \tag{5.72}$$

the controlled solution y of (5.71) belongs to

$$(y, y_t) \in C^s([0, T]; H_0^1(\Omega) \times L^2(\Omega)) \bigcap_{k=0}^{\lfloor s \rfloor} C^k([0, T]; \mathscr{D}(A^{s-k})), \tag{5.73}$$

and, in particular,

$$(y, y_t) \in \bigcap_{k=0}^{\lfloor s \rfloor} C^k([0, T]; H^{s+1-k}(\Omega) \times H^{s-k}(\Omega)). \tag{5.74}$$

Remark that in this case, the time-dependent cut-off function is not needed if χ_ω is assumed to map $\mathscr{D}(A^s)$ to $\mathscr{D}(A^s)$ for all $s > 0$. Note that this requires not only that $\chi_\omega \in H^s(\Omega)$ but also some suitable compatibility conditions on the boundary, that are satisfied for instance when all the normal derivatives of χ_ω on the boundary vanish.

For more details, we refer to our work [35].

5.3.5.2 Boundary Control Operators

Let us assume that

$$\left\|(u^0, u^1)\right\|^2_{H^1_0(\Omega) \times L^2(\Omega)} \le C^2_{obs} \int_0^{T^*} \int_{\partial \Omega} \chi^2_\Gamma |\partial_n u|^2 d\sigma dt \qquad (5.75)$$

for all solutions u of (5.64), where $\chi_\Gamma = \chi_\Gamma(x)$ is a function localized on some part Γ of the boundary $\partial \Omega$.

Then the functional J introduced in (5.23) is now defined on $H^1_0(\Omega) \times L^2(\Omega)$ and reads as

$$J((u^0, u^1)) = \frac{1}{2} \int_0^T \int_{\partial \Omega} \eta(t) \chi_\Gamma(x)^2 |\partial_n u(x, t)|^2 d\Gamma dt + \int_0^1 y^0(x) u^1(x, 0) dx$$

$$- \langle y^1, u^0 \rangle_{H^{-1}(\Omega) \times H^1_0(\Omega)}, \qquad (5.76)$$

where u is the solution of (5.64).

Note that, here again, we have identified $L^2(\Omega)$ with its topological dual. This artificially creates a shift between the spaces X, X^*, and also between $\mathscr{D}(A^j)$ and $\mathscr{D}((A^*)^j) = \mathscr{D}(A^{j+1})$. Besides, this is done in the reverse situation as in the previous paragraph, i.e.:

- The natural space for the controlled trajectory is $X^* = L^2(\Omega) \times H^{-1}(\Omega)$ and therefore the controlled trajectory should lie in the space $\mathscr{D}((A^*)^s) = \mathscr{D}(A^{s-1})$.
- The natural space for the adjoint equation is $X = H^1_0(\Omega) \times L^2(\Omega)$ and therefore the regularity of the trajectory of the adjoint equation should be quantified with the spaces $\mathscr{D}(A^s)$.

Then our results imply the following:

Theorem 5.3.6. *Assume that χ_Γ is compactly supported in $\Gamma \subset \partial \Omega$ and that η is a smooth weight function satisfying (5.21). Also assume (5.75).*

Given any $(y^0, y^1) \in L^2(\Omega) \times H^{-1}(\Omega)$, there exists a unique minimizer (U^0, U^1) of J over $H^1_0(\Omega) \times L^2(\Omega)$. The function

$$V(x, t) = \eta(t) \chi_\Gamma(x) \partial_n U(x, t)_{|\Gamma} \qquad (5.77)$$

is a control function for

$$\begin{cases} y_{tt} - \Delta y = 0, & \text{in } \Omega \times (0, \infty), \\ y = \chi_\Gamma v, & \text{on } \partial\Omega \times (0, \infty), \\ (y(0), y_t(0)) = (y^0, y^1) & \in H_0^1(\Omega) \times L^2(\Omega), \end{cases} \qquad (5.78)$$

with target $(y(T), y_t(T)) = (0, 0)$.

Besides, V can be characterized as the control function which minimizes the $L^2(0, T; dt/\eta; L^2(\Gamma))$-*norm, defined by*

$$\|v\|_{L^2(0,T;dt/\eta;L^2(\Gamma))}^2 = \int_0^T \int_\Gamma |v(x,t)|^2 d\Gamma \frac{dt}{\eta(t)},$$

among all possible controls.

Furthermore, if the function χ_Γ *is smooth, then if* (y^0, y^1) *belongs to* $\mathcal{D}((A^*)^s) = \mathcal{D}(A^{s-1})$ *for some real number* $s \in \mathbb{R}_+$, *the control function V given by* (5.77) *belongs to*

$$V \in H_0^s(0, T; L^2(\Gamma)) \overset{\lfloor s \rfloor}{\underset{k=0}{\cap}} C^k([0, T]; H_0^{s-k-1/2}(\Gamma)) \qquad (5.79)$$

and $(U^0, U^1) \in \mathcal{D}(A^s)$. *In particular, the controlled solution y of* (5.78) *then belongs to*

$$(y, y_t) \in C^s([0, T]; L^2(\Omega) \times H^{-1}(\Omega)) \overset{\lfloor s \rfloor}{\underset{k-0}{\cap}} C^k([0, T]; H^{s-k}(\Omega) \times H^{s-1-k}(\Omega)). \qquad (5.80)$$

5.4 1-d Finite Difference Semidiscretizations

5.4.1 *Orientation*

In Sect. 5.3 we showed how the observability/controllability problem for the constant coefficient wave equation can be solved by Fourier series expansions. We now address the problem of the continuous dependence of the observability constant $C_{obs}(T)$ in (5.41) with respect to finite difference space semidiscretizations as the mesh-size parameter h tends to zero. This problem arises naturally in the numerical implementation of the controllability and observability properties of the continuous wave equation but is of independent interest in the analysis of discrete models for vibrations.

There are several important facts and results that deserve emphasis and that we shall discuss below:

- The observability constant for the semidiscrete model tends to infinity for any T as $h \to 0$. This is related to the fact that the velocity of propagation of solutions tends to zero as $h \to 0$ and the wavelength of solutions is of the same order as the size of the mesh.
- As a consequence of this fact and of the Banach–Steinhaus theorem, there are initial data for the wave equation for which the controls of the semidiscrete models diverge. This proves that one cannot simply rely on the classical convergence (consistency + stability) analysis of the underlying numerical schemes to design algorithms for computing the controls.

However, as we shall explain in Sect. 5.5, one can establish weaker observability results that hold uniformly with respect to $h > 0$. As a consequence, see Sect. 5.6, we will be able to propose numerical methods for which "weak" discrete controls converge.

5.4.2 Finite Difference Approximations

Given $N \in \mathbb{N}$ we define $h = 1/(N+1) > 0$. We consider the mesh $\{x_j = jh, j = 0, \ldots, N+1\}$ which divides $[0, 1]$ into $N + 1$ subintervals $I_j = [x_j, x_{j+1}]$, $j = 0, \ldots, N$.

Consider the following finite difference approximation of the wave (5.39):

$$
\begin{cases}
u_j'' - \dfrac{1}{h^2}\left[u_{j+1} + u_{j-1} - 2u_j\right] = 0, & 0 < t < T,\ j = 1, \ldots, N, \\
u_j(t) = 0, & j = 0,\ N+1,\ 0 < t < T, \\
u_j(0) = u_j^0,\ u_j'(0) = u_j^1, & j = 1, \ldots, N,
\end{cases}
\tag{5.81}
$$

which is a coupled system of N linear differential equations of second order. In (5.81) the function $u_j(t)$ provides an approximation of $u(x_j, t)$ for all $j = 1, \ldots, N$, u being the solution of the continuous wave (5.39). The conditions $u_0 = u_{N+1} = 0$ take account of the homogeneous Dirichlet boundary conditions, and the second order differentiation with respect to x has been replaced by the three-point finite difference. Symbol $'$ denotes differentiation with respect to the time t.

We shall use a vector notation to simplify the expressions. In particular, the column vector

$$
\mathbf{u}_h(t) = \begin{pmatrix} u_1(t) \\ \vdots \\ u_N(t) \end{pmatrix}
\tag{5.82}
$$

will represent the whole set of unknowns of the system. Introducing the matrix

$$A_h = \frac{1}{h^2} \begin{pmatrix} 2 & -1 & 0 & 0 \\ -1 & \ddots & \ddots & 0 \\ 0 & \ddots & \ddots & -1 \\ 0 & 0 & -1 & 2 \end{pmatrix}, \tag{5.83}$$

the system (5.81) reads as follows:

$$\mathbf{u}_h''(t) + A_h \mathbf{u}_h(t) = 0, \quad 0 < t < T; \quad \mathbf{u}_h(0) = \mathbf{u}_h^0, \; \mathbf{u}_h'(0) = \mathbf{u}_h^1. \tag{5.84}$$

The solution $\mathbf{u}_h$ of (5.84) depends also on h, but most often we shall denote it simply by $\mathbf{u}$.

The energy of the solutions of (5.81) is

$$E_h(t) = \frac{h}{2} \sum_{j=0}^{N} \left[|u_j'|^2 + \left| \frac{u_{j+1} - u_j}{h} \right|^2 \right], \tag{5.85}$$

and it is constant in time. It is also a natural discretization of the continuous energy (5.40).

The problem of observability of system (5.81) can be formulated as follows: *To find $T > 0$ and $C_h(T) > 0$ such that*

$$E_h(0) \leq C_h(T)^2 \int_0^T \left| \frac{u_N(t)}{h} \right|^2 dt \tag{5.86}$$

holds for all solutions of (5.81).

Observe that $|u_N/h|^2$ is a natural approximation[5] of $|u_x(1,t)|^2$ for the solution of the continuous system (5.39). Indeed $u_x(1,t) \sim [u_{N+1}(t) - u_N(t)]/h$ and, taking into account that $u_{N+1} = 0$, it follows that $u_x(1,t) \sim -u_N(t)/h$.

System (5.81) is finite-dimensional. Therefore, if observability holds for some $T > 0$, then it holds for all $T > 0$, as we have seen in Sect. 5.3.

Note also that the existence of a constant $C_h(T)$ in (5.86) follows from the equivalence of norms in finite dimensional spaces and the fact that if $\mathbf{u}_h$ is a solution of (5.81) that satisfies $u_N(t) = u_{N+1}(t) = 0$, then $\mathbf{u}_h = 0$. This can be easily seen on (5.81) using an iteration argument.

We are interested mainly in the uniformity of the constant $C_h(T)$ as $h \to 0$. If $C_h(T)$ remains bounded as $h \to 0$, we say that system (5.81) is *uniformly observable* as $h \to 0$. Taking into account that the observability of the limit system (5.39) holds only for $T \geq 2$, it would be natural to expect $T \geq 2$ to be a necessary

---

[5]Here and in what follows u_N refers to the Nth component of the solution $\mathbf{u}$ of the semidiscrete system, which obviously depends also on h.

condition for the uniform observability of (5.81). This is indeed the case but, as we shall see, the condition $T \geq 2$ is far from being sufficient. In fact, *uniform observability fails for all $T > 0$*. In order to explain this fact it is convenient to analyze the spectrum of (5.81).

Let us consider the eigenvalue problem

$$-\frac{1}{h^2}\left[w_{j+1} + w_{j-1} - 2w_j\right] = \lambda w_j, \quad j = 1, \ldots, N; \quad w_0 = w_{N+1} = 0. \tag{5.87}$$

The spectrum can be computed explicitly in this case (Isaacson and Keller [55]):

$$\lambda_h^k = \frac{4}{h^2} \sin^2\left(\frac{k\pi h}{2}\right), \quad k = 1, \ldots, N, \tag{5.88}$$

and the corresponding eigenvectors are

$$\mathbf{w}_h^k = \left(w_{1,h}^k, \ldots, w_{N,h}^k\right)^T : w_{j,h}^k = \sin(k\pi jh), \quad k, j = 1, \ldots, N. \tag{5.89}$$

Obviously, $\lambda_h^k \to \lambda^k = k^2 \pi^2$, as $h \to 0$ for each $k \geq 1$, $\lambda^k = k^2 \pi^2$ being the kth eigenvalue of the continuous wave (5.39). On the other hand we see that the eigenvectors $\mathbf{w}_h^k$ of the discrete system (5.87) coincide with the restriction to the mesh points of the eigenfunctions $w^k(x) = \sin(k\pi x)$ of the continuous wave (5.39).[6]

According to (5.88) we have $\sqrt{\lambda_h^k} = \frac{2}{h} \sin\left(\frac{k\pi h}{2}\right)$, and therefore, in a first approximation, we have

$$\left|\sqrt{\lambda_h^k} - k\pi\right| \sim \frac{k^3 \pi^3 h^2}{24}. \tag{5.90}$$

This indicates that the spectral convergence is uniform only in the range $k \ll h^{-2/3}$, see [91]. Thus, one cannot solve the problem of uniform observability for the semidiscrete system (5.81) as a consequence of the observability property of the continuous wave equation and a perturbation argument with respect to h.

5.4.3 Nonuniform Observability

Multiplying (5.87) by $j(w_{j+1} - w_j)$, one easily obtains (see [53]) the following identity:

---

[6]This is a non generic fact that occurs only for the constant coefficient 1-d problem with uniform meshes.

Lemma 5.4.1. *For any $h > 0$ and any eigenvector of (5.87) associated with the eigenvalue λ,*

$$h \sum_{j=0}^{N} \left| \frac{w_{j+1} - w_j}{h} \right|^2 = \frac{2}{4 - \lambda h^2} \left| \frac{w_N}{h} \right|^2. \tag{5.91}$$

We now observe that the largest eigenvalue λ_h^N of (5.87) is such that $\lambda_h^N h^2 \to 4$ as $h \to 0$ and note the following result on nonuniform observability.

Theorem 5.4.2. *For any $T > 0$,*

$$\lim_{h \to 0} \inf_{\mathbf{u}_h \text{ solution of (5.81)}} \left[\frac{1}{E_h(0)} \left(\int_0^T \left| \frac{u_N}{h} \right|^2 dt \right) \right] = 0. \tag{5.92}$$

Proof (of Theorem 5.4.2). We consider solutions of (5.81) of the form $\mathbf{u}_h = \cos(\sqrt{\lambda_h^N} t) \mathbf{w}_h^N$, where λ_h^N and $\mathbf{w}_h^N$ are the Nth eigenvalue and eigenvector of (5.87), respectively. We have

$$E_h(0) = \frac{h}{2} \sum_{j=0}^{N} \left| \frac{w_{j+1,h}^N - w_{j,h}^N}{h} \right|^2 \tag{5.93}$$

and

$$\int_0^T \left| \frac{u_{N,h}}{h} \right|^2 dt = \left| \frac{w_{N,h}^N}{h} \right|^2 \int_0^T \cos^2 \left(\sqrt{\lambda_h^N} t \right) dt. \tag{5.94}$$

Taking into account that $\lambda_h^N \to \infty$ as $h \to 0$, it follows that

$$\int_0^T \cos^2 \left(\sqrt{\lambda_h^N} t \right) dt \to T/2 \quad \text{as } h \to 0. \tag{5.95}$$

By combining (5.91), (5.93), (5.94) and (5.95), (5.92) follows immediately.   □

It is important to note that the solution we have used in the proof of Theorem 5.4.2 is not the only impediment for the uniform observability inequality to hold.

Indeed, let us consider the following solution of the semidiscrete system (5.81), constituted by the last two eigenvectors:

$$\mathbf{u}_h = \frac{1}{\sqrt{\lambda_h^N}} \left[\exp \left(i \sqrt{\lambda_h^N} t \right) \mathbf{w}_h^N - \frac{w_{N,h}^n}{w_{N,h}^{N-1}} \exp \left(i \sqrt{\lambda_h^{N-1}} t \right) \mathbf{w}_h^{N-1} \right]. \tag{5.96}$$

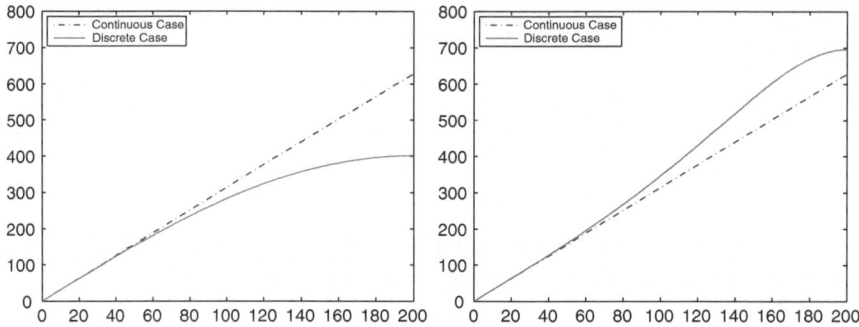

Fig. 5.5 *Left*: Square roots of the eigenvalues in the continuous and discrete cases (*finite difference semidiscretization*). The gaps are clearly independent of k in the continuous case and of order h for large k in the discrete one. *Right*: Dispersion diagram for the piecewise linear finite element space semidiscretization versus the continuous wave equation

This solution is a simple superposition of two monochromatic semidiscrete waves corresponding to the last two eigenfrequencies of the system. The total energy of this solution is of the order 1 (because each of both components has been normalized in the energy norm and the eigenvectors are orthogonal one to each other). However, the trace of its discrete normal derivative is of the order of h in $L^2(0, T)$. This is due to two facts:

- First, the trace of the discrete normal derivative of each eigenvector is very small compared to its total energy.
- Second, and more important, the gap between $\sqrt{\lambda_h^N}$ and $\sqrt{\lambda_h^{N-1}}$ is of the order of h, as is shown in Fig. 5.5, left. The wave packet (5.96) then has a group velocity of the order of h.

    To be more precise, let us compute $|\mathbf{u}_h|^2$, with $\mathbf{u}_h$ as in (5.96):

$$|u_{j,h}(t)|^2 = \frac{1}{\lambda_h^N} \left(\left| w_{j,h}^N - \frac{w_{N,h}^n}{w_{N,h}^{N-1}} w_{j,h}^{N-1} \right|^2 \cos^2 \left(\left(\sqrt{\lambda_h^N} - \sqrt{\lambda_h^{N-1}} \right) \frac{t}{2} \right) \right.$$

$$\left. + \left| \frac{w_{N,h}^n}{w_{N,h}^{N-1}} w_{j,h}^{N-1} + w_{j,h}^N \right|^2 \sin^2 \left(\left(\sqrt{\lambda_h^N} - \sqrt{\lambda_h^{N-1}} \right) \frac{t}{2} \right) \right).$$

By Taylor expansion, the difference between the two frequencies $\sqrt{\lambda_h^N}$ and $\sqrt{\lambda_h^{N-1}}$ is of the order h, and thus we see that the solution is periodic of period of the order of $1/h$.

Note that here, from (5.91), explicit computations yield

$$\left|\frac{u_{N,h}(t)}{h}\right|^2 = \left|\frac{w_{N,h}^N}{h}\right|^2 \frac{4}{\lambda_h^N} \sin^2\left(\left(\sqrt{\lambda_h^N} - \sqrt{\lambda_h^{N-1}}\right)\frac{t}{2}\right)$$

$$= \left(1 - \frac{\lambda_h^N h^2}{4}\right)\sin^2\left(\left(\sqrt{\lambda_h^N} - \sqrt{\lambda_h^{N-1}}\right)\frac{t}{2}\right)$$

$$= \sin^2\left(\frac{\pi h}{2}\right)\sin^2\left(\left(\sqrt{\lambda_h^N} - \sqrt{\lambda_h^{N-1}}\right)\frac{t}{2}\right).$$

Thus, the integral of the square of the normal derivative of $\mathbf{u}_h$ between 0 and T is of order of h^4, where the smallness comes from both the fact that $\sqrt{\lambda_h^N} - \sqrt{\lambda_h^{N-1}} \simeq h$ and (5.91).

High frequency wave packets may be used to show that the observability constant has to blow up at infinite order as $h \to 0$ (see [75, 76]). To do this it is sufficient to proceed as above but combining an increasing number of eigenfrequencies. Actually, Micu in [77] proved that the constant $C_h(T)$ blows up exponentially by means of a careful analysis of the biorthogonal sequences to the family of exponentials $\left\{\exp(i\sqrt{\lambda_h^k}t)\right\}_{k=1,\dots,N}$ as $h \to 0$.

All these high-frequency pathologies are in fact very closely related to the notion of group velocity. We refer to [101, 104] for an in-depth analysis of this notion that we discuss briefly in the context of this example. Since the eigenvectors $\mathbf{w}_h^k$ are sinusoidal functions (see (5.89)) the solutions of the semidiscrete system may be written as linear combinations of complex exponentials (in space-time):

$$\exp\left(\pm i k \pi \left(\frac{\sqrt{\lambda_h^k}}{k\pi}t - x\right)\right).$$

In view of this, we see that each monochromatic wave propagates at a speed

$$\frac{\sqrt{\lambda_h^k}}{k\pi} = \frac{2\sin(k\pi h/2)}{k\pi h} = \left.\frac{\omega(\xi)}{\xi}\right|_{\{\xi = k\pi h\}} = \left. c(\xi)\right|_{\{\xi = k\pi h\}}, \tag{5.97}$$

with $\omega(\xi) = 2\sin(\xi/2)$. This is the so-called *phase velocity*. The velocity of propagation of monochromatic semidiscrete waves (5.97) turns out to be bounded above and below by positive constants, independently of h: $0 < 2/\pi \le c(\xi) \le 1 < \infty$ for all $h > 0, \xi \in [0, \pi]$. Note that $[0, \pi]$ is the relevant range of ξ. Indeed, $\xi = k\pi h$ and $k = 1,\dots,N$, $Nh = 1-h$. This corresponds to frequencies $\zeta = \xi/h$ in $(-\pi/h, \pi/h]$ which is natural due to the sampling of the uniform grid.

But wave packets may travel at a different speed because of the cancellation phenomena we discussed above. The corresponding speed for those semidiscrete

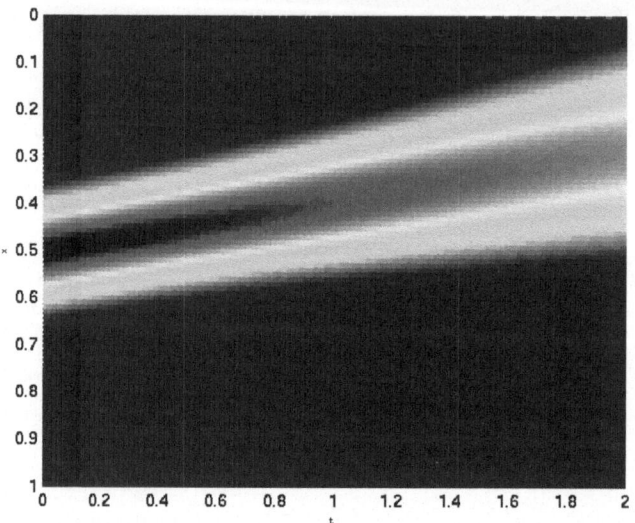

Fig. 5.6 A discrete wave packet and its propagation. In the horizontal axis we represent the time variable, varying between 0 and 2, and the vertical one the space variable x ranging from 0 to 1

wave packets is given by the derivative of $\omega(\cdot)$ (see [101]). At high frequencies ($k \sim N$) the derivative of $\omega(\xi)$ at $\xi = N\pi h = \pi(1-h)$ is of the order of h, the velocity of propagation of the wave packet.

This is illustrated in Fig. 5.6, where we have chosen a discrete initial datum concentrated in space around $x = 0.5$ at $t = 0$ and in frequency at $\zeta \simeq 0.95/h$. As one can see, this discrete wave propagate at a very small velocity.

The fact that the group velocity is of order h is equivalent[7] to the fact that the gap between $\sqrt{\lambda_h^{N-1}}$ and $\sqrt{\lambda_h^{N}}$ is of order h.

According to this analysis, *the group velocity being bounded below is a necessary condition for the uniform observability inequality to hold. Moreover, this is equivalent to a uniform spectral gap condition.*

The convergence property of the numerical scheme guarantees only that the group velocity of numerical waves is the correct one, close to that of the continuous wave equation, for low-frequency wave packets and this is compatible with the high frequency pathologies mentioned above.[8]

---

[7]Defining group velocity as the derivative of ω, i.e., of the curve in the dispersion diagram (see Fig. 5.5), is a natural consequence of the classical properties of the superposition of linear harmonic oscillators with close but not identical phases (see [21]). There is a one-to-one correspondence between the group velocity and the spectral gap which may be viewed as a discrete derivative of this diagram. In particular, when the group velocity decreases, the gap between consecutive eigenvalues also decreases.

[8]Note that in Fig. 5.5, both for finite differences and elements, the semidiscrete and continuous curves are tangent at low frequencies. This is in agreement with the convergence property of the

The careful analysis of this negative example will be useful to design possible remedies, i.e., to propose weaker observability results that would be uniform with respect to the discretization parameter $h > 0$. Actually, all the weak observability results that we shall propose in Sect. 5.5 (and others, see [36] for extensive references and examples) are based, in a way or another, on removing the high-frequency pathologies generated by the numerical scheme under consideration.

As we will see below in the next paragraph, the fact that the observability inequality (5.86) is not uniform with respect to $h > 0$ has an important consequence in controllability: There are some data to be controlled for which the discrete controls diverge.

Remark 5.4.3. According to Fig. 5.5, both finite-difference and finite element methods exhibit a frequency on which the group velocity vanishes. This actually is a generic fact. Indeed, as soon as the discretization method is implemented on a uniform mesh in a symmetric way, the dispersion diagram is given by a continuous function of $\zeta \in (-\pi/h, \pi/h)$ that scales as $\omega(\zeta h)/h$, for some smooth function ω describing the numerical method under consideration. But this function ω can actually be defined for $\zeta \in \mathbb{R}$ as the output of the discrete laplacian when the input is $\exp(i\zeta x)$. Doing that, one easily checks that ω is necessarily 2π-periodic. According to this, if ω is smooth, it necessarily has a critical point in $(-\pi, \pi)$.

Therefore, the existence of waves traveling at zero group velocity is generic with respect to the discretization schemes.

To our knowledge, only the mixed finite element method escapes this pathological fact, but this is so since it corresponds to a non-smooth dispersion relation $\omega(\xi) = 2\tan(\xi/2)$, which is produced by introducing a mass matrix that degenerates at frequency of order π/h where the dispersion relation of the discretization of the laplacian has a critical point. We refer to [17] for a more precise discussion on that particular numerical scheme.

5.4.4 Blow up of Discrete Controls

This section is devoted to analyze the consequences of the negative results on observability obtained in Theorem 5.4.2 at the level of the controllability of the semidiscrete wave (5.98). The finite-dimensional control system reads as follows

$$\begin{cases} y_j'' - \dfrac{1}{h^2}\left[y_{j+1} + y_{j-1} - 2y_j\right] = 0, & 0 < t < T, j = 1, \ldots, N, \\ y_0(0, t) = 0; \; y_{N+1}(1, t) = v(t), & 0 < t < T \\ y_j(0) = y_j^0, \; y_j'(0) = y_j^1, & j = 1, \ldots, N, \end{cases} \tag{5.98}$$

and it is the semidiscrete version of the controlled wave (5.42).

---

numerical scheme under consideration and with the fact that low-frequency wave packets travel essentially with the velocity of the continuous model.

It is easy to see that this semidiscrete system, for all $h > 0$ and all $T > 0$, is exactly controllable because the Kalman rank condition is satisfied. More precisely, for any given $T > 0, h > 0$ and initial data $(\mathbf{y}_h^0, \mathbf{y}_h^1)$, there exists a control $\mathbf{v}_h \in L^2(0, T)$ such that

$$\mathbf{y}_h(T) = \mathbf{y}_h'(T) = 0. \tag{5.99}$$

But, of course, we are interested in the limit process $h \to 0$. In particular, we would like to understand whether, when the initial data in (5.98) are "fixed"[9] to be $(y^0, y^1) \in L^2(0, 1) \times H^{-1}(0, 1)$, the controls $\mathbf{v}_h$ of (5.98) converge in $L^2(0, T)$ as $h \to 0$ to the control of the continuous wave (5.42). The negative results on the observability problem, and the fact that these two problems, observability and controllability are equivalent, see Sect. 5.2, make us predict that, in fact, the convergence of the controls may fail. This is what happens in practice, indeed. In fact for suitable choices of the initial data the controls may diverge as $h \to 0$, whatever $T > 0$ is.

This negative result shows that the discrete approach to numerical control may fail. In other words, *controlling a numerical approximation of a controllable system is not necessarily a good way of computing an approximation of the control of the PDE model.* Summarizing, *the stability and convergence of the numerical scheme for solving the initial-boundary value problem do not guarantee its stability at the level of controllability.*

5.4.4.1 Controllability of the Discrete Schemes

In this section, we prove that the discrete systems (5.98) are exactly controllable for any $h > 0$ and characterize the controls of minimal norm. This actually is a byproduct of (5.86) and Sect. 5.2.1. We only rewrite it in our setting for the convenience of the reader.

Theorem 5.4.4. *For any $T > 0$ and $h > 0$ system (5.98) is exactly controllable. More precisely, for any $(\mathbf{y}_h^0, \mathbf{y}_h^1) \in \mathbb{R}^N \times \mathbb{R}^N$, there exists a control $\mathbf{V}_{hum,h} \in L^2(0, T)$ given by HUM such that the solution of (5.98) satisfies (5.99).*

Moreover, the control $\mathbf{V}_{hum,h}$ of minimal $L^2(0, T)$-norm can be characterized through the minimization of the functional

$$J_h((\mathbf{u}_h^0, \mathbf{u}_h^1)) = \frac{1}{2} \int_0^T \left| \frac{u_N(t)}{h} \right|^2 dt + h \sum_{j=1}^N y_j^0 u_j^1 - h \sum_{j=1}^N y_j^1 u_j^0, \tag{5.100}$$

---

[9]For given initial data (y^0, y^1), the initial data for the controlled semidiscrete system (5.98) are taken to be approximations of (y^0, y^1) on the discrete mesh. The convergence of the controls $\mathbf{v}_h$ in $L^2(0, T)$ is then analyzed for the controls corresponding to these approximate initial data.

in $\mathbb{R}^N \times \mathbb{R}^N$, where $\mathbf{u}_h$ is the solution of the adjoint system (5.81). More precisely, the control $\mathbf{V}_{hum,h}$ is of the form

$$\mathbf{V}_{hum,h}(t) = -\frac{U_N(t)}{h}, \tag{5.101}$$

where $\mathbf{U}_h$ is the solution of the adjoint system (5.81) corresponding to the initial data $(\mathbf{U}_h^0, \mathbf{U}_h^1)$ minimizing the functional J_h.

For each $h > 0$, as explained in Corollary 5.2.7, the control function $\mathbf{V}_{hum,h}$ of minimal $L^2(0, T)$-norm of system (5.98) is given by a linear map $\mathbb{V}_h$ of the initial data $(\mathbf{y}_h^0, \mathbf{y}_h^1)$ and can be written as $\mathbf{V}_{hum,h} = \mathbb{V}_h(\mathbf{y}_h^0, \mathbf{y}_h^1)$.

For convenience, for $h > 0$ we introduce the norms

$$\left\| (\mathbf{u}_h^0, \mathbf{u}_h^1) \right\|_{H_h^1 \times L_h^2}^2 = h \sum_{j=0}^{N} \left[\left(\frac{u_{j+1}^0 - u_j^0}{h} \right)^2 + |u_j^1|^2 \right]$$

and

$$\left\| (\mathbf{y}_h^0, \mathbf{y}_h^1) \right\|_{L_h^2 \times H_h^{-1}} = \sup_{\left\| (\mathbf{u}_h^0, \mathbf{u}_h^1) \right\|_{H_h^1 \times L_h^2} = 1} \left\{ h \sum_{j=1}^{N} y_j^0 u_j^1 - h \sum_{j=1}^{N} u_j^0 y_j^1 \right\}. \tag{5.102}$$

The first one corresponds to the energy of (5.85) and the second one stands for the norm of the space in which the solutions of the controlled semidiscrete system belong to.

In particular, if one extends the discrete functions $(\mathbf{u}_h^0, \mathbf{u}_h^1)$ to continuous ones using Fourier extension, denoted by (u_h^0, u_h^1), the following norms are equivalent:

$$\left\| (\mathbf{u}_h^0, \mathbf{u}_h^1) \right\|_{H_h^1 \times L_h^2}^2 \simeq \left\| (u_h^0, u_h^1) \right\|_{H_0^1 \times L^2}^2.$$

We thus deduce by duality the equivalence between the norms

$$\left\| (\mathbf{y}_h^0, \mathbf{y}_h^1) \right\|_{L_h^2 \times H_h^{-1}} \simeq \left\| (y_h^0, y_h^1) \right\|_{L^2 \times H^{-1}} \tag{5.103}$$

As a simple consequence of the equivalence stated in Theorem 5.2.8, we have

$$\left\| \mathbb{V}_h \right\|_{\mathfrak{L}(L_h^2 \times H_h^{-1}, L^2(0,T))} = \sqrt{2} C_h(T), \tag{5.104}$$

where $C_h(T)$ is the observability constant in (5.86).

By Theorems 5.2.8 and 5.4.2, this indicates that the norms of the discrete control operators blow up when $h \to 0$:

Proposition 5.4.5. *We have*

$$\lim_{h \to 0} \|\mathbb{V}_h\|_{\mathcal{L}(L_h^2 \times H_h^{-1}, L^2(0,T))} = +\infty.$$

Remark 5.4.6. Identity (5.104) indicates that the norm of the discrete controls map blows up when $h \to 0$ at the same rate as $C_h(T)$. In view of the results presented in [77], it blows up with an exponential rate.

As a consequence of Proposition 5.4.5 there are continuous data $(y^0, y^1) \in L^2(0,1) \times H^{-1}(0,1)$ for which the sequence of discrete controls computed on the discrete controlled system (5.98) is not even bounded.

To state our results precisely, we must explain how the continuous data (y^0, y^1) are approximated by discrete ones $(\mathbf{y}_h^0, \mathbf{y}_h^1)$.

For $(y^0, y^1) \in L^2(0,1) \times H^{-1}(0,1)$, with Fourier expansion

$$(y^0, y^1) = \sum_{k=1}^{\infty} (\hat{y}_k^0, \hat{y}_k^1) w^k,$$

we introduce a sequence $(\mathbb{A}_h)_{h>0}$ of discretization operators

$$\mathbb{A}_h : L^2(0,1) \times H^{-1}(0,1) \to \mathbb{R}^N \times \mathbb{R}^N,$$

$$(y^0, y^1) \mapsto (\mathbf{y}_h^0, \mathbf{y}_h^1) = \mathbb{A}_h(y^0, y^1) = \sum_{k=1}^{N} (\hat{y}_k^0, \hat{y}_k^1) \mathbf{w}^k. \tag{5.105}$$

To simplify notations, we will denote similarly by $\mathbb{A}_h(y^0, y^1)$ the discrete functions and their continuous corresponding Fourier extensions.

These operators $\mathbb{A}_h$ map continuous data (y^0, y^1) to discrete ones by truncating the Fourier expansion, and describe a natural relevant discretization process for initial data in $L^2(0,1) \times H^{-1}(0,1)$.

For instance, as one can easily check, for any $(y^0, y^1) \in L^2(0,1) \times H^{-1}(0,1)$,

$$\mathbb{A}_h(y^0, y^1) \xrightarrow[h \to 0]{} (y^0, y^1) \quad \text{in } L^2(0,1) \times H^{-1}(0,1). \tag{5.106}$$

We now prove the following divergence result:

Theorem 5.4.7. *There exists an initial datum* $(y^0, y^1) \in L^2(0,1) \times H^{-1}(0,1)$ *such that the sequence* $(\mathbb{V}_h \circ \mathbb{A}_h(y^0, y^1))_{h>0}$ *is not bounded in* $L^2(0,T)$.

Proof. The proof is by contradiction.

Assume that for all $(y^0, y^1) \in L^2(0,1) \times H^{-1}(0,1)$, the sequence of discrete controls $(\mathbb{V}_h \circ \mathbb{A}_h(y^0, y^1))_{h>0}$ is bounded in $L^2(0,T)$.

Then, applying Banach-Steinhaus Theorem (or the Principle of Uniform Bound-edness) to the operators $(\mathbb{V}_h \circ \mathbb{A}_h)_{h>0}$, there is a constant $C > 0$ such that for all $h > 0$ and $(y^0, y^1) \in L^2(0,1) \times H^{-1}(0,1)$,

$$\left\| \mathbb{V}_h \circ \mathbb{A}_h (y^0, y^1) \right\|_{L^2(0,T)} \le C \left\| (y^0, y^1) \right\|_{L^2 \times H^{-1}}.$$

Due to the particular form of $\mathbb{A}_h$, this implies that for all

$$(\mathbf{y}_h^0, \mathbf{y}_h^1) = \sum_{k=1}^{N} (\hat{y}_{k,h}^0, \hat{y}_{k,h}^1) \mathbf{w}_h^k,$$

we have

$$\left\| \mathbb{V}_h (\mathbf{y}_h^0, \mathbf{y}_h^1) \right\|_{L^2(0,T)} \le C \left\| (\mathbf{y}_h^0, \mathbf{y}_h^1) \right\|_{L_h^2 \times H_h^{-1}}.$$

But this is in contradiction with Proposition 5.4.5 and the equivalence (5.103), which proves the result.   □

Remark 5.4.8. According to Theorem 5.4.7, not only the global cost of controlla-bility diverges, but there exist specific initial data such that its cost diverges. This is a direct consequence of the Principle of Uniform Boundedness. As we indicated above here we refer to the cost of controlling the sequence of discrete initial data $(\mathbf{y}_h^0, \mathbf{y}_h^1)$ built specifically from the initial data (y^0, y^1) by truncating Fourier series.

But the approximation $\mathbb{A}_h$ of the initial data can be defined differently as well, and the result will remain true. For instance, we may take discrete averages of the continuous data over intervals centered on the mesh-points $x_j = jh$. Of course, in what concerns y^1, we have to be particularly careful since the fact that it belongs to $H^{-1}(0,1)$ allows only doing averages against test functions in $H_0^1(0,1)$. The use of these test functions can be avoided by first, taking a smooth approximation of y^1 in $H^{-1}(0,1)$ and then taking averages.

Remark 5.4.9. This lack of convergence of the semidiscrete controls $\mathbf{V}_{hum,h}$ towards the continuous one V can be understood easily. Indeed, as we have shown above, the semidiscrete system, even in the absence of controls, generates a lot of spurious high frequency oscillations. The control $\mathbf{V}_{hum,h}$ of the semidiscrete system (5.98) has to take all these spurious components into account. When doing this it gets further and further away from the true control V of the continuous wave (5.42), as the numerical experiments in the following section illustrate.

5.4.5   Numerical Experiments

In this section, we describe some numerical experiments showing both the insta-bility of the numerical controls for suitable initial data to be controlled. These simulations were performed by Alejandro Maass Jr. using MATLAB.

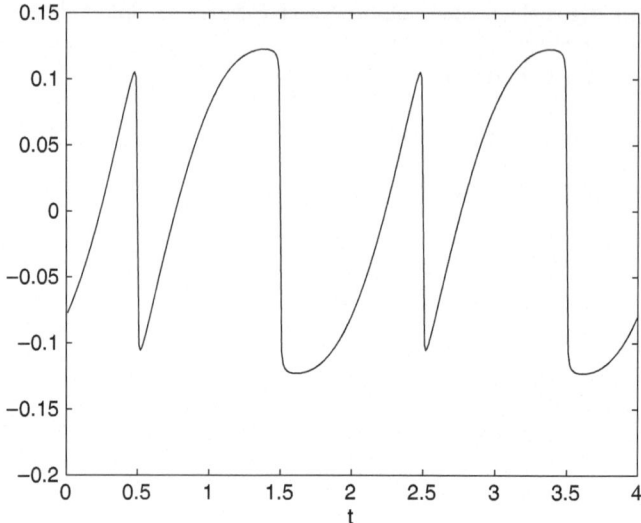

Fig. 5.7 Plot of the continuous control corresponding to the initial data (y^0, y^1) in Fig. 5.8

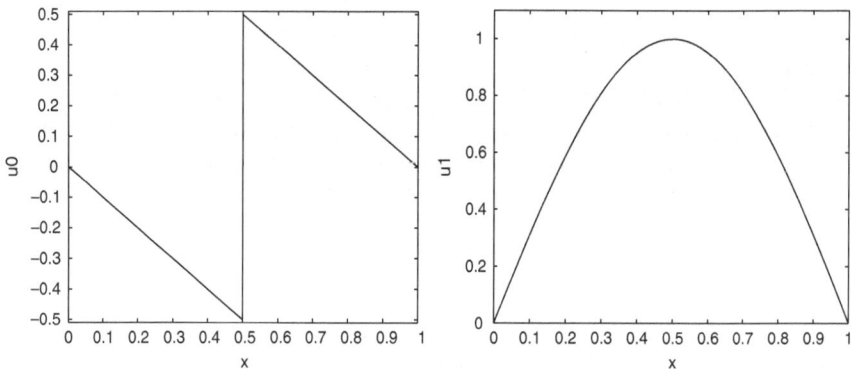

Fig. 5.8 Plot of the initial datum to be controlled: *Left*, the position y^0. *Right*, the velocity y^1

We consider the wave equation in time $T = 4$ on the space interval $(0, 1)$. This suffices for the boundary control of the continuous wave equation for which the minimal time is $T = 2$, see Proposition 5.3.2.

Given an initial datum to be controlled, for instance the one plotted in Fig. 5.8, we can then compute explicitly the control of the continuous equation.

The control function can then be computed explicitly using Fourier series, see Sect. 5.3.3. In Fig. 5.7 we present its plot.

The control can also be computed explicitly by using D'Alembert formula. This also explains the form of the control in Fig. 5.7, right, which looks very much like the superposition of the initial data to be controlled.

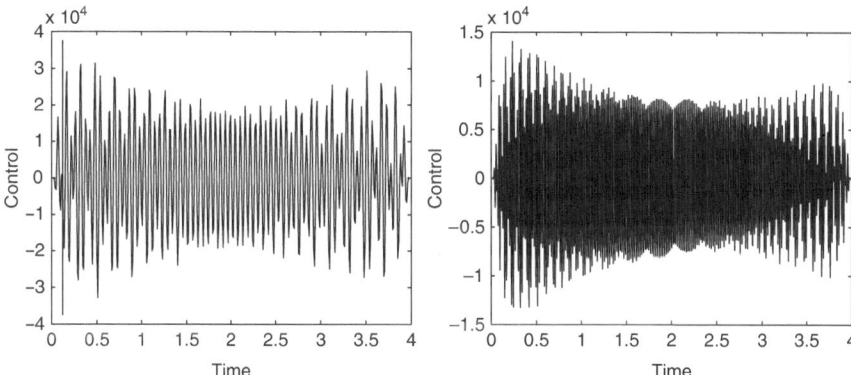

Fig. 5.9 Divergent evolution of the discrete exact controls when the number N of mesh-points increases. *Left*: the number of mesh points is $N = 50$. *Right*: $N = 150$. In both cases, we plot the control obtained after 500 iterations of the conjugate gradient algorithm for the minimization of J_h

We now consider the finite-difference semidiscrete approximation of the wave equation by finite-differences. We then compute the exact control of the semidiscrete system (5.98) for several values of N.

Of course, in practice, we do not deal with the space semidiscrete adjoint (5.81) but rather with fully discrete approximations. In our experiments we employ the centered discretization in time with time-step $\Delta t = 0.5\,h$, which, of course, guarantees the convergence of the scheme.

Following the discrete approach for numerical control, we compute the controls for the resulting fully discrete system. This is done minimizing the corresponding time-discrete version of the functional J_h in (5.100) using a conjugate gradient algorithm. It turns out that the number of iterations needed for convergence is huge. We stop the conjugate gradient algorithm after 500 iterations. The obtained results are plot in Fig. 5.9 for $N = 50$ and $N = 150$. Increasing the number of iterations would not change significantly the shape of the obtained controls. Note that they are very far from the shape of the actual control above. This is a clear evidence of the divergence of the discrete procedure to compute an effective numerical approximation of the control by controlling the approximate discrete dynamics. This is due to the very weak observability of the corresponding discrete system which makes the coercivity of the corresponding J_h functional to be very weak. This produces two effects. First, the descent algorithms are very slow and, second, the norm of the minimizers is huge. This is what we see in these numerical experiments.

It is also very surprising that the conjugate gradient method needs so many iterations whereas it minimizes a functional on a finite-dimensional space of dimension $2N$. Indeed, it is well-known that the conjugate gradient algorithm yields the exact minimizer after K iterations, where K is the size of dimension of the space we are working in, hence, in our case $K = 2N$. Then the functional is very ill-conditioned and the numerical errors cannot be negligible and prevent the conjugate gradient algorithm from converging in $2N$ iterations.

The descent iterative method does converge in 500 iterations when the number of mesh points is less than $N \leq 44$. But the controls one obtains when doing that are very similar to those plotted in Fig. 5.9.

5.5 Remedies for High-Frequency Pathologies

In the previous section we have shown that the discrete wave equations are not uniformly (with respect to the space mesh size h) observable, whatever the time $T > 0$ is.

We have mentioned that this is due to high-frequency spurious waves. In this section, we show that, when employing convenient filtering mechanisms, ruling out the high frequency components, one can recover uniform observability inequalities. At this point it is important to observe that the high-frequency pathologies cannot be avoided by simply taking, for instance, a different approximation of the discrete normal derivative since the fact that the group velocity vanishes is due to the numerical approximation scheme itself and, therefore, cannot be compensated by suitable boundary measurements. One has really to take care of the spurious high frequency solutions that the numerical scheme generates.

5.5.1 Fourier Filtering

We introduce a Fourier filtering mechanism that consists in eliminating the high frequency Fourier components and restricting the semidiscrete wave equation under consideration to the subspace of solutions generated by the Fourier components corresponding to the eigenvalues $\lambda \leq \gamma h^{-2}$ with $0 < \gamma < 4$ or with indices $0 < j < \delta h^{-1}$ with $0 < \delta < 1$. In this subspace the observability inequality becomes uniform. Note that these classes of solutions correspond to taking projections of the complete solutions by cutting off all frequencies with $\sqrt{\gamma} h^{-1} < \zeta < 2h^{-1}$.

The following classical result due to Ingham in the theory of nonharmonic Fourier series (see Ingham [54] and Young [105]) is useful for proving the uniform observability of filtered solutions.

Theorem 5.5.1 (Ingham [56]). *Let $\{\mu_k\}_{k \in \mathbb{Z}}$ be a sequence of real numbers such that $\mu_{k+1} - \mu_k \geq \sigma > 0$ for all $k \in \mathbb{Z}$. Then for any $T > 2\pi/\sigma$ there exists a positive constant $C(T, \sigma) > 0$ depending only on T and σ such that*

$$\frac{1}{C(T,\sigma)^2} \sum_{k \in \mathbb{Z}} |a_k|^2 \leq \int_0^T \left| \sum_{k \in \mathbb{Z}} a_k e^{i\mu_k t} \right|^2 dt \leq C(T,\sigma)^2 \sum_{k \in \mathbb{Z}} |a_k|^2 \qquad (5.107)$$

for all sequences of complex numbers $\{a_k\} \in \ell^2$.

Remark 5.5.2. Ingham's inequality can be viewed as a generalization of the orthogonality property of trigonometric functions we used to prove the observability of the 1-d wave equation in Sect. 5.3, known as Paserval's identity.

Ingham's inequality allows showing that, as soon as the gap condition is satisfied, there is uniform observability provided the time is large enough.

All these facts confirm that a suitable cutoff or filtering of the spurious numerical high frequencies may be a cure for these pathologies.

Let us now describe the basic *Fourier filtering mechanism* in more detail. We recall that solutions of (5.81) can be developed in Fourier series as follows:

$$\mathbf{u}_h(t) = \sum_{k=1}^{N} \left(a_k \cos\left(\sqrt{\lambda_h^k}\, t\right) + \frac{b_k}{\sqrt{\lambda_h^k}} \sin\left(\sqrt{\lambda_h^k}\, t\right) \right) \mathbf{w}_h^k,$$

where a_k, b_k are the Fourier coefficients of the initial data, i.e., $\mathbf{u}_h^0 = \sum_{k=1}^{N} a_k \mathbf{w}_h^k$, $\mathbf{u}_h^1 = \sum_{k=1}^{N} b_k \mathbf{w}_h^k$.

Given $s > 0$, we introduce the following classes of solutions of (5.81):

$$\mathscr{C}_h(s) = \left\{ \mathbf{u}_h(t) = \sum_{\lambda_h^k \leq s} \left(a_k \cos\left(\sqrt{\lambda_h^k}\, t\right) + \frac{b_k}{\sqrt{\lambda_h^k}} \sin\left(\sqrt{\lambda_h^k}\, t\right) \right) \mathbf{w}_h^k \right\}, \quad (5.108)$$

in which the high frequencies corresponding to the indices $j > \lfloor \delta(N+1) \rfloor$ have been cut off. As a consequence of Ingham's inequality and the analysis of the gap of the spectra of the semidiscrete systems we have the following result.[10]

Theorem 5.5.3 (see [53]). *For any $\gamma \in (0, 4)$ there exists $T(\gamma) > 0$ such that for all $T > T(\gamma)$ there exists $C = C(T, \gamma) > 0$ such that*

$$\frac{1}{C^2} E_h(0) \leq \int_0^T \left| \frac{u_N(t)}{h} \right|^2 dt \leq C^2 E_h(0) \qquad (5.109)$$

for every solution $\mathbf{u}_h$ of (5.81) in the class $\mathscr{C}_h(\gamma/h^2)$ and for all $h > 0$. Moreover, the minimal time $T(\gamma)$ for which (5.109) holds is such that $T(\gamma) \to 2$ as $\gamma \to 0$ and $T(\gamma) \to \infty$ as $\gamma \to 4$.

Remark 5.5.4. Theorem 5.5.3 guarantees the uniform observability in each class $\mathscr{C}_h(\gamma/h^2)$ for all $0 < \gamma < 4$, provided the time T is larger than $T(\gamma)$.

---

[10]These results may also be obtained using discrete multiplier techniques (see [53] and [32] for an improved version with a sharp estimate of the time $T(\delta)$).

The last statement in the theorem shows that when the filtering parameter γ tends to zero, i.e., when the solutions under consideration contain fewer and fewer frequencies, the time for uniform observability converges to $T = 2$, which is the corresponding one for the continuous equation. This is in agreement with the observation that the group velocity of the low-frequency semidiscrete waves coincides with the velocity of propagation in the continuous model.

By contrast, when the filtering parameter increases, i.e., when the solutions under consideration contain more and more frequencies, the time of uniform control tends to infinity. This is in agreement and explains further the negative result showing that, in the absence of filtering, there is no finite time T for which the uniform observability inequality holds.

The proof of Theorem 5.5.3 below provides an explicit estimate on the minimal observability time in the class $\mathscr{C}_h(\gamma/h^2)$: $T(\gamma) = 2/\sqrt{1 - \gamma/4}$.

Remark 5.5.5. In the context of the numerical computation of the boundary control for the wave equation the need of an appropriate filtering of the high frequencies was observed by Glowinski [40] and further investigated numerically by Asch and Lebeau in [2].

Let us now briefly sketch the proof of Theorem 5.5.3. The easiest one relies on the explicit representation of the solutions in $\mathscr{C}_h(\gamma/h^2)$ and the application of Ingham's theorem. This can be made possible since for all k with $\lambda_h^k \leq \gamma h^{-2}$,

$$\sqrt{\lambda_h^{k+1}} - \sqrt{\lambda_h^k} \geq \pi \cos(k\pi h/2) \geq \pi \sqrt{1 - \gamma/4}, \text{ as explicit computations yield.}$$

Another proof can be derived using the so-called discrete multiplier identity: for all solutions $\mathbf{u}_h$ of (5.81),

$$TE_h(0) + X_h(t)\Big|_0^T = \frac{1}{2} \int_0^T \left|\frac{u_N(t)}{h}\right|^2 dt + \frac{h^3}{4} \sum_{j=0}^N \int_0^T \left|\frac{u'_{j+1} - u'_j}{h}\right|^2 dt, \quad (5.110)$$

with

$$X_h(t) = h \sum_{j=1}^N jh \left(\frac{u_{j+1} - u_{j-1}}{2h}\right) u'_j. \quad (5.111)$$

Using (5.110) and straightforward bounds on the time boundary term X_h and on the extra term

$$\frac{h^3}{4} \sum_{j=0}^N \int_0^T \left|\frac{u'_{j+1} - u'_j}{h}\right|^2 dt, \quad (5.112)$$

one will be able to prove Theorem 5.5.3 in any time $T > 2/(1 - \gamma/4)$, see [53]. However, using more refined estimates on these terms, one can recover the observability time $T(\gamma) = 2/\sqrt{1 - \gamma/4}$, see [32].

Let us also note that the time $T(\gamma) = 2/\sqrt{1-\gamma/4}$ is sharp. More precisely, when $T < T(\gamma)$, there is no uniform observability results in the class $\mathscr{C}_h(\gamma/h^2)$ since $T(\gamma)$ is the time corresponding to the minimum group velocity within the class $\mathscr{C}_h(\gamma/h^2)$. But the proof is technically more involved and is beyond the scope of these notes. We refer to [72] and [36] for detailed proofs.

5.5.2 A Two-Grid Algorithm

Glowinski and Li in [42] introduced a two-grid algorithm that makes it possible to compute efficiently the control of the continuous model. The method was further developed by Glowinski in [40].

The relevance and impact of using two grids can be easily understood in view of the analysis of the 1-d semidiscrete equation developed in the previous paragraph.

In (5.88) we have seen that all the eigenvalues of the semidiscrete system satisfy $\lambda \leq 4/h^2$. We have also seen that the observability inequality becomes uniform when one considers solutions involving eigenvectors corresponding to eigenvalues $\lambda \leq \gamma/h^2$, with $\gamma < 4$, see Theorem 5.5.3.

The key idea of this two-grid filtering mechanism consists in using two grids: one, the computational one in which the discrete wave equations are solved, with step size h and a coarser one of size $2h$. In the fine grid, the eigenvalues satisfy the sharp upper bound $\lambda \leq 4/h^2$. And the coarse grid will "select" half of the eigenvalues, the ones corresponding to $\lambda \leq 2/h^2$. This indicates that in the fine grid the solutions obtained in the coarse one would behave very much as filtered solutions.

To be more precise, let $N \in \mathbb{N}$ be an odd number, and still consider the semidiscrete wave (5.81). We then define the class

$$\mathscr{V}_h = \left\{ (\mathbf{u}_h^0, \mathbf{u}_h^1) \in \mathbb{R}^N \times \mathbb{R}^N, \quad u_{2j+1}^\ell = \frac{u_{2j}^\ell + u_{2j+2}^\ell}{2}, \right.$$

$$\left. j \in \{0, \ldots, (N-1)/2\}, \ell \in \{0,1\} \right\}. \quad (5.113)$$

The idea of Glowinski and Li is then to consider initial data lying in this space, which can be easily described, as we said, in the physical space.

Formally, the oscillations in the coarse mesh that correspond to the largest eigenvalues $\lambda \simeq 4\sin(\pi/4)^2/h^2$, in the finer mesh are associated to eigenvalues in the class of filtered solutions with parameter $\gamma = 4\sin(\pi/4)^2 = 2$. Formally, this corresponds to a situation where the observability inequality is uniform for $T > 2/\sqrt{1-\gamma/4} = 2\sqrt{2}$.

The following holds:

Theorem 5.5.6. *For $N \in \mathbb{N}$ an odd integer and $T > 2\sqrt{2} + 2h$, for any initial data $(\mathbf{u}_h^0, \mathbf{u}_h^1) \in \mathcal{V}_h$, the solution $\mathbf{u}_h$ of (5.81) satisfies:*

$$E_h(0) \leq \frac{2}{T - 2\sqrt{2} - 2h} \int_0^T \left| \frac{u_N}{h} \right|^2 dt. \qquad (5.114)$$

Theorem 5.5.6 has been obtained recently in [36] using the multiplier identity (5.110) and careful estimates on each term in this identity. This approach yields the most explicit estimate on the observability constant for bi-grid techniques.

This issue has also been studied theoretically in the article [83] using the multiplier techniques in 1-d (but getting an observation time $T > 4$), and later in [50] in 2d using a dyadic decomposition argument. The time has later been improved in 1-d to $T > 2\sqrt{2}$ using Ingham techniques in [71], loosing track of the observability constants.

Theorem 5.5.6 justifies the efficiency of the two-grid algorithm for computing the control of the continuous wave equation, as we shall derive more explicitly in Sect. 5.6.

This method was introduced by Glowinski [40] in the context of the full finite difference (in time) and finite element space discretization in 2D. It was then further developed in the framework of finite differences by M. Asch and G. Lebeau in [2], where the Geometric Control Condition for the wave equation in different geometries was tested numerically.

5.5.3 Tychonoff Regularization

Glowinski, Li, and Lions [43] proposed a Tychonoff regularization technique that allows one to recover the uniform (with respect to the mesh size) coercivity of the functional that one must minimize to get the controls in the HUM approach. The method was tested to be efficient in numerical experiments.

In the context of observability Tychonoff regularization corresponds to relaxing the boundary observability inequality by adding an extra observation, distributed everywhere in the domain and at the right scale so that it asymptotically vanishes as h tends to zero but is strong enough to capture the energy of the pathological high frequency components. The corresponding observability inequality is as follows:

$$E_h(0) \leq C(T)^2 \left[\int_0^T \left| \frac{u_N(t)}{h} \right|^2 dt + h^3 \sum_{j=0}^N \int_0^T \left| \frac{u_{j+1}' - u_j'}{h} \right|^2 dt \right]. \qquad (5.115)$$

The following holds:

Theorem 5.5.7 ([100]). *For any time $T > 2$, there exists a constant $C(T)$ such that, for all $h > 0$, inequality (5.115) holds for all solutions $\mathbf{u}_h$ of (5.81). Furthermore, $C(T)^2$ can be taken to be $2/(T-2)$.*

In (5.115) we have the extra term (5.112) that has already been encountered in the multiplier identity (5.110). By inspection of the solutions of (5.81) in separated variables it is easy to understand why this added term is a suitable one to reestablish the uniform observability property. Indeed, consider the solution of the semidiscrete system $\mathbf{u}_h = \exp\left(\pm i \sqrt{\lambda_h^k} t\right) \mathbf{w}_h^k$. The extra term we have added is of the order of $h^2 \lambda_h^k E_h(0)$. Obviously this term is negligible as $h \to 0$ for the low-frequency solutions (for k fixed) but becomes relevant for the high-frequency ones when $\lambda_h^k \sim 1/h^2$. Accordingly, when inequality (5.86) fails, i.e., for the high-frequency solutions, the extra term in (5.115) reestablishes the uniform character of the estimate with respect to h. It is important to emphasize that both terms are needed for (5.115) to hold. Indeed, (5.112) by itself does not suffice since its contribution vanishes as $h \to 0$ for the low-frequency solutions.

We do not give the proof of Theorem 5.5.7, which is an easy consequence of the discrete multiplier identity (5.110)–(5.111).

5.5.4   Space Semidiscretizations of the 2D Wave Equations

In this section we briefly discuss the results in [111] on the space finite difference semidiscretizations of the 2D wave equation in the square $\Omega = (0, \pi) \times (0, \pi)$ of $\mathbb{R}^2$:

$$\begin{cases} u_{tt} - \Delta u = 0 & \text{in} \quad Q = \Omega \times (0, T), \\ u = 0 & \text{on} \quad \partial \Omega \times (0, T), \\ u(x, 0) = u^0(x), \ u_t(x, 0) = u^1(x) & \text{in} \quad \Omega. \end{cases} \tag{5.116}$$

Obviously, the fact that classical finite differences provide divergent results for 1-d problems in what concerns observability and controllability indicates that the same should be true in two dimensions as well. This is indeed the case. However, the multidimensional case exhibits some new features and deserves additional analysis, in particular in what concerns filtering techniques. Given $\left(u^0, u^1\right) \in H_0^1(\Omega) \times L^2(\Omega)$, system (5.116) admits a unique solution $u \in C\left([0, T]; H_0^1(\Omega)\right) \cap C^1 \left([0, T]; L^2(\Omega)\right)$. Moreover, the energy

$$E(t) = \frac{1}{2} \int_\Omega \left[|u_t(x, t)|^2 + |\nabla u(x, t)|^2 \right] dx \tag{5.117}$$

remains constant, i.e.,

$$E(t) = E(0) \quad \forall 0 < t < T. \tag{5.118}$$

Let Γ denote a subset of the boundary of Ω constituted by two consecutive sides, for instance,

$$\Gamma = \{(x_1, \pi) : x_1 \in (0, \pi)\} \cup \{(\pi, x_2) : x_2 \in (0, \pi)\}. \tag{5.119}$$

It is well known (see [68, 69]) that for $T > 2\sqrt{2}\pi$ there exists $C_{obs}(T) > 0$ such that

$$E(0) \leq C_{obs}(T)^2 \int\limits_0^T \int\limits_\Gamma \left| \frac{\partial u}{\partial n} \right|^2 d\sigma dt \tag{5.120}$$

holds for every finite-energy solution of (5.116).

We can now address the standard five-point finite difference space semidiscretization scheme for the 2-d wave equation.

As in one dimension we may perform a complete description of both the continuous solutions and those of the semidiscrete systems in terms of Fourier series. One can then deduce the following:

• The semidiscrete system is observable for all time T and mesh size h.
• The observability constant $C_h(T)$ blows up as h tends to 0 because of the spurious high-frequency numerical solutions.
• The uniform (with respect to h) observability property may be reestablished by a suitable filtering of the high frequencies.

However, filtering needs to be implemented more carefully in the multi-dimensional case.

Indeed, the upper bound on the spectrum of the semidiscrete system in two dimensions is $8/h^2$ but it is not sufficient to filter by a constant $0 < \gamma < 8$, i.e., to consider solutions that do not contain the contribution of the high frequencies $\lambda > \gamma \, h^{-2}$, to guarantee uniform observability.

In fact, one has to filter by means of a constant $0 < \gamma < 4$. This is due to the existence of solutions corresponding to high-frequency oscillations in one direction and very slow oscillations in the other. Roughly speaking, one needs to filter efficiently in both space directions, and this requires taking $\gamma < 4$ (see [111]).

In order to better understand the necessity of filtering and getting sharp observability times it is convenient to adopt the approach of [72, 73] based on the use of discrete Wigner measures. The symbol of the semidiscrete system for solutions of wavelength h is

$$\tau^2 - 4 \left(\sin^2(\xi_1/2) + \sin^2(\xi_2/2) \right) \tag{5.121}$$

and can be easily obtained as in the von Neumann analysis of the stability of numerical schemes by taking the Fourier transform of the semidiscrete equation: the continuous one in time and the discrete one in space.[11]

Note that in the symbol in (5.121) the parameter h disappears. This is due to the fact that we are analyzing the propagation of waves of wavelength of the order of h.

The bicharacteristic rays are then defined as follows:

$$\begin{cases} x_j'(s) = -2\sin(\xi_j/2)\cos(\xi_j/2) = -\sin(\xi_j), & j = 1,2, \\ t'(s) = \tau, \\ \xi_j'(s) = 0, & j = 1,2, \\ \tau'(s) = 0. \end{cases} \tag{5.122}$$

on the characteristic set $\tau^2 - 4(\sin^2(\xi_1/2) + \sin^2(\xi_2/2)) = 0$.

It is interesting to note that the rays are still straight lines, as for the constant coefficient wave equation, since the coefficients of the equation and the numerical discretization are both constant. We see, however, that in (5.122) the velocity of propagation changes with respect to that of the continuous wave equation.

Let us now consider initial data for this Hamiltonian system with the following particular structure: x^0 is any point in the domain Ω, the initial time $t_0 = 0$, and the initial microlocal direction (τ^*, ξ^*) is such that

$$(\tau^*)^2 = 4\left(\sin^2(\xi_1^*/2) + \sin^2(\xi_2^*/2)\right). \tag{5.123}$$

Note that the last condition is compatible with the choice $\xi_1^* = 0$ and $\xi_2^* = \pi$ together with $\tau^* = 2$. Thus, let us consider the initial microlocal direction $\xi_2^* = \pi$ and $\tau^* = 2$. In this case the ray remains constant in time, $x(t) = x^0$, since, according to the first equation in (5.122), x_j' vanishes both for $j = 1$ and $j = 2$. Thus, the projection of the ray over the space x does not move as time evolves. This ray never reaches the exterior boundary $\partial\Omega$ where the equation evolves and excludes the possibility of having a uniform boundary observability property. More precisely, this construction allows one to show that, as $h \to 0$, there exists a sequence of solutions of the semidiscrete problem whose energy is concentrated in any finite time interval $0 \le t \le T$ as much as one wishes in a neighborhood of the point x^0.

This example corresponds to the case of very slow oscillations in the space variable x_1 and very rapid ones in the x_2-direction, and it can be ruled out, precisely, by taking the filtering parameter $\gamma < 4$. In view of the structure of the Hamiltonian system, it is clear that one can be more precise when choosing the space of filtered solutions. Indeed, it is sufficient to exclude by filtering the rays that do not propagate at all to guarantee the existence of a minimal velocity of propagation (see Fig. 5.10).

---

[11]This argument can be easily adapted to the case where the numerical approximation scheme is discrete in both space and time by taking discrete Fourier transforms in both variables.

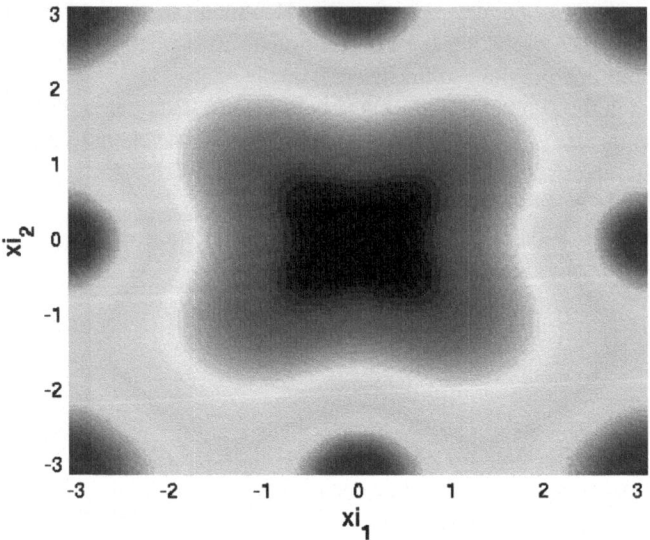

Fig. 5.10 Level set representation of the group velocity as a function of the frequency $(h\xi_1, h\xi_2) \in (-\pi, \pi)$. In *red*, the points where the group velocity is 1, which is the velocity of propagation of continuous waves. In *blue*, the points where the group velocity is close to zero. When, by means of a filtering method the blue areas are removed, the velocity of propagation of rays is uniformly bounded from below

Roughly speaking, this suffices for the observability inequality to hold uniformly in h for a sufficiently large time [72, 73].

This ray approach makes it possible to conjecture the optimal uniform observability time depending on the class of filtered solutions under consideration. The optimal time is the one that all characteristic rays entering in the class of filtered solutions need to reach the controlled region. This constitutes the discrete version of the GCC for the continuous wave equation. Moreover, if the filtering is done so that the wavelength of the solutions under consideration is of an order strictly less than h, then one recovers the classical observability result for the constant coefficient continuous wave equation with the optimal observability time.

5.5.5 A More General Result

Here, we describe the most general result available in the literature for uniform observability of space semidiscrete wave equations.

This concerns the finite-element discretization of (5.59) observed through some subdomain ω. Let us emphasize from the beginning that the results presented in that section hold under the Geometric Control Condition for (Ω, ω, T), whatever the dimension is and under very mild assumptions on the finite-element discretization under consideration.

In the following, to simplify the presentation, we focus on the constant coefficient wave equation:

$$\begin{cases} u_{tt} - \Delta u = 0, & \text{in } \Omega \times (0, T), \\ u = 0, & \text{on } \partial\Omega \times (0, T), \\ u(0) = u^0, \quad u_t(0) = u^1, & \text{in } \Omega \end{cases} \tag{5.124}$$

observed through $\chi_\omega u_t$ on $\omega \times (0, T)$.

The corresponding observability inequality is

$$\left\| \nabla u^0 \right\|_{L^2(\Omega)}^2 + \left\| u^1 \right\|_{L^2(\Omega)}^2 \leq C_{obs}^2 \int_0^T \left\| \chi_\omega u_t(t) \right\|_{L^2(\Omega)}^2 \, dt. \tag{5.125}$$

Let us now describe the finite element method we use to discretize (5.124).

Consider $(V_h)_{h>0}$ a sequence of vector spaces of finite dimension n_h that embed V_h into $L^2(\Omega)$ using a linear morphism $\rho_h : V_h \to L^2$. For each $h > 0$, the inner product $\langle \cdot, \cdot \rangle_{L^2}$ in L^2 induces a structure of Hilbert space for V_h endowed by the scalar product $\langle \cdot, \cdot \rangle_h = \langle \rho_h \cdot, \rho_h \cdot \rangle_{L^2}$. We assume that for each $h > 0$, the vector space $\rho_h(V_h)$ is a subspace of $\mathcal{D}((-\Delta_D)^{1/2}) = H_0^1(\Omega)$. We thus define the linear operator $A_{0h} : V_h \to V_h$ by

$$\langle A_{0h}\boldsymbol{\phi}_h, \boldsymbol{\psi}_h \rangle_h = \langle \nabla \rho_h \boldsymbol{\phi}_h, \nabla \rho_h \boldsymbol{\psi}_h \rangle_{L^2}, \quad \forall (\boldsymbol{\phi}_h, \boldsymbol{\psi}_h) \in V_h^2. \tag{5.126}$$

The operator A_{0h} defined in (5.126) obviously is self-adjoint and positive definite. Formally, definition (5.126) implies that

$$A_{0h} = (\nabla \rho_h)^* \nabla \rho_h. \tag{5.127}$$

This operator A_{0h} corresponds to the finite element discretization of $-\Delta_D$, the Laplace operator with Dirichlet boundary conditions. System (5.124) is then discretized into

$$\mathbf{u}_h'' + A_{0h}\mathbf{u}_h = 0, \quad \mathbf{u}_h(0) = \mathbf{u}_h^0 \in V_h, \quad \mathbf{u}_h'(0) = \mathbf{u}_h^1 \in V_h. \tag{5.128}$$

In this context, for all $h > 0$, the observation operator naturally becomes $\chi_\omega \rho_h u_h'(t)$.

We now make precise the assumptions we have, usually, on ρ_h, and which will be needed in our analysis. For this, we introduce the adjoint of ρ_h from V_h endowed with the scalar product of $\langle A_{0h}^{1/2} \cdot, A_{0h}^{1/2} \cdot \rangle_h$ to $\mathcal{D}(A_0^{1/2}) = H_0^1(\Omega)$ endowed with the scalar product $\langle \nabla \cdot, \nabla \cdot \rangle_{L^2}$.

One easily checks that $\rho_h^* \rho_h = Id_{V_h}$. Besides, the embedding ρ_h describes the finite element approximation we have chosen. In particular, the vector space $\rho_h(V_h)$

approximates, in the sense given hereafter, the space $\mathcal{D}(A_0^{1/2}) = H_0^1(\Omega)$: There exist $\theta > 0$ and $C_0 > 0$, such that for all $h > 0$,

$$\begin{cases} \left\| \nabla(\rho_h \rho_h^* - I)u \right\|_{L^2(\Omega)} \leq C_0 \left\| \nabla u \right\|_{L^2(\Omega)}, \quad \forall u \in H_0^1(\Omega), \\ \left\| \nabla(\rho_h \rho_h^* - I)u \right\|_{L^2(\Omega)} \leq C_0 h^\theta \left\| -\Delta u \right\|_{L^2(\Omega)}, \quad \forall u \in H^2 \cap H_0^1(\Omega). \end{cases} \tag{5.129}$$

Note that in many applications, estimates (5.129) are satisfied for $\theta = 1$. This is in particular true when discretizing on uniformly regular meshes (see [92]).

We will not discuss convergence results for the numerical approximation schemes presented here, which are classical under assumption (5.129), and which can be found for instance in the textbook [92].

In view of the previous results, it is natural to restrict ourselves to filtered initial data. For all $h > 0$, since A_{0h} is a self adjoint positive definite matrix, the spectrum of A_{0h} is given by a sequence of positive eigenvalues

$$0 < \lambda_h^1 \leq \lambda_h^2 \leq \cdots \leq \lambda_h^{n_h} \tag{5.130}$$

and normalized (in V_h) eigenvectors $(\mathbf{w}_h^k)_{1 \leq k \leq n_h}$. For any s, we can now define, for each $h > 0$, the filtered space (to be compared with (5.108))

$$\mathscr{C}_h(s) = \left\{ \mathbf{u}_h = \sum_{\lambda_h^k \leq s} \left(a_k \cos\left(\sqrt{\lambda_h^k} t\right) + \frac{b_k}{\sqrt{\lambda_h^k}} \sin\left(\sqrt{\lambda_h^k} t\right) \right) \mathbf{w}_h^k \right\}.$$

We are now in position to state the following results:

Theorem 5.5.8 ([27]). *Assume that the maps $(\rho_h)_{h>0}$ satisfy property (5.129) and that (ω, Ω, T) satisfies the Geometric Control Condition, i.e. that system (5.124) is exactly observable.*

Then there exist $\varepsilon > 0$, a time T^* and a positive constant C_{obs} such that, for any $h \in (0, 1)$, any solution of (5.128) lying in $\mathscr{C}_h(\varepsilon/h^\theta)$ satisfies

$$\left\| \nabla \rho_h \mathbf{u}_h^0 \right\|_{L^2(\Omega)}^2 + \left\| \rho_h \mathbf{u}_h^1 \right\|_{L^2(\Omega)}^2 \leq C_{obs}^2 \int_0^{T^*} \left\| \chi_\omega \rho_h \mathbf{u}_h'(t) \right\|_{L^2(\Omega)}^2 \, dt. \tag{5.131}$$

Note in particular that this yields the same results as the one obtained in [90] in a 1-d framework and generalizes it to any dimension.

The proof of this Theorem combines, essentially, the observability inequality of the continuous wave equation and sharp estimates on the convergence of the numerical scheme towards the continuous model. Roughly speaking, one needs to build the subspace of initial data so that numerical solutions are uniformly close to the continuous ones so that they inherit the observability properties of the later.

The interest of this result is that it holds in any space dimension and in a very general Galerkin approximation setting. To our knowledge, [27] and the companion paper [29] are the first ones in which this kind of results are presented with such a degree of generality.

The proof of this statement can be derived using resolvent estimates [16, 79] (see also [88] for a similar estimate) but this method does not yield sharp estimates on the observability time. Hence T^* in Theorem 5.5.8 may be much larger than the time for which (ω, Ω, T) satisfies GCC and the one could expect to be sharp in view of the analysis of the dispersion diagram of the numerical scheme.

Note also that (5.131) holds within a class of functions that are much more filtered than in Theorem 5.5.3. The later holds up to the critical scale within subclasses of the form $\mathscr{C}_h(\gamma / h^2)$, $\gamma < 4$. Whether the result in Theorem 5.5.8 is true or not in these optimal subclasses is an interesting open problem. Note, in any case, that Theorem 5.5.8 holds in a much more general setting, where new phenomena could occur. Even in 1-d, for the finite element method on non-uniform meshes, whether Theorem 5.5.8 can be improved or not is an open problem.

5.6 Convergence Results

The goal of this section is to describe a general approach to show the convergence of the discrete controls, obtaining convergence rates, from the observability results presented in the previous section.

5.6.1 A General Procedure for the Convergence of the Discrete Controls

In this section, we describe the setting in which we are working, and present the main ideas.

Let A be a skew-adjoint operator $A : \mathscr{D}(A) \subset X \to X$ with compact resolvent and dense domain, and B be an admissible control operator $B \in \mathfrak{L}(\mathscr{U}, X_{-1})$.

We assume that the continuous system (5.13) is controllable in some time $T > 0$.

Now, we approximate the continuous model (5.13) by a sequence of finite-dimensional systems

$$\mathbf{x}'_h = A_h \mathbf{x}_h + B_h \mathbf{v}_h, \quad t \geq 0, \qquad \mathbf{x}_h(0) = \mathbf{x}^0_h \in X_h, \qquad (5.132)$$

where (A_h, B_h) is a sequence of finite-dimensional approximations of the operators (A, B) respectively, where for each $h > 0$, A_h is a skew-adjoint operator defined on a finite dimensional space X_h embedded into X, and B_h is defined on a vector space $\mathscr{U}_h$ that embeds into the Hilbert space $\mathscr{U}$ with values in X_h.

We consider the embedding $\rho_h : X_h \to X$, which provides an Hilbert structure on X_h by $\|\cdot\|_h = \|\rho_h(\cdot)\|_X$.

To simplify the presentation, we further assume that B_h is simply given by $\rho_h^* B$, where B is the continuous control operator, so that $\mathscr{U}_h$ simply coincides with $\mathscr{U}$. Otherwise, similar ideas can be applied, see for instance Sect. 5.6.2.3.

We also assume that the spaces X_h fill the space X as $h \to 0$ in a sense that will be made precise below. Of course, the finite difference or the finite-element approximation schemes for the wave equation fit into this setting, and a more precise description can be made in these cases.

We have already seen in Sect. 5.4.4 that, for the finite-difference method, the discrete controls fulfilling the control requirement $\mathbf{x}_h(T) = 0$ may blow up as $h \to 0$, due to the fact that observability properties do not hold uniformly with respect to the discretization parameter $h > 0$.

However, we have seen in Sect. 5.5 that weak observability results can be shown to hold uniformly with respect to the discretization parameter $h > 0$, provided suitable filtering mechanisms are implemented. To be more precise, we assume that there exist a positive constant C_{obs} and a time T such that, for all $h > 0$,

$$\left\|\varphi_h^T\right\|_h^2 \leq C_{obs}^2 \int_0^T \eta(t) \left\|B_h^* \varphi_h(t)\right\|_{\mathscr{U}}^2 \, dt, \quad \forall \varphi_h^T \in \mathfrak{C}_h, \tag{5.133}$$

where $\mathfrak{C}_h$ is a subspace of X_h, η is a smooth function with values in $[0,1]$, vanishing for $t \notin [0,T]$ and equals to 1 on some non trivial subset of $[0,T]$, similarly as in (5.21), and φ_h is the solution of the adjoint system

$$\varphi_h' = A_h \varphi_h, \quad t \in (0,T), \qquad \varphi_h(T) = \varphi_h^T. \tag{5.134}$$

We now consider the HUM-type functional J_h, defined for $\varphi_h^T \in \mathfrak{C}_h$ by

$$J_h(\varphi_h^T) = \frac{1}{2} \int_0^T \eta(t) \left\|B_h^* \varphi_h(t)\right\|_{\mathscr{U}}^2 \, dt + \langle \mathbf{x}_h^0, \varphi_h(0)\rangle_h. \tag{5.135}$$

Using the same arguments as in Theorem 5.2.1 and Corollary 5.2.6, one easily checks that:

Theorem 5.6.1. *Assume that* (5.133) *holds with constants C and T independent of $h > 0$.*

Let $h > 0$ and $\mathbf{x}_h^0 \in X_h$. Then the functional J_h in (5.135) *is continuous, strictly convex and coercive on $\mathfrak{C}_h$ and it admits a unique minimizer $\Phi_h^T \in \mathfrak{C}_h$. Then, setting $\mathbf{V}_h = \eta(t) B_h^* \Phi_h$, where Φ_h is the solution of* (5.134) *with initial data Φ_h^T, the solution $\mathbf{x}_h$ of* (5.132) *satisfies*

$$\forall \varphi_h^T \in \mathfrak{C}_h, \quad \langle \varphi_h^T, \mathbf{x}_h(T)\rangle_h = 0, \tag{5.136}$$

or equivalently $x_h(T) \in \mathfrak{C}_h^\perp$.

Besides, this is the only control $\mathbf{V}_h \in L^2(0, T; \mathscr{U})$ *such that the corresponding* $\mathbf{x}_h$ *satisfies* (5.136) *and which is of the form* $\mathbf{V}_h = \eta B_h^* \boldsymbol{\varphi}_h$ *for some* $\boldsymbol{\varphi}_h$ *solution of* (5.134) *with* $\boldsymbol{\varphi}_h(T) \in \mathfrak{C}_h$.

Moreover,

$$\frac{1}{C_{obs}^2} \left\| \Phi_h^T \right\|_h^2 \leq \int_0^T \left\| \mathbf{V}_h(t) \right\|_{\mathscr{U}}^2 \frac{dt}{\eta(t)} = \int_0^T \eta(t) \left\| B_h^* \Phi_h(t) \right\|_{\mathscr{U}}^2 dt \leq C_{obs}^2 \left\| \mathbf{x}_h^0 \right\|_h^2.$$

(5.137)

The following two questions arise now naturally:

- *Convergence.* Given $\mathbf{x}_h^0$ that converge (weakly or strongly) to x^0 in X as $h \to 0$ (in a sense to be made precise), can we show that the discrete controls $\mathbf{V}_h$ converge to V, the continuous control corresponding to x^0 for (5.13)?
- *Convergence rates.* Can we furthermore give a convergence rate for the convergence of $\mathbf{V}_h$ towards V?

These two questions will be investigated below in this very general setting. Of course, getting such results requires a more precise knowledge of the numerical schemes under consideration.

We shall then present a general frame on which, under suitable hypotheses that should then be carefully verified in each situation, the convergence will be proved with convergence rates.

5.6.1.1 Convergence

To derive the convergence of the discrete controls $\mathbf{V}_h$ given by Theorem 5.6.1, we need the following hypotheses, that should be verified in each situation:

Hypothesis #1. *For* $\varphi^T \in \cap_{s>0} \mathscr{D}(A^s)$ *and* φ *be the corresponding solution of* (5.14), *there exists a sequence of functions* $\varphi_h^T \in \mathfrak{C}_h$ *such that, if* $\boldsymbol{\varphi}_h$ *denotes the corresponding solution of* (5.134),

$$\rho_h \boldsymbol{\varphi}_h(0) \xrightarrow[h \to 0]{} \varphi(0) \quad in \ X$$

(5.138)

$$B_h^* \boldsymbol{\varphi}_h \xrightarrow[h \to 0]{} B^* \varphi \quad in \ L^2(0, T; \mathscr{U}).$$

(5.139)

Hypothesis #1 looks like a classical result of convergence of the numerical methods under consideration. This is indeed the case, except for the fact that the approximations of φ^T are searched within the restricted subspace $\mathfrak{C}_h$ of X_h. This in practice requires proving the convergence of suitable projections of the numerical approximations.

We also need the following assumption:

Hypothesis #2. For $\varphi_h^T \in X_h$ and $\varphi^T \in X$ such that

$$\rho_h \varphi_h^T \underset{h \to 0}{\rightharpoonup} \varphi^T \quad in\ X \quad and \quad \sup_h \| B_h^* \varphi_h(t) \|_{L^2(0,T;U)} < \infty, \tag{5.140}$$

denoting by φ_h and φ respectively the solutions of (5.134) and (5.14) with initial data φ_h^T and φ^T respectively,

$$\rho_h \varphi_h \underset{h \to 0}{\rightharpoonup} \varphi \quad in\ L^2(0,T;X) \tag{5.141}$$

$$B_h^* \varphi_h \underset{h \to 0}{\rightharpoonup} B^* \varphi \quad in\ L^2(0,T;\mathcal{U}) \tag{5.142}$$

$$\rho_h \varphi_h(0) \underset{h \to 0}{\rightharpoonup} \varphi(0) \quad in\ X. \tag{5.143}$$

The statements in Hypothesis #2 typically hold for classical numerical approximation schemes.

Under these two main hypotheses we get the following result:

Theorem 5.6.2. Let $x^0 \in X$ and $\mathbf{x}_h^0 \in X_h$ be such that $\rho_h \mathbf{x}_h^0$ weakly converges to x^0 in X as $h \to 0$.

We further assume that Hypotheses #1 and #2 hold true.

Then the discrete controls $\mathbf{V}_h$ given by Theorem 5.6.1 weakly converge to V given by Proposition 5.2.11 in $L^2(0,T;dt/\eta;\mathcal{U})$ as $h \to 0$.

Moreover, if $\rho_h \mathbf{x}_h^0$ strongly converge to x^0, $\mathbf{V}_h$ strongly converge to V in the norm of $L^2(0,T;dt/\eta;\mathcal{U})$ (hence in the $L^2(0,T;\mathcal{U})$-norm as well) as $h \to 0$.

Proof. The proof of Theorem 5.6.2 is divided into several steps.

Step 1. Extraction of a weakly convergent sequence of controls. From Theorem 5.6.1, the sequence $\mathbf{V}_h$ is bounded in $L^2(0,T;dt/\eta;\mathcal{U})$. Hence, up to extraction of a subsequence, the controls $\mathbf{V}_h$ weakly converge to some function v in $L^2(0,T;dt/\eta;\mathcal{U})$.

Step 2. Any weak accumulation point of $\mathbf{V}_h$ is a control function for (5.13). The Euler–Lagrange equation satisfied by the minimizer Φ_h^T of J_h in (5.135) is the following one:

$$\forall \varphi_h^T \in \mathfrak{C}_h, \quad \int_0^T \langle \mathbf{V}_h(t), B_h^* \varphi_h \rangle_{\mathcal{U}}\, dt + \langle \mathbf{x}_h^0, \varphi_h(0) \rangle_h = 0. \tag{5.144}$$

Let us then take $\varphi^T \in \cap_{s>0} \mathscr{D}(A^s)$. Using Hypothesis #1, we obtain a sequence $\varphi_h^T \in \mathfrak{C}_h$ such that the strong convergences (5.139)–(5.138) hold. Further using that

$$\langle \mathbf{x}_h^0, \varphi_h(0) \rangle_h = \langle \rho_h \mathbf{x}_h^0, \rho_h \varphi_h(0) \rangle_X,$$

and passing to the limit in (5.144), we obtain that for all $\varphi^T \in \cap_{s>0} \mathscr{D}(A^s)$,

$$\int_0^T \langle v(t), B^*\varphi \rangle_{\mathscr{U}} \, dt + \langle x^0, \varphi(0) \rangle_X = 0. \tag{5.145}$$

By density, this also holds true for all $\varphi^T \in X$. From (5.20), this implies that v is a control function for (5.13).

Step 3. Any weak accumulation point v of $\mathbf{V}_h$ can be written as $v = \eta B^*\varphi$ for some φ solution of the adjoint system (5.14). For all $h > 0$, $\mathbf{V}_h = \eta B_h^* \Phi_h$, where Φ_h is the solution of (5.134) with initial data Φ_h^T, and $\mathbf{V}_h$ and $\rho_h \Phi_h^T$ are bounded, respectively, in $L^2(0, T; dt/\eta; \mathscr{U})$ and X, due to (5.137). Thus, up to subsequence, $\rho_h \Phi_h^T$ weakly converge in X to some φ^T. Thus, from Hypothesis #2, $v = \eta B^*\varphi$, where φ is the solution of (5.14) corresponding to φ^T.

Step 4. Any weak accumulation point of $\mathbf{V}_h$ is the control V given by Proposition 5.2.11. This follows from the uniqueness of the control functions that can be written $\eta B^*\varphi$ for some φ solution of (5.14) (see Proposition 5.2.11).

Hence there is only one weak accumulation point for the sequence $(\mathbf{V}_h)$, which coincides with the control V given by Proposition 5.2.11. Therefore, the sequence $(\mathbf{V}_h)$ weakly converges to V in $L^2(0, T; dt/\eta; \mathscr{U})$ as $h \to 0$.

Step 5. Strong convergence when $\rho_h \mathbf{x}_h^0$ strongly converges to x^0. In view of the weak convergence property from Step 4, we only need to prove the convergence of the $L^2(0, T; dt/\eta; \mathscr{U})$-norms of $\mathbf{V}_h$ as $h \to 0$.

But, from (5.144) applied to Φ_h^T ($\in \mathfrak{C}_h$),

$$\|\mathbf{V}_h\|_{L^2(0,T;dt/\eta;\mathscr{U})}^2 = \int_0^T \eta(t) \|B_h^* \Phi_h(t)\|_{\mathscr{U}}^2 \, dt = -\langle \rho_h \mathbf{x}_h^0, \rho_h \Phi_h(0) \rangle_X. \tag{5.146}$$

On the other hand, $V = \eta B^*\Phi$, where Φ is given by Proposition 5.2.11. From (5.20) applied to $\varphi^T = \Phi^T$, we obtain

$$\|V\|_{L^2(0,T;dt/\eta;\mathscr{U})}^2 = \int_0^T \eta(t) \|B^* \Phi(t)\|_{\mathscr{U}}^2 \, dt = -\langle x_0, \Phi(0) \rangle_X. \tag{5.147}$$

Now, using Step 3 and Hypothesis #2, $\rho_h \Phi_h^T$ weakly converges to some φ^T in X which is such that $V = \eta B^*\varphi$. From the observability inequality (5.22), $\varphi \equiv \Phi$, the one corresponding to the minimizer of the functional J in (5.23). Hence $\rho_h \Phi_h^T$ weakly converges in X to Φ^T. Applying again Hypothesis #2, $\rho_h \Phi_h(0)$ weakly converges to $\Phi(0)$ in X as $h \to 0$.

Passing to the limit, $\langle \rho_h \mathbf{x}_h^0, \rho_h \Phi_h(0) \rangle_X$ converges to $\langle x_0, \Phi(0) \rangle_X$ as $h \to 0$, and then passing to the limit in (5.146) and using (5.147), the $L^2(0, T; dt/\eta; \mathscr{U})$-norms of $\mathbf{V}_h$ converge to the $L^2(0, T; dt/\eta; \mathscr{U})$-norm of V.

This concludes the proof of the theorem.   □

Note that this method of proof is not new (see, for instance, [53]) and it has been shown to be robust and efficient, whatever the discretization scheme or the weak observability properties under consideration are.

However, this approach did not seem to be sufficient to get convergence rates for the discrete controls. The main reason is that it was not known, with this degree of generality, that smooth initial data to be controlled yield smooth controls. As we have explained above, this holds true in a broad abstract setting, but only when the cut-off function in time $\eta(t)$ is introduced or when the control operator is bounded, i.e. $B \in \mathfrak{L}(\mathscr{U}, X)$. Then, using Theorem 5.2.12, we will be in conditions to prove also convergence rates.

5.6.1.2 Convergence Rates

To prove convergence rates for the discrete controls towards the continuous ones, it is necessary, as is standard in numerical analysis, to assume some smoothness on the initial data. One then needs to make sure that the numerical schemes approximating the PDE model have suitable convergence rates that we will then transfer to the controls. In the following Hypothesis #3 we require this property to be fulfilled.

Hypothesis #3. *There exist $s_1 > 0$ and a constant $\theta_1 > 0$ such that for all $\varphi^T \in \mathscr{D}(A^{s_1})$, one can find a sequence of functions $\boldsymbol{\varphi}_h^T \in \mathfrak{C}_h$ such that the corresponding solutions $\boldsymbol{\varphi}_h$ of (5.134) satisfy, for $h > 0$,*

$$\sup_{t \in (0,T)} \left(\|\rho_h \boldsymbol{\varphi}_h - \varphi\|_X \right) + \|B^*(\rho_h \boldsymbol{\varphi}_h - \varphi)\|_{L^2(0,T;\mathscr{U})} \leq C h^{\theta_1} \|\varphi^T\|_{\mathscr{D}(A^{s_1})}, \quad (5.148)$$

where φ is the solution of (5.14) with initial data φ^T.

Note that Hypothesis #3 is a stronger version of Hypothesis #1. It always holds with X_h instead of $\mathfrak{C}_h$ for convergent numerical approximation schemes. As we shall see, in specific examples, similar results hold within the classes $\mathfrak{C}_h$ as assumed in Hypothesis #3.

Also note that when B^* is bounded, estimate (5.148) is implied by the weaker one:

$$\sup_{t \in (0,T)} \|\rho_h \boldsymbol{\varphi}_h - \varphi\|_X \leq C h^{\theta_1} \|\varphi^T\|_{\mathscr{D}(A^{s_1})}. \quad (5.149)$$

We also need a similar convergence assumption for the controlled equation:

Hypothesis #4. *There exist $s_2 > 0$ and a constant θ_2 such that for all $x^0 \in \mathscr{D}(A^{s_2})$ and $\Phi^T \in \mathscr{D}(A^{s_2})$, setting $\mathbf{x}_h^0 = \rho_h^* x^0$, $v = \eta B^* \Phi$ where Φ is the solution of (5.14) with initial data Φ^T and $\mathbf{v}_h \in L^2(0, T; \mathscr{U})$, the corresponding solutions $\mathbf{x}_h$ and x of (5.132) and (5.13) respectively satisfy:*

$$\|\rho_h \mathbf{x}_h(T) - x(T)\|_X \leq C h^{\theta_2} \left(\|x^0\|_{\mathscr{D}(A^{s_2})} + \|\Phi^T\|_{\mathscr{D}(A^{s_2})} \right) + C \|\mathbf{v}_h - v\|_{L^2(0,T;\mathscr{U})}. \tag{5.150}$$

Note that Hypothesis #4 looks like a classical convergence result for numerical methods. The fact that the source term is given as $\eta B^* \Phi$ is needed to guarantee that the controlled trajectory x lies in a smooth space, and in particular that this is a strong solution, see Corollary 5.2.13 and Sect. 5.3.3.

We are now in position to state our main result:

Theorem 5.6.3. *Assume that Hypotheses #3 and #4 hold.*

Let $s = \max\{s_1, s_2\}$ and $\theta = \min\{\theta_1, \theta_2\}$.

Then, for any $x^0 \in \mathscr{D}(A^s)$, setting $\mathbf{x}_h^0 = \rho_h^* x^0$, the discrete controls $\mathbf{V}_h$ given by Theorem 5.6.1 converge to the control V given by Proposition 5.2.11 and

$$\|\mathbf{V}_h - V\|_{L^2(0,T;dt/\eta;\mathscr{U})} \leq C h^{\theta} \|x^0\|_{\mathscr{D}(A^s)}. \tag{5.151}$$

Proof. The proof is divided into several steps.

Step 1. The continuous control is smooth. Let $x^0 \in \mathscr{D}(A^s)$. From Theorem 5.2.12, the weighted HUM method yields a control $V(t) = \eta(t) B^* \Phi(t)$, computed by Proposition 5.2.11 where Φ is the solution of (5.14) corresponding to the minimizer Φ^T of the functional J in (5.23), which is smooth:

$$\|\Phi^T\|_{\mathscr{D}(A^s)} \leq C \|x^0\|_{\mathscr{D}(A^s)}.$$

Step 2. An approximate control. Since $\Phi^T \in \mathscr{D}(A^s)$, by Hypothesis #3, one can approximate Φ by a sequence $\tilde{\Phi}_h$ of solutions of the discrete (5.134) with initial data $\tilde{\Phi}_h^T \in \mathfrak{C}_h$ such that

$$\|B^*(\rho_h \tilde{\Phi}_h - \Phi)\|_{L^2(0,T;\mathscr{U})} \leq C h^{\theta} \|\Phi^T\|_{\mathscr{D}(A^s)} \leq C h^{\theta} \|x^0\|_{\mathscr{D}(A^s)}.$$

Hence, setting

$$\tilde{\mathbf{v}}_h(t) = \eta(t) B_h^* \tilde{\Phi}_h(t), \tag{5.152}$$

$\tilde{\mathbf{v}}_h$ satisfies

$$\|\tilde{\mathbf{v}}_h - V\|_{L^2(0,T;\eta;\mathscr{U})} \leq C h^{\theta} \|x^0\|_{\mathscr{D}(A^s)}. \tag{5.153}$$

Then, using Hypothesis #4, we get that the solution $\tilde{\mathbf{x}}_h$ of

$$\tilde{\mathbf{x}}_h' = A_h \tilde{\mathbf{x}}_h + B_h \tilde{\mathbf{v}}_h, \quad t \geq 0, \qquad \tilde{\mathbf{x}}_h(0) = \mathbf{x}_h^0,$$

satisfies

$$\|\tilde{\mathbf{x}}_h(T)\|_h \leq C h^{\theta} \|x^0\|_{\mathscr{D}(A^s)}.$$

Step 3. An exact discrete control. From Theorem 5.6.1, there exists a control function $\hat{\mathbf{v}}_h \in L^2(0, T; \mathscr{U})$ such that the function $\mathbf{w}_h$ solution of

$$\mathbf{w}_h' = A_h \mathbf{w}_h + B_h \hat{\mathbf{v}}_h, \quad t \geq 0, \qquad \mathbf{w}_h(0) = 0,$$

satisfies

$$\forall \boldsymbol{\varphi}_h^T \in \mathfrak{C}_h, \quad \langle \mathbf{w}_h(T) + \tilde{\mathbf{x}}_h(T), \boldsymbol{\varphi}_h^T \rangle_h = 0.$$

Besides, from Theorem 5.6.1, this can be done with a control function $\hat{\mathbf{v}}_h \in L^2(0, T; \mathscr{U})$ that can be written $\hat{\mathbf{v}}_h = \eta B_h^* \zeta_h$ for ζ_h solution of (5.134) with initial data $\boldsymbol{\zeta}_h^T \in \mathfrak{C}_h$, and with

$$\|\hat{\mathbf{v}}_h\|_{L^2(0,T;dt/\eta;\mathscr{U})} \leq C \|\tilde{\mathbf{x}}_h(T)\|_h \leq C h^{\theta} \|x^0\|_{\mathscr{D}(A^s)}. \tag{5.154}$$

Hence $\tilde{\mathbf{v}}_h + \hat{\mathbf{v}}_h$ is a control for (5.132) (in the sense of (5.136)).

Step 4. Identification of the controls. From the uniqueness of the discrete controls that can be written as $\eta B_h^* \boldsymbol{\varphi}_h$ with $\boldsymbol{\varphi}_h^T \in \mathfrak{C}_h$ stated in Theorem 5.6.1, $\mathbf{V}_h = \tilde{\mathbf{v}}_h + \hat{\mathbf{v}}_h$.

Hence, from (5.153)–(5.154),

$$\|V - \mathbf{V}_h\|_{L^2(0,T;dt/\eta;\mathscr{U})} \leq \|V - \tilde{\mathbf{v}}_h\|_{L^2(0,T;dt/\eta;\mathscr{U})} + \|\hat{\mathbf{v}}_h\|_{L^2(0,T;dt/\eta;\mathscr{U})}$$

$$\leq C h^{\theta} \|x^0\|_{\mathscr{D}(A^s)}.$$

This completes the proof of the theorem.   □

The approach presented above is very general and can be applied in many situations. Below, we shall explain how it yields convergence results from the weak observability results stated in Sect. 5.5.

Remark 5.6.4. We refer to the recent work [20] for approximation results based on the continuous approach. In that approach the approximate controls are not built as controls for an approximate discrete dynamics but rather discretizing an iterative algorithm leading to convergence at the continuous level, but necessarily to the control of minimal norm. Note also that the method developed in [20] only converges for initial data to be controlled lying in $\mathscr{D}(A^{3/2})$ (the proofs in [20] focus on the finite element methods for the wave equation, for which this space is the natural one), but does not a priori converge when the initial data to be controlled only lie in X. The discrete approach we develop here provides both, convergence results in the optimal class of initial data and convergence rates for smooth data.

Remark 5.6.5. In a first reading, the fact that the proof of convergence of the discrete controls does not require the convergence of the controlled equations might seem surprising. Indeed, Hypotheses #1, #2 and #3 refer only to the adjoint (5.134)–(5.14) and only Hypothesis #4 directly refers to the convergence of the controlled equation.

But the convergence properties of the adjoint (5.134) towards the continuous one (5.14) in Hypotheses #1, #2 and #3 also yield convergence results for the discrete controlled system (5.132)–(5.13) since their solutions are defined by transposition, taking scalar products with the solutions of the adjoint system.

5.6.2 Controllability Results

In this section we apply the above procedure for deriving convergence rates for numerical controls in various relevant examples.

Before going further, let us emphasize that the problem of boundary control, as the internal control problem above, corresponds to a case in which the energy space is not identified with its dual, as it is done in the previous paragraph. This fact creates a shift in the functional spaces below. We made the choice of presenting the abstract theory in the reflexive case with the identification between X and its dual for the sake of simplicity.

More precisely, in the case of the boundary controllability of the wave equation, the adjoint (5.39) lies in $X = H_0^1(0, 1) \times L^2(0, 1)$, whereas the controlled (5.42) is solved in the space $X^* = L^2(0, 1) \times H^{-1}(0, 1)$.

Note in particular that the wave semigroup is an isometry in both spaces X and X^*, and thus the only difference with respect to the presentation above is that the identification between X and its dual is not done.

Hence, Hypotheses #1, #2, #3 should be checked in the energy space $H_0^1(0, 1) \times L^2(0, 1)$, whereas Hypothesis #4, that refers to the convergence of the continuous controlled equation towards (5.42), should be proved in the space $L^2(0, 1) \times H^{-1}(0, 1)$.

5.6.2.1 Filtering Methods

Based on Theorem 5.5.3, we can set $\mathfrak{C}_h = \mathscr{C}_h(\gamma/h^2)$ with $\gamma \in (0, 4)$. Note that, here $\mathscr{C}_h(\gamma/h^2)$ refers to the space in which the trajectories $\mathbf{u}_h$, solutions of (5.81), live. Of course, this can be identified with the set of data such that for some $t \in (0, T)$

(and then for all $t \in (0, T)$), $(\mathbf{u}_h(t), \mathbf{u}'_h(t))$ belongs to the vector space spanned by the first eigenvectors $\mathbf{w}_h^k$ corresponding to the eigenvalues $\lambda_h^k \leq \gamma/h^2$.

In that case, the control requirement (5.136) for solutions $\mathbf{y}_h$ of (5.98) becomes:

$$\forall \mathbf{u}_h \in \mathscr{C}_h(\gamma/h^2), \quad h \sum_{j=1}^{N} y_j(T) u'_j(T) - h \sum_{j=1}^{N} y'_j(T) u_j(T) = 0, \qquad (5.155)$$

or, equivalently,

$$\pi_{\mathscr{C}_h(\gamma/h^2)} \mathbf{y}_h(T) = 0, \text{ and } \pi_{\mathscr{C}_h(\gamma/h^2)} \mathbf{y}'_h(T) = 0, \qquad (5.156)$$

where $\pi_{\mathscr{C}_h(\gamma/h^2)}$ denotes the orthogonal projection of $L_h^2(0, 1)$ on the vector space spanned by the eigenfunctions $\mathbf{w}_h^k$ corresponding to eigenvalues $\lambda_h^k \leq \gamma/h^2$.

Fix now $\gamma \in (0, 4)$, and $T > T(\gamma)$ given by Theorem 5.5.3. Introduce $\delta > 0$ such that $T > T(\gamma) + 2\delta$. Let η be a smooth function of time such that

$$\eta : \mathbb{R} \to [0, 1], \quad \eta(t) = \begin{cases} 1 \text{ on } [\delta, T(\gamma) + \delta], \\ 0 \text{ on } \mathbb{R} \setminus (0, T). \end{cases} \qquad (5.157)$$

According to the analysis done in the previous section, it is then natural to introduce the following functional

$$J_h(\mathbf{u}_h) = \frac{1}{2} \int_0^T \eta(t) \left| \frac{u_N(t)}{h} \right|^2 dt + h \sum_{j=1}^{N} y_j^0 u_j^1 - h \sum_{j=1}^{N} y_j^1 u_j^0, \qquad (5.158)$$

for $\mathbf{u}_h \in \mathscr{C}_h(\gamma/h^2)$.

Then, similarly as in Theorem 5.4.4, we have:

Theorem 5.6.6. *Let* $\gamma \in (0, 4)$ *and* $T > T(\gamma)$ *given by Theorem 5.5.3.*

For all $h > 0$ *system* (5.98) *is controllable in the sense of* (5.155) *(or, equivalently,* (5.156)).

More precisely, for any $(\mathbf{y}_h^0, \mathbf{y}_h^1) \in \mathbb{R}^N \times \mathbb{R}^N$, *there exists a control* $\mathbf{V}_h \in L^2(0, T; dt/\eta)$ *such that the solution of* (5.98) *satisfies* (5.155).

Moreover, the control $\mathbf{V}_h$ *of minimal* $L^2(0, T; dt/\eta)$-*norm fulfilling* (5.155) *can be characterized through the minimization (over* $\mathscr{C}_h(\gamma/h^2)$) *of the functional* J_h *in* (5.158) *as*

$$\mathbf{V}_h(t) = -\eta(t) \frac{U_N(t)}{h}, \qquad (5.159)$$

where $\mathbf{U}_h$ *is the minimizer of* J_h *in* (5.158) *over* $\mathscr{C}_h(\gamma/h^2)$.

Here, the difference with the situation in Theorem 5.4.4 is that discrete systems are observable within the space $\mathscr{C}_h(\gamma/h^2)$, uniformly with respect to the discretization parameter $h > 0$. This allows to deduce that the discrete controls $\mathbf{V}_h$ given by Theorem 5.6.6 are bounded.

One should then prove that the Hypotheses #1 and #2 hold in this case, to obtain a convergence result. In this case, they take the following form:

Lemma 5.6.7 ([32, 53]). *Let* $(u^0, u^1) \in C_0^\infty(0, 1)^2$ *and u be the corresponding solution of* (5.39). *Then there exists a sequence of functions* $\mathbf{u}_h \in \mathscr{C}_h(\gamma/h^2)$ *such that*

$$\left(\rho_h \mathbf{u}_h(0), \rho_h \mathbf{u}_h'(0)\right) \xrightarrow[h \to 0]{} (u(0), u'(0)) \quad \text{in } H_0^1(0, 1) \times L^2(0, 1) \quad (5.160)$$

$$-\frac{u_{N,h}}{h} \xrightarrow[h \to 0]{} \partial_x u(1, t) \quad \text{in } L^2(0, T), \quad (5.161)$$

where ρ_h *is the continuous extension of the discrete function* $\mathbf{u}_h$ *by Fourier series.*

In other words, Hypothesis #1 is satisfied in this case. Corresponding to Hypothesis #2, we have:

Lemma 5.6.8 ([36, 53]). *Let* $(\mathbf{u}_h^0, \mathbf{u}_h^1)$ *be discrete functions and* $(u^0, u^1) \in H_0^1(0, 1) \times L^2(0, 1)$ *such that*

$$(\rho_h \mathbf{u}_h^0, \rho_h \mathbf{u}_h^1) \xrightarrow[h \to 0]{} (u^0, u^1) \quad \text{in } H_0^1(0, 1) \times L^2(0, 1) \quad (5.162)$$

and

$$\sup_h \left\| \frac{u_{N,h}(t)}{h} \right\|_{L^2(0,T)} < \infty. \quad (5.163)$$

Then, denoting by $\mathbf{u}_h$ *and u respectively the solutions of* (5.81) *and* (5.39) *with initial data* $(\mathbf{u}_h^0, \mathbf{u}_h^1)$ *and* (u^0, u^1) *respectively, we have*

$$(\rho_h \mathbf{u}_h, \rho_h \mathbf{u}_h') \rightharpoonup_{h \to 0} (u, u') \quad \text{in } L^2(0, T; H_0^1(0, 1) \times L^2(0, 1)) \quad (5.164)$$

$$-\frac{u_{N,h}}{h} \rightharpoonup_{h \to 0} \partial_x u(1, t) \quad \text{in } L^2(0, T) \quad (5.165)$$

$$(\rho_h \mathbf{u}_h^0, \rho_h \mathbf{u}_h^1) \rightharpoonup_{h \to 0} (u^0, u^1) \quad \text{in } H_0^1(0, 1) \times L^2(0, 1). \quad (5.166)$$

Here, again, ρ_h *denotes the continuous extension operator of discrete functions by Fourier series.*

In other words, Hypothesis #2 is satisfied in this case.

Note that, due to the multiplier identity (5.110), one easily checks that (5.163) is a consequence of (5.162). Indeed, weakly convergent sequences are bounded, and (5.110) immediately yields an uniform admissibility result for the discrete wave equation (5.81).

We refer to [32, 36, 53] for the proof of Lemmas 5.6.7–5.6.8.

Accordingly, based on the convergence result in Theorem 5.6.2, we get

Theorem 5.6.9. *Within the setting of Theorem 5.6.6, given* $(y^0, y^1) \in L^2(0,1) \times H^{-1}(0,1)$ *and a sequence of discrete initial data* $(\mathbf{y}_h^0, \mathbf{y}_h^1)$ *such that* $(\rho_h \mathbf{y}_h^0, \rho_h \mathbf{y}_h^1)$ *weakly converges to* (y^0, y^1) *in* $L^2(0,1) \times H^{-1}(0,1)$, *the discrete controls* $\mathbf{V}_h$ *provided by Theorem 5.6.6 weakly converge in* $L^2(0,T; dt/\eta)$ *to* V, *the control provided by Theorem 5.3.6, as* $h \to 0$.

Besides, if the discrete initial data $(\mathbf{y}_h^0, \mathbf{y}_h^1)$ *are such that* $(\rho_h \mathbf{y}_h^0, \rho_h \mathbf{y}_h^1)$ *strongly converges to* (y^0, y^1) *in* $L^2(0,1) \times H^{-1}(0,1)$, *then the discrete controls* $\mathbf{V}_h$ *strongly converge to* V *in* $L^2(0,T; dt/\eta)$ *as* $h \to 0$.

It is then natural to address the issue of the convergence rates for the discrete controls $\mathbf{V}_h$ given by Theorem 5.6.6. For this to be done, as we have said, it is sufficient to derive the order of convergence for the discrete wave equation, and, more precisely, to check that Hypotheses #3 and #4 hold.

The following result is proved in [32]:

Proposition 5.6.10 ([32]). *Let* $(u^0, u^1) \in H^2 \cap H_0^1(0,1) \times H_0^1(0,1)$. *Then there exists a constant* $C = C(T)$ *independent of* (u^0, u^1) *and a sequence* $(\mathbf{u}_h^0, \mathbf{u}_h^1) \in \mathscr{C}_h(1/h^{4/3})$ *of initial data such that for all* $h > 0$,

$$\left\| (\rho_h \mathbf{u}_h^0, \rho_h \mathbf{u}_h^1) - (u^0, u^1) \right\|_{H_0^1 \times L^2} \le C h^{2/3} \left\| (u^0, u^1) \right\|_{H^2 \cap H_0^1 \times H_0^1} \tag{5.167}$$

and the solutions u *of* (5.39) *with initial data* (u^0, u^1) *and* $\mathbf{u}_h$ *of* (5.81) *with initial data* $(\mathbf{u}_h^0, \mathbf{u}_h^1)$ *satisfy, for all* $h > 0$,

$$\sup_{t \in [0,T]} \left\| (\rho_h \mathbf{u}_h(t), \rho_h \mathbf{u}_h'(t)) - (u(t), u'(t)) \right\|_{H_0^1 \times L^2}$$

$$\le C h^{2/3} \left\| (u^0, u^1) \right\|_{H^2 \cap H_0^1 \times H_0^1}, \tag{5.168}$$

$$\left\| \frac{u_{N,h}(\cdot)}{h} + u_x(1, \cdot) \right\|_{L^2(0,T)} \le C h^{2/3} \left\| (u^0, u^1) \right\|_{H^2 \cap H_0^1 \times H_0^1}, \quad h > 0. \tag{5.169}$$

Moreover,

$$\sup_{t \in [0,T]} \left\| (\rho_h \mathbf{u}_h(t), \rho_h \mathbf{u}_h'(t)) \right\|_{H^2 \cap H_0^1 \times H_0^1} \le C \left\| (u^0, u^1) \right\|_{H^2 \cap H_0^1 \times H_0^1}, \tag{5.170}$$

$$\left\| \frac{u_{N,h}(\cdot)}{h} \right\|_{H^1(0,T)} \le C \left\| (u^0, u^1) \right\|_{H^2 \cap H_0^1 \times H_0^1}, \quad h > 0. \tag{5.171}$$

Note that Proposition 5.6.10 is proved by taking the Fourier series decomposition of the continuous solution u of (5.39) and truncating it at the best order, which turns out to be $\lambda_h^k \simeq 1/h^{4/3}$. This might be surprising since it introduces powers of the form $h^{2/3}$ for the rate of convergence of the numerical scheme. But, actually, this strategy is optimal, as explained in [91]. This is due to the fact that

$$\sqrt{\lambda_h^k} - k\pi = \frac{2}{h} \sin\left(\frac{k\pi h}{2}\right) - k\pi \simeq \frac{\pi^3}{24} k^3 h^2,$$

which is small within the range of k such that $k \lesssim h^{-2/3}$, hence corresponding to $\lambda_h^k \lesssim h^{-4/3}$.

Also note that ρ_h denotes the Fourier extension of the discrete solutions. Hence it is smooth and one can take the $H^2(0, 1)$ norms of these continuous approximations as required in the statement above.

Finally, let us emphasize that Proposition 5.6.10 is well-known except for what concerns the convergence of the normal derivatives on the boundary. In particular, our approach strongly uses the uniform hidden regularity property given by the multiplier identity (5.110).

Once this is done, we are in position to state the following counterpart of Hypothesis #4:

Theorem 5.6.11 ([32]). *Let* $(y^0, y^1) \in H_0^1(0, 1) \times L^2(0, 1)$ *and* $v \in H_0^1(0, T)$ *and denote by* y *the corresponding solution of* (5.42).

Consider a sequence of initial data $(\mathbf{y}_h^0, \mathbf{y}_h^1)$ *and control functions* $\mathbf{v}_h \in L^2(0, T)$ *and denote by* $\mathbf{y}_h$ *the corresponding solution of* (5.98). *Then there exists a positive constant* C *independent of* $h > 0$ *such that*

$$\| (\rho_h y_h(T), \rho_h y_h'(T)) - (y(T), y'(T)) \|_{L^2 \times H^{-1}}$$

$$\leq C h^{2/3} \left\{ \| (y^0, y^1) \|_{H_0^1 \times L^2} + \| v \|_{H_0^1(0,T)} \right\}$$

$$+ \| (\rho_h \mathbf{y}_h^0, \rho_h \mathbf{y}_h^1) - (y^0, y^1) \|_{L^2 \times H^{-1}} + C \| \mathbf{v}_h - v \|_{L^2(0,T)}. \quad (5.172)$$

The details of the proof of Theorem 5.6.11 will be given in [32].

This is slightly more subtle than Proposition 5.6.10 at least for two reasons:

- To give a precise definition of the solution of the wave equation with initial data in $L^2(0, 1) \times H^{-1}(0, 1)$ with a boundary data $v \in L^2(0, T)$, one needs to introduce the concept of solutions in the sense of transposition, i.e. based on the duality with solutions u of equations similar to (5.39) lying in the energy space $H_0^1(0, 1) \times L^2(0, 1)$, and to use hidden regularity results that show that $u_x(1, t) \in L^2(0, T)$, see [68].
- One should then use the explicit convergence results stated in Proposition 5.6.10, and in particular the one on the normal derivative (5.169).

Then, using Proposition 5.6.10 and Theorems 5.6.11 and 5.6.3, we get:

Theorem 5.6.12. *Let* $(y^0, y^1) \in H_0^1(0, 1) \times L^2(0, 1)$ *and consider a sequence of discrete initial data* $(\mathbf{y}_h^0, \mathbf{y}_h^1)$ *such that* $(\rho_h \mathbf{y}_h^0, \rho_h \mathbf{y}_h^1)$ *strongly converges to* (y^0, y^1) *in* $L^2(0, 1) \times H^{-1}(0, 1)$.

Let $\gamma \in (0, 4)$ and $T > T(\gamma)$. Then the controls $\mathbf{V}_h$ given by Theorem 5.6.6 strongly converge to V in $L^2(0, T; dt/\eta)$, where V is the control given by Theorem 5.3.6 corresponding to (y^0, y^1).

Besides, there exists a constant C such that for all $h > 0$,

$$\| \mathbf{V}_h - V \|_{L^2(0,T;dt/\eta)} \leq C h^{2/3} \| (y^0, y^1) \|_{H_0^1 \times L^2}$$
$$+ C \| (\rho_h \mathbf{y}_h^0, \rho_h \mathbf{y}_h^1) - (y^0, y^1) \|_{L^2 \times H^{-1}} . \quad (5.173)$$

In particular, choosing $(\mathbf{y}_h^0, \mathbf{y}_h^1)$ such that for some C independent of $h > 0$,

$$\left\| (\rho_h \mathbf{y}_h^0, \rho_h \mathbf{y}_h^1) - (y^0, y^1) \right\|_{L^2 \times H^{-1}} \leq C h^{2/3} \left\| (y^0, y^1) \right\|_{H_0^1 \times L^2}, \quad (5.174)$$

one immediately gets

$$\| \mathbf{V}_h - V \|_{L^2(0,T;dt/\eta)} \leq C h^{2/3} \left\| (y^0, y^1) \right\|_{H_0^1 \times L^2} . \quad (5.175)$$

To our knowledge, this is the first result on the order of convergence for the discrete controls obtained in Theorem 5.6.6.

Let us also emphasize that the convergence results stated in (5.174) are satisfied when taking as discrete initial data the restriction to the mesh points of the orthogonal projections in $L^2(0, 1)$ or $H_0^1(0, 1)$ on the vector space spanned by the functions $(w^k(x) = \sin(k\pi x))_{1 \leq k \leq N}$. Of course, other interpolation operators can be considered for which assumption (5.174) is satisfied.

Remark 5.6.13. The observability results in classes of filtered solutions stated in Sect. 5.5.4 and obtained in [111] for the semidiscrete finite-difference approximations of the multi-dimensional wave equation, also yield similar convergence estimates with proofs that follow line to line those above. We do not write down the details here for the sake of conciseness.

The results stated in Theorem 5.5.8 [27] do not apply in the context of boundary controllability, but rather when the control is distributed inside the domain. In that case one does not need to use transposition methods since solutions are defined in a classical manner and this can be done by standard energy and semigroup methods (see Theorem 5.3.5). Consequently, the needed convergence results are more classical. But still, to our knowledge, a rigorous proof of the fact that Hypothesis #3 holds in that case is still missing.

Of course, despite of this, Hypotheses #1 and #2 hold and follow from classical convergence results for the finite element methods, see [4]. Therefore, one can prove the counterpart of Theorem 5.6.9 in that case, see [27] for details.

5.6.2.2 The Bi-grid Technique

The methods above can also be used to obtain convergence results and convergence rates for the two-grid filtering technique.

In this case $\mathfrak{C}_h = \mathscr{V}_h$, where $\mathscr{V}_h$ is given by (5.113). We are then precisely in the same setting as the one in Sect. 5.6.1.

Based on the observability result stated in Theorem 5.5.6, using Theorem 5.6.1 we obtain:

Theorem 5.6.14. *Let $T > 2\sqrt{2}$ and η be as in (5.157) with $T(\gamma)$ replaced by $2\sqrt{2}$. Let $(\mathbf{y}_h^0, \mathbf{y}_h^1)$ be discrete initial data.*

Then introduce the functional J_h defined as in (5.158) for $\mathbf{u}_h$ solution of (5.81) such that $(\mathbf{u}_h(T), \mathbf{u}_h'(T)) \in \mathscr{V}_h$. This functional has a unique minimizer $\mathbf{U}_h$ solution of (5.81) with $(\mathbf{U}_h(T), \mathbf{U}_h'(T)) \in \mathscr{V}_h$, among the space of solutions $\mathbf{u}_h$ such that $(\mathbf{u}_h(T), \mathbf{u}_h'(T)) \in \mathscr{V}_h$.

Then $\mathbf{V}_h$ defined as in (5.159) is a control function for which the solution $\mathbf{y}_h$ of (5.98) satisfies

$$\forall (\mathbf{u}_h^{0,T}, \mathbf{u}_h^{1,T}) \in \mathscr{V}_h, \quad h \sum_{j=1}^N y_j(T) u_j^{1,T} - h \sum_{j=1}^N y_j'(T) u_j^{0,T} = 0. \qquad (5.176)$$

Moreover, $\mathbf{V}_h$ is the control of minimal $L^2(0,T;dt/\eta)$ norm for which the corresponding solution of (5.98) satisfies the control requirement (5.176). It is also the only control satisfying (5.176) that can be written as in (5.159) for a solution $\mathbf{u}_h$ of (5.81) with $(\mathbf{u}_h(T), \mathbf{u}_h'(T)) \in \mathscr{V}_h$.

Now, using Theorem 5.6.2, Lemma 5.6.8 and an easy variant of Lemma 5.6.7 left to the reader, one can then prove the following:

Theorem 5.6.15 ([36]). *Within the setting of Theorem 5.6.14, given $(y^0, y^1) \in L^2(0,1) \times H^{-1}(0,1)$ and a sequence of discrete initial data $(\mathbf{y}_h^0, \mathbf{y}_h^1)$ such that $(\rho_h \mathbf{y}_h^0, \rho_h \mathbf{y}_h^1)$ weakly converges to (y^0, y^1) in $L^2(0,1) \times H^{-1}(0,1)$, the discrete controls $\mathbf{V}_h$ provided by Theorem 5.6.14 weakly converge in $L^2(0,T;dt/\eta)$ to V, the control provided by Theorem 5.3.6, as $h \to 0$.*

Besides, if the discrete initial data $(\mathbf{y}_h^0, \mathbf{y}_h^1)$ are such that $(\rho_h \mathbf{y}_h^0, \rho_h \mathbf{y}_h^1)$ strongly converge to (y^0, y^1) in $L^2(0,1) \times H^{-1}(0,1)$, the discrete controls $\mathbf{V}_h$ strongly converge to V in $L^2(0,T;dt/\eta)$ as $h \to 0$.

To go further, one should then prove a variant of Proposition 5.6.10 for the solutions $\mathbf{u}_h$ of the discrete wave equation (5.81) such that $(\mathbf{u}_h(T), \mathbf{u}_h'(T)) \in \mathscr{V}_h$. One way of doing that is to take the discrete solutions given by (5.6.10), which belong to $\mathscr{C}_h(1/h^{4/3})$ and to add to them high-frequency components so that $(\mathbf{u}_h(T), \mathbf{u}_h'(T)) \in \mathscr{V}_h$. Doing this, one can check that the high-frequency components that have been added that way are small and do not modify the estimates in Proposition 5.6.10.

Note that, of course, these approximations will not belong anymore to $\mathscr{C}_h(1/h^{4/3})$ but it does not matter for our purpose.

Then, using Theorem 5.6.2, Theorem 5.6.11 and this slightly modified variant of Proposition 5.6.10 where we further imposed on the discrete data the condition

$(\mathbf{u}_h(T), \mathbf{u}'_h(T)) \in \mathscr{V}_h$, one can obtain convergence rates for the convergence of the discrete controls:

Theorem 5.6.16 ([36]). *Let* $(y^0, y^1) \in H_0^1(0,1) \times L^2(0,1)$ *and consider a sequence of discrete initial data* $(\mathbf{y}_h^0, \mathbf{y}_h^1)$ *such that* $(\rho_h \mathbf{y}_h^0, \rho_h \mathbf{y}_h^1)$ *strongly converges to* (y^0, y^1) *in* $L^2(0,1) \times H^{-1}(0,1)$.

Let $T > 2\sqrt{2}$. *Then the controls* $\mathbf{V}_h$ *given by Theorem 5.6.14 strongly converge to* V *in* $L^2(0,T; dt/\eta)$, *where* V *is the control given by Theorem 5.3.6 corresponding to* (y^0, y^1).

Besides, there exists a constant C *such that for all* $h > 0$, *estimate* (5.173) *holds. In particular, choosing* $(\mathbf{y}_h^0, \mathbf{y}_h^1)$ *such that for some* C *independent of* $h > 0$, (5.174) *is satisfied, one immediately gets* (5.175).

The proof can be found in [36] but, again, it follows the general theory developed in Sect. 5.6.1.

5.6.2.3 Tychonoff Regularization

The Tychonoff regularization is of slightly different nature since, in agreement with Theorem 5.5.7, one has to reinforce the observation operator by adding an extra observation, distributed everywhere in the discrete grid, so that observability holds uniformly on the mesh-size parameter for all solutions. In view of this, the applied control mechanism has to be reinforced as well, adding an extra control distributed everywhere in the domain. However, this added control will vanish as $h \to 0$ and the methods of Sect. 5.6.1 will apply to show the convergence towards the limit control of the leading term. There are however some minor modifications to be introduced with respect to the abstract functional setting provided in Sect. 5.6.1 that we describe below.

Let η be as in (5.157) with $T(\gamma)$ replaced by 2.

First, we introduce the functional $\hat{J}_h$ defined for $(\mathbf{u}_h^0, \mathbf{u}_h^1) \in \mathbb{R}^N \times \mathbb{R}^N$ by:

$$\hat{J}_h(\mathbf{u}_h^0, \mathbf{u}_h^1) = \frac{1}{2} \int_0^T \eta(t) \left| \frac{u_N(t)}{h} \right|^2 dt + \frac{h^3}{4} \sum_{j=0}^N \int_0^T \eta(t) \left| \frac{u'_{j+1} - u'_j}{h} \right|^2 dt$$

$$+ h \sum_{j=1}^N y_j^0 u_j^1 - h \sum_{j=1}^N y_j^1 u_j^0, \quad (5.177)$$

where $\mathbf{u}_h$ is the solution of the adjoint system (5.81) with initial datum $(\mathbf{u}_h^0, \mathbf{u}_h^1)$.

Using this functional and based on Theorem 5.5.7, we get the following:

Theorem 5.6.17. *Set* $T > 2$, *and consider an initial datum* $(\mathbf{y}_h^0, \mathbf{y}_h^1) \in \mathbb{R}^N \times \mathbb{R}^N$.

For each $h > 0$, *the functional* $\hat{J}_h$ *in* (5.177) *has a unique minimizer* $(\mathbf{U}_h^0, \mathbf{U}_h^1)$. *Then, setting*

$$\begin{cases} \mathbf{V}_h(t) = -\eta(t)\dfrac{U_N(t)}{h} \\ G_{j,h}(t) = \dfrac{\eta(t)}{2h^2}\left(U'_{j+1} - 2U'_j + U'_{j-1}\right), \quad j = 1,\ldots,N, \end{cases} \tag{5.178}$$

the solution $\mathbf{y}_h$ of

$$\begin{cases} y''_j - \dfrac{1}{h^2}\left[y_{j+1} + y_{j-1} - 2y_j\right] = h^2 G'_{j,h}, & 0 < t < T,\ j = 1,\ldots,N \\ y_0(t) = 0;\ y_{N+1}(t) = \mathbf{V}_h(t), & 0 < t < T \\ y_j(0) = y_j^0,\ y'_j(0) = y_j^1, & j = 1,\ldots,N, \end{cases} \tag{5.179}$$

satisfies the control requirement

$$(\mathbf{y}_h(T), \mathbf{y}'_h(T)) = (0,0). \tag{5.180}$$

Theorem 5.6.17 shows how the Tychonoff regularization modifies the control problem. It introduces a control everywhere in the domain, that weakly converges to zero. This is of course compatible with our analysis, which states the existence of high-frequency spurious solutions which do not propagate and therefore can not be controlled from the boundary. Therefore, if one wants to satisfy the strong control requirement (5.180), one needs to introduce a control everywhere in the domain. But this control can be built in such a way that it vanishes when $h \to 0$.

Note that Theorem 5.5.7 gives a lot more of information, and in particular the following one:

Proposition 5.6.18. *Under the assumptions of Theorem 5.6.17, there exists a constant $C(T)$ independent of $h > 0$ such that*

$$\|\mathbf{V}_h\|_{L^2(0,T)} + h\,\|\rho_h \mathbf{G}_h\|_{L^2(0,T;H^{-1}(0,1))} + h^2\,\|\rho_h \mathbf{G}_h\|_{L^\infty(0,T;L^2(0,1))}$$
$$\leq C(T)\,\left\|(\rho_h \mathbf{y}_h^0, \rho_h \mathbf{y}_h^1)\right\|_{L^2(0,1)\times H^{-1}(0,1)}. \tag{5.181}$$

We now state the following counterparts of Lemmas 5.6.7 and 5.6.8:

Lemma 5.6.19 ([36]). *In the setting of Lemma 5.6.7 (with $\gamma = 4$ so that no filtering is implemented), we further have*

$$h^2 \rho_h \Delta_h \mathbf{u}'_h \xrightarrow[h\to 0]{} 0 \quad \text{in } L^2((0,T)\times(0,1)). \tag{5.182}$$

Lemma 5.6.20 ([36]). *In the setting of Lemma 5.6.8, we further have*

$$h^2 \rho_h \Delta_h \mathbf{u}'_h \xrightharpoonup[h\to 0]{} 0 \quad \text{in } L^2((0,T)\times(0,1)). \tag{5.183}$$

Based on Proposition 5.6.18, Lemmas 5.6.19–5.6.20 and using the same ideas as in Theorem 5.6.2, one gets the following:

Theorem 5.6.21 ([36]). *Within the setting of Theorem 5.6.17, given* $(y^0, y^1) \in L^2(0, 1) \times H^{-1}(0, 1)$ *and a sequence of discrete initial data* $(\mathbf{y}_h^0, \mathbf{y}_h^1)$ *such that* $(\rho_h \mathbf{y}_h^0, \rho_h \mathbf{y}_h^1)$ *weakly converges to* (y^0, y^1) *in* $L^2(0, 1) \times H^{-1}(0, 1)$, *then the discrete controls* $(\mathbf{V}_h, \mathbf{G}_h)$ *provided by Theorem 5.6.17 weakly converge in the following sense:*

$$
\begin{aligned}
\mathbf{V}_h \underset{h \to 0}{\rightharpoonup} V, &\quad \text{in } L^2(0, T; dt/\eta), \\
h^2 \rho_h \mathbf{G}_h \underset{h \to 0}{\rightharpoonup} 0, &\quad \text{in } L^2((0, T) \times (0, 1)),
\end{aligned}
\tag{5.184}
$$

where V is the control provided by Theorem 5.3.6, as $h \to 0$.

Besides, if the discrete initial data $(\mathbf{y}_h^0, \mathbf{y}_h^1)$ *are such that* $(\rho_h \mathbf{y}_h^0, \rho_h \mathbf{y}_h^1)$ *strongly converges to* (y^0, y^1) *in* $L^2(0, 1) \times H^{-1}(0, 1)$, *then the discrete controls* $(\mathbf{V}_h, h^2 \rho_h \mathbf{G}_h)$ *strongly converge to* $(V, 0)$ *in* $L^2(0, T; dt/\eta) \times L^2((0, T) \times (0, 1))$ *as* $h \to 0$.

One can even follow the proof of Theorem 5.6.3 to obtain convergence rates for the discrete controls. For doing that, inspecting the proof of Theorem 5.6.11, we need the following for the convergence of the equations of (5.81) to (5.39):

Proposition 5.6.22 ([36]). *In the setting of Lemma 5.6.8, we further have*

$$
\sup_t \left\| h \rho_h \Delta_h \mathbf{u}_h'(t) \right\|_{L^2(0,1)} \leq C h \left\| (u^0, u^1) \right\|_{H^2 \cap H_0^1 \times H_0^1}.
\tag{5.185}
$$

The proof of this additional estimate is easy: Basically, it uses that $h \nabla_h$ are uniformly bounded with norm smaller than 2, and then

$$
\left\| h \rho_h \Delta_h \mathbf{u}_h' \right\|_{L^2(0,1)} \lesssim \left\| \rho_h \mathbf{u}_h' \right\|_{H_0^1} \lesssim \left\| (\rho_h \mathbf{u}_h^0, \rho_h \mathbf{u}_h^1) \right\|_{H^2 \cap H_0^1 \times H_0^1}.
$$

We also need to be able to give an estimate on the controlled equation, which is mainly the one in Theorem 5.6.11 except that an internal control in $H^{-1}(0, T; L^2(0, 1))$ has been added. When the distributed source terms are in $L^2(0, T; L^2(0, 1))$ convergence results in the energy space are classical and can be found, for instance, in [4]. One can easily deal with source terms in $H^{-1}(0, T; L^2(0, 1))$ integrating the equations in time, and working in the space $L^2(0, 1) \times H^{-1}(0, 1)$.

Hence we can derive the following result:

Theorem 5.6.23 ([36]). *Within the setting of Theorem 5.6.21, let* $(y^0, y^1) \in H_0^1(0, 1) \times L^2(0, 1)$ *and consider a sequence of discrete initial data* $(\mathbf{y}_h^0, \mathbf{y}_h^1)$ *such that* $(\rho_h \mathbf{y}_h^0, \rho_h \mathbf{y}_h^1)$ *strongly converge to* (y^0, y^1) *in* $L^2(0, 1) \times H^{-1}(0, 1)$.

Let $T > 2$. *Then the controls* $(\mathbf{V}_h, h^2 \mathbf{G}_h)$ *given by Theorem 5.6.17 strongly converge to* $(V, 0)$ *in* $L^2(0, T; dt/\eta) \times L^2((0, T) \times (0, 1))$, *where V is the control given by Theorem 5.3.6 corresponding to* (y^0, y^1).

Besides, there exists a constant C such that for all h > 0,

$$\| \mathbf{V}_h - V \|_{L^2(0,T;dt/\eta)} + \| h^2 \rho_h \mathbf{G}_h \|_{L^2((0,T)\times(0,1))}$$

$$\leq Ch^{2/3} \| (y^0, y^1) \|_{H_0^1 \times L^2} + \| (\rho_h \mathbf{y}_h^0, \rho_h \mathbf{y}_h^1) - (y^0, y^1) \|_{L^2 \times H^{-1}} . \quad (5.186)$$

In particular, if (5.174) is satisfied, we get

$$\|\mathbf{V}_h - V \|_{L^2(0,T;dt/\eta)} + \|h^2 \rho_h \mathbf{G}_h \|_{L^2((0,T)\times(0,1))} \leq Ch^{2/3} \|(y^0, y^1)\|_{H_0^1 \times L^2} .$$
$$(5.187)$$

The precise proofs will be given in [36], but here again, they rely on the same ideas as for Theorem 5.6.3. Indeed it consists in using that the minimizer (U^0, U^1) of the continuous HUM functional is smooth. Therefore, one can approximate it with a known error term by a discrete solution $(\tilde{\mathbf{U}}_h^0, \tilde{\mathbf{U}}_h^1)$ of (5.81), which corresponds to some approximate controls $(\tilde{\mathbf{v}}_h, h^2 \tilde{\mathbf{g}}_h)$ defined by (5.178) with $\tilde{\mathbf{U}}_h$ instead of $\mathbf{U}_h$. One should then correct this error, and this can be done with small controls using the observability result in Proposition 5.6.18. We finally conclude by the uniqueness of controls $(\mathbf{v}_h, \mathbf{g}_h)$ that can be written as (5.178) for some solution $\mathbf{u}_h$ of (5.81).

5.6.3 Numerical Experiments

In this section, our goal is to illustrate the convergence results proven above. We focus on the study of the filtering method, the others being very similar.

We first consider the case in which the initial datum to be controlled lies in $L^2(0, 1) \times H^{-1}(0, 1)$: $y^0(x) = x^2$ for $x \in (0, 1/2)$, $y^0(x) = -(1 - x)^2$ for $x \in (1/2, 1)$ and $y^1 \equiv 0$ (see Fig. 5.11).

We then represent in Fig. 5.12 the control functions for various choices of N. Note that here, due to the weight function in time, the explicit expression of the control that is given through the minimization of the functional J in (5.76) is not available anymore.

Here, the wave equation is discretized in time, with a CFL condition $\Delta t = 0.5h$. The filtering parameter is taken to be $\gamma = 1$. The function η is chosen such that: $\eta = 1$ for $t \in (0.4, 3.6)$. On $t \in (0, 0.4)$, $\eta(t)$ is a polynomial of order 3 so that $\eta(0) = \eta'(0) = \eta(1) = \eta'(1)$ and $\eta(0.4) = 1$, and we choose it in a similar way in $(3.6, 4)$. Of course, η is not C^∞ smooth but only C^1, but this would be enough for our purpose. With these choices, the time of control $T = 4$ suffices to control the fully discrete dynamics.

As one can see, the controls in Fig. 5.12 exhibit some kind of Gibbs phenomenon close to the discontinuities of the control.

Let us now present similar numerical results, but for an initial datum to be controlled in $H_0^1(0, 1) \times L^2(0, 1)$. Now, (y^0, y^1) are chosen such that: $y^0 = 0$ and

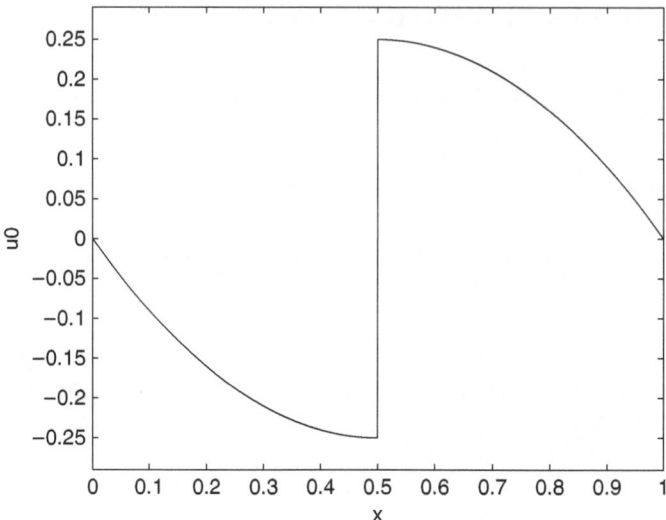

Fig. 5.11 The initial position y^0 to be controlled

y^1 is the discontinuous triangular function in Fig. 5.13. The analytic expression of y^1 is $y^1(x) = -x$ for $x \in (0, 1/2)$ and $y^1(x) = 1 - x$ in $(1/2, 1)$.

This corresponds to an initial datum to be controlled in $H_0^1(0, 1) \times L^2(0, 1)$. Therefore, we should expect better convergence properties as before.

We present in Fig. 5.14 the controls computed for that initial data and for several values of N. One can see that there, the controls in Fig. 5.14 seem to be smoother than the ones in Fig. 5.12. This is of course consistent with our analysis which states that:

- The smoothness of the continuous control corresponds to the smoothness of the initial datum to be controlled.
- The discrete controls converge towards the continuous one.

To conclude our analysis, we illustrate our results on the rate of convergence of the discrete controls. For that to be done, we take as reference control the one carefully computed for some large reference system size N_{ref}. Using this accurately computed control $V_{N_{ref}}$, we compute the norm of $V_N - V_{N_{ref}}$ for various $N \leq N_{ref}$. The rate of convergence of V_N towards $V_{N_{ref}}$ should give a realistic estimate of the convergence rate of the discrete controls towards the continuous one. In log–log scales, this yields Fig. 5.15.

The linear interpolations of the obtained curves have slope -1.04 when controlling $(0, y^1)$ with y^1 as in Fig. 5.13 and slope -0.34 when controlling $(y^0, 0)$ with y^0 as in Fig. 5.11.

The fact that, for $(0, y^1)$ with y^1 as in Fig. 5.13, the rate is much better than the expected rate $-2/3$ predicted by Theorem 5.6.12 comes from the fact that the initial

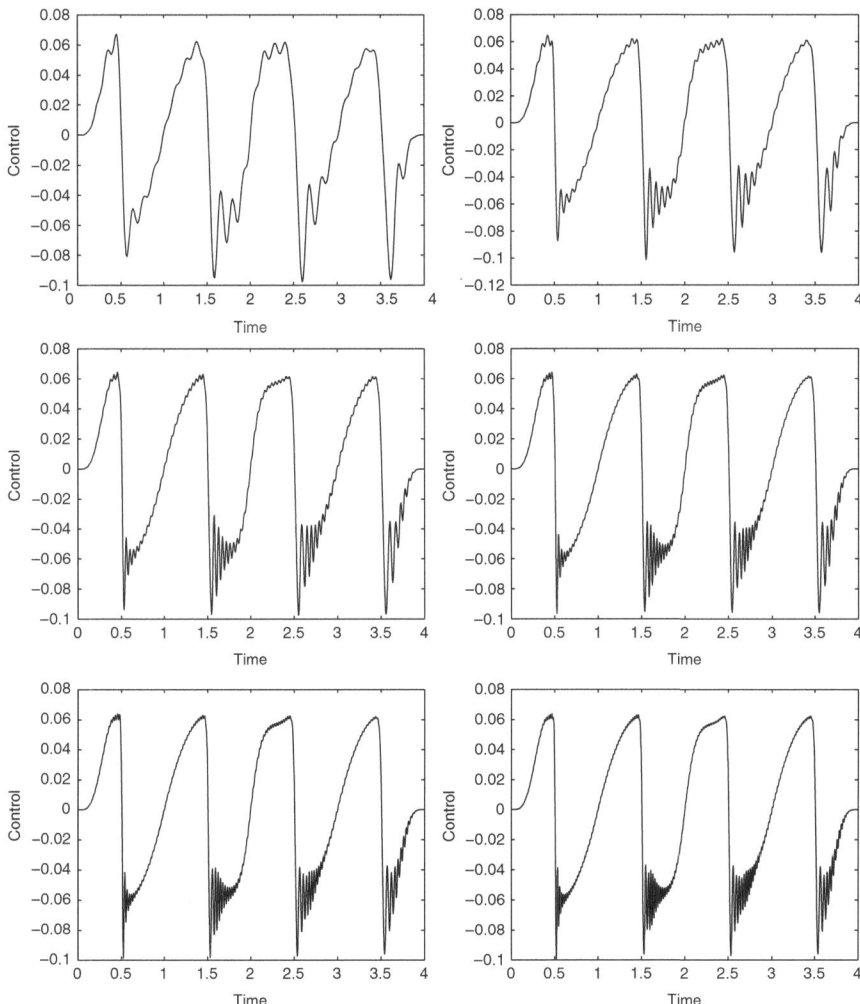

Fig. 5.12 Discrete controls computed for the initial datum $(y^0, 0)$ with y^0 as in Fig. 5.11, for different values of N, under the CFL condition $\Delta t = 0.5h$, in time $T = 4$ and with a filtering parameter $\gamma = 1$. From *left to right* and *top to bottom*: $N = 50$, $N = 100$, $N = 150$, $N = 200$, $N = 250$ and $N = 300$

datum to be controlled $(0, y^1)$, with y^1 as in Fig. 5.13, lies not only in $H_0^1(0, 1) \times L^2(0, 1)$ but in $H_0^s(0, 1) \times H_0^{s-1}(0, 1)$ for all $s < 3/2$. This gain of $1/2^-$ derivative with respect of the energy space explain the faster convergence rate as we shall explain below.

Similarly, $(y^0, 0)$ with y^0 as in Fig. 5.11, lies not only in $L^2(0, 1) \times H^{-1}(0, 1)$ but also in $H_0^s(0, 1) \times H_0^{s-1}(0, 1)$ for all $s < 1/2$, thus explaining why the controls seem to converge with a rate of the order of $1/3$.

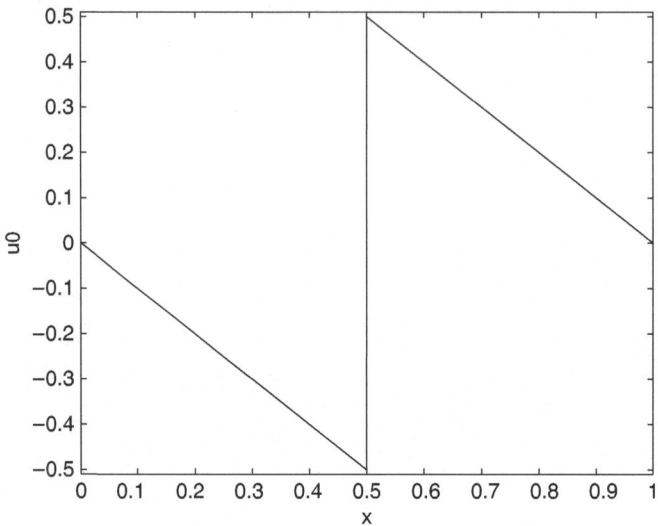

Fig. 5.13 The initial velocity y^1 to be controlled

In fact, the numerical approximations of the controls converge to that of the limit system with rates corresponding to the class of regularity of the initial data under consideration. Actually, following the proofs of [32, 36], if the initial data to be controlled lye in $H_0^s(0, 1) \times H_0^{s-1}(0, 1)$ for $s \in (0, 3/2)$ (above the value $s = 3/2$, more compatibility boundary conditions are required), the convergence rate is of the order of $h^{2s/3}$. This is completely consistent with the numerical simulations in Fig. 5.15 since the theory then predicts a convergence rate of order $h^{1/3}$ for $s = 1/2$ and of h for $s = 3/2$, to be compared with the rates $h^{0.34}$ and $h^{1.04}$ found in Fig. 5.15. For the proof of these more general convergence rates results it suffices, in fact, to prove the analogs of Theorems 5.6.11–5.6.12 in the spaces of the corresponding regularity and convergence rates.

5.7 Further Comments and Open Problems

5.7.1 Further Comments

1. *Time-discrete and fully discrete approximations.* In these notes, we have addressed the problem of the convergence of the controls for space semidiscrete approximations of the wave equation as the mesh-size goes to zero. But one can go further and discretize in time these space semidiscrete approximations to obtain fully discrete approximation schemes. This time-discretization adds further spurious high-frequency waves and, consequently, extra difficulties to the fulfillment of the

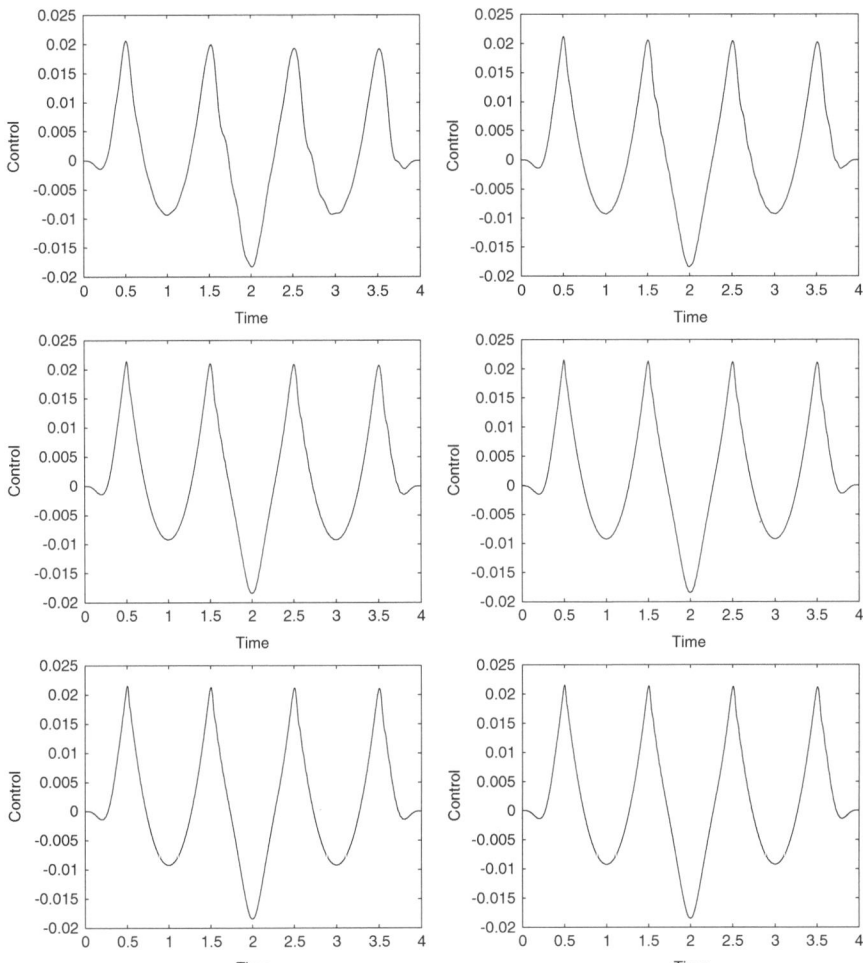

Fig. 5.14 Discrete controls computed for the initial datum $(0, y^1)$ with y^1 as in Fig. 5.13, for different values of N, under the CFL condition $\Delta t = 0.5\,\text{h}$, in time $T = 4$ and with a filtering parameter $\gamma = 1$. From *left to right* and *top to bottom*: $N = 50$, $N = 100$, $N = 150$, $N = 200$ $N = 250$ and $N = 300$

observability inequalities. This is so since the time-discretization process deforms the spectrum and the dispersion relation of the system.

This added numerical dispersion effect has been studied more precisely in [31] for abstract conservative systems (see also [107] for a study of a time discrete and space continuous wave equation) using resolvent type estimates [16, 79, 88]. The interest of the method developed there is that it completely decouples the effects of the space discretization process from the ones originating from the time discretization. Again, the main results can be stated as follows: removing high-

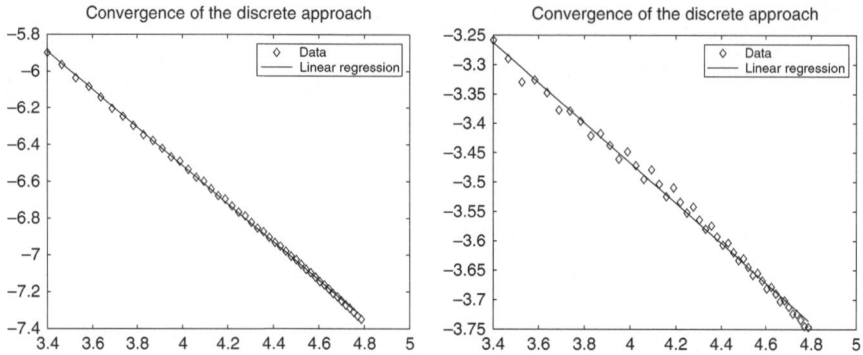

Fig. 5.15 Graph of $\log(\|V_N - V_{300}\|)$ as a function of $\log(N)$ for $N \in (30, 120)$: *left*, for the discrete controls computed for the initial datum $(0, y^1)$ with y^1 as in Fig. 5.13, the linear interpolant has slope -1.04; *right*, for the discrete controls computed for the initial datum $(y^0, 0)$ with y^0 as in Fig. 5.11, the linear interpolant has slope -0.34

frequency solutions, one can get uniform observability properties, where, here, uniformity is referred to space and time discretization parameters. Spurious waves appear at frequencies of the order of $1/(\Delta t)$, where Δt is the time discretization parameter [31]. On the other hand, the added filtering that the time-discretization processes require can be avoided through suitable CFL type conditions on the space and time discretization parameters. These results are sharp, as it has been shown explicitly in [107].

However, the results in [31] do not provide any precise estimate on the time needed to guarantee the uniform observability inequality. This is a drawback of the method developed in [31], which is based on resolvent estimates.

To overcome this drawback, more recently in [36], we have developed a discrete transmutation technique, inspired on previous works, in particular by Miller [79,80], which establishes a connection between solutions of the time continuous systems and the time-discrete ones. This approach yields explicit estimates on the time needed to guarantee uniform observability results.

The approach developed in Sect. 5.6.1 also applies in the context of fully discrete schemes and also yields convergence results for the corresponding discrete controls with explicit convergence rates based on the existing results on the convergence of the fully discrete systems towards the continuous one.

2. *Other space discretization methods.* In these notes, we have mainly considered the 1-d wave equation discretized using finite differences and we have proved that their observability and controllability properties fail to be uniform as the mesh-size parameters tend to zero. This turns out to occur for most numerical methods. In particular, this is also the case for the finite element method, see [53], among others.

However, there are some schemes that enjoy uniform observability properties, but they seem to be very rare. This is the case for instance for the mixed finite element method [17, 18, 28, 41]. For these schemes observability and controllability properties are uniform, without any need of filtering, and the discrete controls converge towards the continuous ones. But this discretization method has an important drawback: Its CFL type condition for stability has the form $\Delta t \leq h^2$, where Δt is the time discretization parameter. This is in contrast with the above methods which only require $\Delta t \leq h$.

3. *Stabilization and discretization.* As already noticed in [96], the theory of stabilization and observation/control are strongly linked.

This connection has been made even more precise in [46], showing that the damped wave system

$$
\begin{cases}
z_{tt} - \Delta z + \chi_\omega z_t = 0 & \text{in} \quad Q, \\
z = 0 & \text{on} \quad \Sigma, \\
z(x,0) = z^0(x), z_t(x,0) = z^1(x) & \text{in} \quad \Omega.
\end{cases}
\tag{5.188}
$$

is exponentially stable, in the sense that there exist a constant C and a strictly positive constant $\mu > 0$ such that for all initial data $(z^0, z^1) \in H_0^1(\Omega) \times L^2(\Omega)$ and for all $t \geq 0$,

$$
E(t) \leq C e^{-\mu t} E(0),
$$

if and only if the wave system (5.64) is observable through ω.

This result can be easily extended to an abstract framework, provided the damping and control operators are bounded.

In the context of stabilization of waves one often considers boundary damping operators. They turn out to be unbounded perturbations of the conservative semi-group and, therefore, the equivalence of stabilization of the damped system and the observation of the conservative one does not apply. We refer to [1] for results in that direction.

Going back to the problem of stabilization by distributed damping as above, and in what concerns the numerical approximation issues, our understanding of the lack of observability for space semidiscrete systems (and fully discrete ones, see Comment #1 above) suggests that similar pathologies may arise making the decay properties of the corresponding semidiscrete or fully discrete systems not to be uniform. That is indeed the case. As a by byproduct of the lack of uniform observability for (5.64), the apparently most natural discretizations of (5.188) are not exponentially uniformly stable, see e.g. [81, 90, 99]. Again, this is due to high-frequency phenomena and spurious solutions coming from the numerical schemes under consideration. One shall then add a numerical viscosity term everywhere in the domain to damp out efficiently these spurious waves. This is the idea that has been developed in [99] for the 1 and 2-d wave equation and then later formalized in a much more general form in [33, 34].

The possible use of two-grid filtering techniques to ensure uniform decay properties is an interesting subject that requires further analysis. Of course, one of the main difficulties is related to the fact that the property of being of two-grid form is not preserved along the dissipative dynamics.

4. *Other models.* Let us also point out that many control results exist for other type of models, such as Schrödinger equation [67], beam equation [89], where similar ideas as the one presented above can be applied, even if of course, each case presents some specificity and should be handled carefully.

The convergence properties of controls for discrete heat equations has also been developed lately in [8–10, 30, 61, 70]. The later works [8, 9] are based on Carleman estimates for discrete elliptic operators, which require important technical developments.

5.7.2   Open Problems

Problem 5.7.1. Semilinear Wave Equations. We have studied the convergence of the discrete controls for linear wave equations, and we have described the difficulties encountered because of the spurious high-frequency solutions and how to remedy them.

Of course, the same questions arise in the context of semilinear wave equations, even with globally Lipschitz nonlinearities, a case that has been handled for instance in [113]. Most often the nonlinear problems are addressed by means of a fixed point argument together with a careful analysis of the control properties of the linearized system. One of the main difficulties that appears when doing that is to estimate the dependence of the observability constants on the (t, x)-depending potentials of the linearized equation. This can be handled using sidewise energy estimates (but this works only in 1-d), multipliers or Carleman estimates [26, 37, 109, 110], thus yielding various constraints on the growth of the non-linearity at infinity for the controllability property to hold. This kind of results guarantees the controllability of the nonlinear system for all initial data in an uniform time.

But one can relax the control problem, analyzing it locally, for small data. Local results, together with exponential convergence ones obtained by means of suitable damping mechanisms, allow showing that, eventually, every initial data can be controlled to zero but on a time that depends on the size of the initial data and that may tend to infinity when the norm of the data tend to infinity. Local results can be proved for nonlinearities growing at infinity in a superlinear manner. When using energy methods, however, one needs to impose growth conditions at infinity. More recently, using dispersive estimates (see [24, 25]), the class of nonlinearities for which this kind of results holds has been extended to cover the range of nonlinearities that can be handled for the well-posedness of the Cauchy problem in the energy space by means of Strichartz inequalities. We refer to the survey article [108] for a discussion of this issue.

The extension of the numerical analysis we have developed and presented in this article to this semilinear setting is a widely open problem. In [115], the adaptation of the two-grid technique to globally Lipschitz nonlinearities is presented, together with some open problems and directions of research.

There is also plenty to be done to adapt the numerical analysis techniques presented here to super-critical exponents since the theory of dispersive estimates for linear discrete waves is also difficult matter in itself. We refer to in [49–51] for the first results in that direction in the context of Schrödinger equations.

The same problems arise in the context of many other nonlinear PDE, for instance: semilinear Schrödinger equations [62], KdV equations [95], semilinear heat equations [26, 112], etc.

Problem 5.7.2. Non-uniform meshes. In applications, one usually deals with non-uniform meshes for finite element methods. But the Fourier analysis methods we have developed here can not be applied in that setting. Roughly speaking, the only existing result in this direction is the one presented in Theorem 5.5.8 ([27]), ensuring that, when filtering the high frequencies at the scale $1/\sqrt{h}$, uniform observability holds. But on uniform meshes, the critical scale is $1/h$. An in depth analysis is needed in order to explain what is the behavior of numerical waves in this intermediate range for frequencies in between $1/\sqrt{h}$ and $1/h$.

The issue is even open in 1-d. For instance, it would worth identifying the class of quasi-uniform meshes for which the $1/h$ filtering scale suffices.

In this context, the article [28] is worth mentioning: There, it has been proved that, for the mixed finite element method in 1-d on non-uniform meshes, uniform observability properties hold under some mild restrictions on the mesh. This is based on the very nature of the mixed finite element discretization which allows to compute explicitly the spectrum of the discrete equations and then to apply Fourier analysis techniques.

Note that this issue can also be related to the observability properties of the wave equation with variable coefficients in uniform meshes. For the continuous 1-d wave equation the assumption on the BV regularity of the coefficients is sharp (see [19]). Adapting the numerical analysis results presented in this paper to that setting is a challenging open problem.

Problem 5.7.3. Uniform control of the low frequencies. In [77] it has been proved that, in 1-d, for initial data having only a finite number of Fourier components, the discrete controls are uniformly bounded and converge as $h \rightarrow 0$ towards the control of the wave equation. This result has been proved using moment problem techniques. The article [77] provides explicit estimates on the bi-orthogonal functions depending both on the frequency and the mesh-size parameters and in particular yields uniform estimates in the case in which only a finite number of frequencies are involved. This analysis is limited by now to 1-d problems. The extension of this result to multi-dimensional problems, even in the case of the unit square observed from two consecutive boundaries, is a challenging and interesting open problem.

Problem 5.7.4. Wigner measures. In [72, 73], Macià adapted Wigner measures to study the propagation of the singularities of waves in a discrete setting on uniform meshes of the whole space (see Problem 5.7.2). Roughly speaking, to any sequence of solutions of the discrete wave equation one associates a measure living on the space and frequency variables that is constant along the bicharacteristic flow of the Hamiltonian corresponding to the wave process under consideration. This Wigner measure has some interesting features. In particular, when considering sequences that weakly converge to zero in L^2, the Wigner measure describes the possible lack of strong convergence very accurately.

But this theory is still to be developed more completely to handle, for instance, boundary conditions and non-uniform meshes or to adapt the notion of polarization introduced in [15] to the discrete setting.

Problem 5.7.5. Numerical methods using randomness. When discretizing one dimensional hyperbolic systems of conservation laws, one can use the so-called Glimm's random choice method.

This idea, originally developed in [39], has even been used to prove existence of solutions for one dimensional hyperbolic systems.

A natural question then is the following one: Can we use Glimm's random choice method to obtain convergent sequences of discrete controls? So far, this issue is widely open. The only contribution we are aware of is [22], which states that, for the corresponding discrete 1-d wave equation, with an excellent probability, uniform observability holds. Here, excellent probability means with a probability greater than $\exp(-C(T)(\Delta t / h)^2/(\Delta t))$, where Δt is the time discretization parameter, and $C(T)$ is a strictly positive constant when $T > 2$.

But of course, this first result should be further developed, in particular for conservation laws. Also, one could try to extend Glimm's idea to higher dimensions and derive numerical schemes for the 2-d wave equation with some random effects that could help on the obtention of discrete observability properties.

Problem 5.7.6. Inverse Problems. The literature on inverse problems for hyperbolic equations is wide. We refer, for instance, to the works of Bukgheim and Klibanov [13] and the books [12, 56, 58, 59] (and the references therein) for a presentation of the state of the art in that field. For what concerns the acoustic wave equation, we can also refer to the works of [52, 86].

Roughly speaking, the problem is that of determining the properties of a medium by making boundary measurements on the waves propagating in it.

To illustrate the kind of problems that arise in this field and their intrinsic complexity let us consider the example of the 1d wave equation in which the velocity of propagation c is a positive unknown constant:

$$u_{tt} - c^2 u_{xx} = 0, \quad 0 < x < 1, 0 < t < T, \quad u(0,t) = u(1,t) = 0, \quad 0 < t < T. \tag{5.189}$$

One could then consider the problem of determining the velocity c out of boundary measurements $u_x(1, t)$ for $t \in \mathbb{R}$.

In this continuous setting, using the time periodicity of solutions with time period $2/c$, one could determine the value of c in terms of the periodicity of the boundary measurement. But, of course, this cannot be applied in the discrete setting since the discrete versions of (5.189) generate a lot of spurious high-frequency waves that travel at any velocity between 0 and c, thus breaking down the periodicity properties of continuous waves.

Hence, even on that simple example, the convergence of the solutions of the discrete inverse problems towards those of the continuous one is not so obvious and very unlikely to hold. Of course, on more intricate examples, the situation will become even worse. Generally speaking, the problem of solving discrete inverse problems and passing to the limit as the mesh-size parameter tends to zero is widely open.

Note that these questions are also of interest for what concerns the so-called Calderón problem, which consists, in the elliptic setting, on identifying the electrical conductivity of a medium by the knowledge of the so-called Dirichlet to Neumann map (or voltage to current map), see [103]. There again, to our knowledge, convergence issues for numerical approximation schemes have not been analyzed.

Problem 5.7.7. Unique continuation for discrete waves. For the continuous wave equation in a bounded domain, it is well-known that if the solution vanishes in some open subset during a certain amount of time (which shall be large enough and depends on the whole geometry of the set ω where the solution vanishes and the domain Ω where the equation holds), then the solution is identically zero everywhere. For the constant coefficient wave equation this is a consequence of Holmgren's uniqueness theorem, see [48].

Such result is not true for the discrete wave equation, as an explicit counterexample by O. Kavian shows (mentioned in [114]): In the unit square, when discretizing the Laplacian on a uniform grid using the usual 5-points finite-difference discretization, there exists a concentrated eigenvalue, alternating between 1 and -1 on the diagonal, and taking the null value 0 outside. This corresponds to the eigenvalue $4/h^2$, where h is the mesh-size, hence to a very high eigenfunction. Of course, this makes the discrete version of the unique continuation property above to be false. However, one could expect this uniqueness property to be true within a class of filtered solutions. This is indeed the case, as it has been recently proved in [9].

But the same can be said about the quantitative versions of the uniqueness theorem above that are by now well known in the continuous setting (see among others, [63,85,93,94]). These results consist in weak observability estimates for the continuous wave equation when no geometric condition is fulfilled.

When no geometric condition is fulfilled, such weak observability estimates for discrete wave equations are so far completely unknown, but we expect this to be reachable using suitable discrete versions of the Carleman inequalities, the preliminary results by [9] and the so-called Fourier–Bros–Iagoniltzer transform [85,94].

Problem 5.7.8. Waves on networks. Several important applications require the understanding of the propagation of waves into networks, and their control theoretical properties. Even in the continuous setting, this question is intricate since the propagation of the waves in a network depend strongly on its geometrical and topological properties. In particular, when the network includes a closed loop, some resonant effects may appear. We refer to [23] (and to the references therein) for a precise description of the state of the art in this field, updated in the recent survey [116].

Hence, when discretizing these models, understanding the propagation, observation and control properties of discrete waves propagate into networks, becomes a complex topic that is widely open. Some preliminary results have been obtained in [11] on a star shaped network of three strings controlled from the exterior nodes. But there is still an important gap between the understanding of the observability properties of the waves on networks in the discrete and continuous frameworks.

Problem 5.7.9. Hybrid parabolic/hyperbolic systems. In these notes we focused on the classical wave equation and its semidiscrete approximation schemes, but in many applications the relevant models are much more complex.

A classical example is given by the system of linear thermoelasticity, whose null-controllability properties have been derived in [65]. This system is composed of one parabolic type equation coupled with an hyperbolic one. In [65], it is proved that the system of linear thermoelasticity is null-controllable when the Geometric Control Condition is satisfied, which of course comes from the hyperbolic nature of the underlying wave equation.

When discretizing such equations, in view of the results developed above, it is natural to expect that the discrete controllability properties may fail to be uniform. But this should be discussed more precisely, because of the coupling with the parabolic component that may strongly influence the dynamics.

Acknowledgements When preparing the last version of this manuscript we were supported by Alejandro Maas Jr., internship student from the Universidad Técnica Federico Santa María (UTFSM), Chile, visiting BCAM for two months early 2011. He contributed to improve our plots and also to run the numerical experiments we present here. We express our gratitude to him for his efficient and friendly help. This work was supported by the ERC Advanced Grant FP7–246775 NUMERIWAVES, the Grant PI2010–04 of the Basque Government, the ESF Research Networking Program OPTPDE and Grant MTM2008–03541 of the MICINN, Spain. The first author acknowledges the hospitality and support of the Basque Center for Applied Mathematics where part of this work was done.

References

1. K. Ammari, M. Tucsnak, Stabilization of second order evolution equations by a class of unbounded feedbacks. ESAIM Contr. Optim. Calc. Var. **6**, 361–386 (2001)
2. M. Asch, G. Lebeau, Geometrical aspects of exact boundary controllability for the wave equation—a numerical study. ESAIM Contr. Optim. Calc. Var. **3**, 163–212 (1998)
3. D. Auroux, J. Blum, Back and forth nudging algorithm for data assimilation problems. C. R. Math. Acad. Sci. Paris **340**(12), 873–878 (2005)
4. G.A. Baker, J.H. Bramble, Semidiscrete and single step fully discrete approximations for second order hyperbolic equations. RAIRO Anal. Numér. **13**(2), 75–100 (1979)
5. C. Bardos, F. Bourquin, G. Lebeau, Calcul de dérivées normales et méthode de Galerkin appliquée au problème de contrôlabilité exacte. C. R. Acad. Sci. Paris Sér. I Math. **313**(11), 757–760 (1991)

6. C. Bardos, G. Lebeau, J. Rauch, Sharp sufficient conditions for the observation, control and stabilization of waves from the boundary. SIAM J. Contr. Optim. **30**(5), 1024–1065 (1992)
7. T.Z. Boulmezaoud, J.M. Urquiza, On the eigenvalues of the spectral second order differentiation operator and application to the boundary observability of the wave equation. J. Sci. Comput. **31**(3), 307–345 (2007)
8. F. Boyer, F. Hubert, J. Le Rousseau, Discrete carleman estimates for elliptic operators and uniform controllability of semi-discretized parabolic equations. J. Math. Pures Appl. (9) **93**(3), 240–276 (2010)
9. F. Boyer, F. Hubert, J. Le Rousseau, Discrete carleman estimates for elliptic operators in arbitrary dimension and applications,. SIAM J. Contr. Optim. **48**, 5357–5397 (2010)
10. F. Boyer, F. Hubert, J. Le Rousseau, Uniform null-controllability properties for space/time-discretized parabolic equations. Numer. Math. **118**(4), 601–661 (2011)
11. U. Brauer, G. Leugering, On boundary observability estimates for semi-discretizations of a dynamic network of elastic strings. Contr. Cybern. **28**(3), 421–447 (1999); Recent advances in control of PDEs.
12. A.L. Bughgeim, *Volterra Equations and Inverse Problems*, Inverse and Ill-posed Problems Series. (VSP, Utrecht, 1999)
13. A.L. Bukhgeïm, M.V. Klibanov, Uniqueness in the large of a class of multidimensional inverse problems. Dokl. Akad. Nauk SSSR **260**(2), 269–272 (1981)
14. N. Burq, P. Gérard, Condition nécessaire et suffisante pour la contrôlabilité exacte des ondes. C. R. Acad. Sci. Paris Sér. I Math. **325**(7), 749–752 (1997)
15. N. Burq, G. Lebeau, Mesures de défaut de compacité, application au système de Lamé. Ann. Sci. École Norm. Sup. (4) **34**(6), 817–870 (2001)
16. N. Burq, M. Zworski, Geometric control in the presence of a black box. J. Amer. Math. Soc. **17**(2), 443–471 (2004)
17. C. Castro, S. Micu, Boundary controllability of a linear semi-discrete 1-d wave equation derived from a mixed finite element method. Numer. Math. **102**(3), 413–462 (2006)
18. C. Castro, S. Micu, A. Münch, Numerical approximation of the boundary control for the wave equation with mixed finite elements in a square. IMA J. Numer. Anal. **28**(1), 186–214 (2008)
19. C. Castro, E. Zuazua, Concentration and lack of observability of waves in highly heterogeneous media. Arch. Ration. Mech. Anal. **164**(1), 39–72 (2002)
20. N. Cîndae, S. Micu, M. Tucsnak, An approximation method for exact controls of vibrating systems. SIAM J. Control Optim. 1283–1305 (2011)
21. G.C. Cohen, *Higher-Order Numerical Methods for Transient Wave Equations*. Scientific Computation. (Springer, Berlin, 2002); With a foreword by R. Glowinski.
22. J.-M. Coron, S. Ervedoza, O. Glass, Uniform observability estimates for the 1-d discretized wave equation and the random choice method. Compt. Rendus Math. **347**(9–10), 505–510 (2009)
23. R. Dáger, E. Zuazua, *Wave Propagation, Observation and Control in 1-d Flexible Multi-structures, Mathématiques & Applications (Berlin)*, vol. 50. (Springer, Berlin, 2006)
24. B. Dehman, G. Lebeau, Analysis of the HUM control operator and exact controllability for semilinear waves in uniform time. SIAM J. Contr. Optim. **48**(2), 521–550 (2009)
25. B. Dehman, G. Lebeau, E. Zuazua, Stabilization and control for the subcritical semilinear wave equation. Ann. Sci. École Norm. Sup. (4) **36**(4), 525–551 (2003)
26. T. Duyckaerts, X. Zhang, E. Zuazua, On the optimality of the observability inequalities for parabolic and hyperbolic systems with potentials. Ann. Inst. H. Poincaré Anal. Non Linéaire **25**(1), 1–41 (2008)
27. S. Ervedoza, Spectral conditions for admissibility and observability of wave systems: Applications to finite element schemes. Numer. Math. **113**(3), 377–415 (2009)
28. S. Ervedoza, Observability properties of a semi-discrete 1D wave equation derived from a mixed finite element method on nonuniform meshes. ESAIM Contr. Optim. Calc. Var. **16**(2), 298–326 (2010)
29. S. Ervedoza, Admissibility and observability for Schrödinger systems: Applications to finite element approximation schemes. Asymptot. Anal. **71**(1–2), 1–32 (2011)

30. S. Ervedoza, J. Valein, On the observability of abstract time-discrete linear parabolic equations. Rev. Mat. Complut. **23**(1), 163–190 (2010)
31. S. Ervedoza, C. Zheng, E. Zuazua, On the observability of time-discrete conservative linear systems. J. Funct. Anal. **254**(12), 3037–3078 (2008)
32. S. Ervedoza, E. Zuazua, On the numerical approximation of controls for waves, Springer Briefs in Mathematics, to appear.
33. S. Ervedoza, E. Zuazua, *Uniform exponential decay for viscous damped systems*, vol. 78, In: Advances in Phase Space Analysis of Partial Differential Equations, Progress in Nonlinear Differential Equations and Their Applications, vol. 78 (Birkhäuser Boston Inc., Boston, MA, 2009) pp. 95–112
34. S. Ervedoza, E. Zuazua, Uniformly exponentially stable approximations for a class of damped systems. J. Math. Pure. Appl. **91**, 20–48 (2009)
35. S. Ervedoza, E. Zuazua, A systematic method for building smooth controls for smooth data. Discrete Contin. Dyn. Syst. Ser. B **14**(4), 1375–1401 (2010)
36. S. Ervedoza, E. Zuazua, *Propagation, observation and numerical approximations of waves*. Book in preparation.
37. X. Fu, J. Yong, X. Zhang, Exact controllability for multidimensional semilinear hyperbolic equations. SIAM J. Contr. Optim. **46**(5), 1578–1614 (2007)
38. P. Gérard, Microlocal defect measures. Comm. Part. Differ. Equat. **16**(11), 1761–1794 (1991)
39. J. Glimm, Solutions in the large for nonlinear hyperbolic systems of equations. Comm. Pure Appl. Math. **18**, 697–715 (1965)
40. R. Glowinski, Ensuring well-posedness by analogy: Stokes problem and boundary control for the wave equation. J. Comput. Phys. **103**(2), 189–221 (1992)
41. R. Glowinski, W. Kinton, M.F. Wheeler, A mixed finite element formulation for the boundary controllability of the wave equation. Int. J. Numer. Meth. Eng. **27**(3), 623–635 (1989)
42. R. Glowinski, C.H. Li, On the numerical implementation of the Hilbert uniqueness method for the exact boundary controllability of the wave equation. C. R. Acad. Sci. Paris Sér. I Math. **311**(2), 135–142 (1990)
43. R. Glowinski, C.H. Li, J.-L. Lions, A numerical approach to the exact boundary controllability of the wave equation. I. Dirichlet controls: description of the numerical methods. Japan J. Appl. Math. **7**(1), 1–76 (1990)
44. R. Glowinski, J.-L. Lions, J. He, *Exact and Approximate Controllability for Distributed Parameter Systems, Encyclopedia of Mathematics and Its Applications*, vol. 117, (Cambridge University Press, Cambridge, 2008); A numerical approach.
45. P. Grisvard, Contrôlabilité exacte des solutions de l'équation des ondes en présence de singularités. J. Math. Pure. Appl. (9) **68**(2), 215–259 (1989)
46. A. Haraux, Une remarque sur la stabilisation de certains systèmes du deuxième ordre en temps. Port. Math. **46**(3), 245–258 (1989)
47. L.F. Ho, Observabilité frontière de l'équation des ondes. C. R. Acad. Sci. Paris Sér. I Math. **302**(12), 443–446 (1986)
48. L. Hörmander, *Linear Partial Differential Operators*. Die Grundlehren der mathematischen Wissenschaften, Bd. 116. (Academic Press, New York, 1963)
49. L.I. Ignat, E. Zuazua, Dispersive properties of a viscous numerical scheme for the Schrödinger equation. C. R. Math. Acad. Sci. Paris **340**(7), 529–534 (2005)
50. L.I. Ignat, E. Zuazua, A two-grid approximation scheme for nonlinear Schrödinger equations: dispersive properties and convergence. C. R. Math. Acad. Sci. Paris **341**(6), 381–386 (2005)
51. L.I. Ignat, E. Zuazua, Numerical dispersive schemes for the nonlinear Schrödinger equation. SIAM J. Numer. Anal. **47**(2), 1366–1390 (2009)
52. O.Y. Imanuvilov, M. Yamamoto, Determination of a coefficient in an acoustic equation with a single measurement. Inverse Probl. **19**(1), 157–171 (2003)
53. J.A. Infante, E. Zuazua, Boundary observability for the space semi discretizations of the 1-d wave equation. Math. Model. Num. Ann. **33**, 407–438 (1999)
54. A.E. Ingham, Some trigonometrical inequalities with applications to the theory of series. Math. Z. **41**(1), 367–379 (1936)

55. E. Isaacson, H.B. Keller, *Analysis of Numerical Methods*, (Wiley, New York, 1966)
56. V. Isakov, *Inverse Problems for Partial Differential Equations, Applied Mathematical Sciences*, 2nd edn. vol. 127, (Springer, New York, 2006)
57. J. Klamka, *Controllability of Dynamical Systems, Mathematics and Its Applications (East European Series)*. vol. 48, (Kluwer, Dordrecht, 1991)
58. M.V. Klibanov, Inverse problems and Carleman estimates. Inverse Probl. **8**(4), 575–596 (1992)
59. M.V. Klibanov, A. Timonov, *Carleman Estimates for Coefficient Inverse Problems and Numerical Applications*. Inverse and Ill-posed Problems Series. (VSP, Utrecht, 2004)
60. V. Komornik, A new method of exact controllability in short time and applications. Ann. Fac. Sci. Toulouse Math. (5) **10**(3), 415–464 (1989)
61. S. Labbé, E. Trélat, Uniform controllability of semidiscrete approximations of parabolic control systems. Syst. Contr. Lett. **55**(7), 597–609 (2006)
62. C. Laurent, Global controllability and stabilization for the nonlinear Schrödinger equation on some compact manifolds of dimension 3. SIAM J. Math. Anal. **42**(2), 785–832 (2010)
63. G. Lebeau, Contrôle analytique. I. Estimations a priori. Duke Math. J. **68**(1), 1–30 (1992)
64. G. Lebeau, M. Nodet, Experimental study of the HUM control operator for linear waves. Exp. Math. **19**(1), 93–120 (2010)
65. G. Lebeau, E. Zuazua, Null-controllability of a system of linear thermoelasticity. Arch. Ration. Mech. Anal. **141**(4), 297–329 (1998)
66. E.B. Lee, L. Markus, *Foundations of Optimal Control Theory*, 2nd edn. ed. by Robert E. Krieger (Melbourne, FL, 1986)
67. L. León, E. Zuazua, Boundary controllability of the finite-difference space semi-discretizations of the beam equation. ESAIM Contr. Optim. Calc. Var. **8**, 827–862 (2002); A tribute to J. L. Lions.
68. J.-L. Lions, *Contrôlabilité Exacte, Stabilisation et Perturbations de Systèmes Distribués. Tome 1. Contrôlabilité exacte*, vol. RMA 8. (Masson, Paris, 1988)
69. J.-L. Lions, Exact controllability, stabilization and perturbations for distributed systems. SIAM Rev. **30**(1), 1–68 (1988)
70. A. López, E. Zuazua, *Some New Results Related to the Null Controllability of the 1-d Heat Equation*, In: Séminaire sur les Équations aux Dérivées Partielles, 1997–1998, (École Polytech., Palaiseau, 1998) p. Exp. No. VIII, 22.
71. P. Loreti, M. Mehrenberger, An ingham type proof for a two-grid observability theorem. ESAIM: COCV **14**(3), 604–631 (2008)
72. F. Macià, *The Effect of Group Velocity in the Numerical Analysis of Control Problems for the Wave Equation*, In: Mathematical and numerical aspects of wave propagation—WAVES 2003, (Springer, Berlin, 2003) pp. 195–200
73. F. Macià, Wigner measures in the discrete setting: High-frequency analysis of sampling and reconstruction operators. SIAM J. Math. Anal. **36**(2), 347–383 (2004)
74. F. Macià, E. Zuazua, On the lack of observability for wave equations: a Gaussian beam approach. Asymptot. Anal. **32**(1), 1–26 (2002)
75. A. Marica, Propagation and dispersive properties for the discontinuous Galerkin and higher order finite element approximations of the wave and Schrödinger equations, Ph D Thesis, Universidad Autónoma de Madrid, 2010
76. A. Marica, E. Zuazua, Localized solutions for the finite difference semi-discretization of the wave equation. C. R. Math. Acad. Sci. Paris **348**(11–12), 647–652 (2010)
77. S. Micu, Uniform boundary controllability of a semi-discrete 1-D wave equation. Numer. Math. **91**(4), 723–768 (2002)
78. S. Micu, E. Zuazua, *An Introduction to the Controllability of Partial Differential Equations*, ed. by T. Sari, Collection Travaux en Cours Hermannin Quelques Questions de Théorie du Contrôle, pp. 67–150 (2005)
79. L. Miller, Controllability cost of conservative systems: Resolvent condition and transmutation. J. Funct. Anal. **218**(2), 425–444 (2005)

80. L. Miller, The control transmutation method and the cost of fast controls. SIAM J. Contr. Optim. **45**(2), 762–772 (2006)
81. A. Münch, A.F. Pazoto, Uniform stabilization of a viscous numerical approximation for a locally damped wave equation. ESAIM Contr. Optim. Calc. Var. **13**(2), 265–293 (2007)
82. M. Negreanu, A.-M. Matache, C. Schwab, Wavelet filtering for exact controllability of the wave equation. SIAM J. Sci. Comput. **28**(5), 1851–1885 (2006)
83. M. Negreanu, E. Zuazua, Convergence of a multigrid method for the controllability of a 1-d wave equation. C. R. Math. Acad. Sci. Paris **338**(5), 413–418 (2004)
84. A. Osses, A rotated multiplier applied to the controllability of waves, elasticity, and tangential Stokes control. SIAM J. Contr. Optim. **40**(3), 777–800 (2001)
85. K.D. Phung, *Waves, Damped Wave and Observation*, ed. by Ta-Tsien Li, Yue-Jun Peng, Bo-Peng Rao. Some Problems on Nonlinear Hyperbolic Equations and Applications, Series in Contemporary Applied Mathematics CAM 15, 2010.
86. J.-P. Puel, M. Yamamoto, On a global estimate in a linear inverse hyperbolic problem. Inverse Probl. **12**(6), 995–1002 (1996)
87. J.V. Ralston, Solutions of the wave equation with localized energy. Comm. Pure Appl. Math. **22**, 807–823 (1969)
88. K. Ramdani, T. Takahashi, G. Tenenbaum, M. Tucsnak, A spectral approach for the exact observability of infinite-dimensional systems with skew-adjoint generator. J. Funct. Anal. **226**(1), 193–229 (2005)
89. K. Ramdani, T. Takahashi, M. Tucsnak, Semi-discrétisation en espace du problème de la stabilisation interne de l'équation des poutres. ESAIM Proc. **18**, 48–56 (2007)
90. K. Ramdani, T. Takahashi, M. Tucsnak, Uniformly exponentially stable approximations for a class of second order evolution equations—application to LQR problems. ESAIM Contr. Optim. Calc. Var. **13**(3), 503–527 (2007)
91. J. Rauch, On convergence of the finite element method for the wave equation. SIAM J. Numer. Anal. **22**(2), 245–249 (1985)
92. P.-A. Raviart, J.-M. Thomas, *Introduction à l'analyse Numérique des Équations aux Dérivées Partielles*, Collection Mathématiques Appliquées pour la Maitrise. [Collection of Applied Mathematics for the Master's Degree]. (Masson, Paris, 1983)
93. L. Robbiano, Théorème d'unicité adapté au contrôle des solutions des problèmes hyperboliques. Comm. Part. Differ. Equat. **16**(4–5), 789–800 (1991)
94. L. Robbiano, Fonction de coût et contrôle des solutions des équations hyperboliques. Asymptotic Anal. **10**(2), 95–115 (1995)
95. L. Rosier, B.-Y. Zhang, Control and stabilization of the Korteweg-de Vries equation: recent progresses. J. Syst. Sci. Complex **22**(4), 647–682 (2009)
96. D.L. Russell, Controllability and stabilizability theory for linear partial differential equations: recent progress and open questions. SIAM Rev. **20**(4), 639–739 (1978)
97. T.I. Seidman, J. Yong, How violent are fast controls? II. Math. Contr. Signals Syst. **9**(4), 327–340 (1996)
98. E.D. Sontag, *Mathematical Control Theory, Texts in Applied Mathematics*. vol. 6, 2nd edn. (Springer, New York, 1998); Deterministic finite-dimensional systems.
99. L.R. Tcheugoué Tebou, E. Zuazua, Uniform boundary stabilization of the finite difference space discretization of the 1d wave equation. Adv. Comput. Math. **26**(1–3), 337–365 (2007)
100. L.R. Tcheugoué Tébou, E. Zuazua, Uniform exponential long time decay for the space semi-discretization of a locally damped wave equation via an artificial numerical viscosity. Numer. Math. **95**(3), 563–598 (2003)
101. L.N. Trefethen, Group velocity in finite difference schemes. SIAM Rev. **24**(2), 113–136 (1982)
102. M. Tucsnak, G. Weiss, *Observation and Control for Operator Semigroups Birkäuser Advanced Texts*, vol. 11, (Springer, Basel, 2009)
103. G. Uhlmann, *Developments in Inverse Problems Since Calderón's Foundational Paper*, In Harmonic Analysis and Partial Differential Equations. (Chicago, IL, 1996), Chicago Lectures in Mathematics, (Chicago, IL, 1999) pp. 295–345

104. R. Vichnevetsky, J.B. Bowles, *Fourier Analysis of Numerical Approximations of Hyperbolic Equations, SIAM Studies in Applied Mathematics*. vol. 5, (SIAM, Philadelphia, PA 1982); With a foreword by G. Birkhoff.

105. R.M. Young, *An Introduction to Nonharmonic Fourier Series*, 1st edn. (Academic Press, San Diego, CA, 2001)

106. X. Zhang, Explicit observability estimate for the wave equation with potential and its application. R. Soc. Lond. Proc. Ser. A Math. Phys. Eng. Sci. **456**(1997), 1101–1115 (2000)

107. X. Zhang, C. Zheng, E. Zuazua, Exact controllability of the time discrete wave equation. *Discrete and Continuous Dynamical Systems*, (2007)

108. X. Zhang, E. Zuazua, *Exact Controllability of the Semi-linear Wave Equation*, Unsolved Problems in Mathematical Systems and Control Theory, (Princeton University Press, Princeton, 2004), pp. 173–178

109. E. Zuazua, Exact controllability for the semilinear wave equation. J. Math. Pures Appl. (9) **69**(1), 1–31 (1990)

110. E. Zuazua, Exact controllability for semilinear wave equations in one space dimension. Ann. Inst. H. Poincaré Anal. Non Linéaire **10**(1), 109–129 (1993)

111. E. Zuazua, Boundary observability for the finite-difference space semi-discretizations of the 2-D wave equation in the square. J. Math. Pure. Appl. (9) **78**(5), 523–563 (1999)

112. E. Zuazua, *Some Results and Open Problems on the Controllability of Linear and Semilinear Heat Equations*, In Carleman Estimates and Applications to Uniqueness and Control Theory (Cortona, 1999), Progress in Nonlinear Differential Equations and Their Applications, vol. 46, (Birkhäuser Boston, Boston, MA, 2001), pp. 191–211

113. E. Zuazua, Controllability of partial differential equations and its semi-discrete approximations. Discrete Contin. Dyn. Syst. **8**(2), 469–513 (2002); Current developments in partial differential equations (Temuco, 1999)

114. E. Zuazua, Propagation, observation, and control of waves approximated by finite difference methods. SIAM Rev. **47**(2), 197–243 (2005)

115. E. Zuazua, *Control and Numerical Approximation of the Wave and Heat Equations*, In International Congress of Mathematicians. vol. 3, European Mathematical Society, (Zürich, 2006), pp. 1389–1417

116. E. Zuazua, Control and stabilization of waves on 1-d networks, "Traffic flow on networks", B. Piccoli and M. Rascle, eds., Lecture Notes in Mathematics- C.I.M.E. Foundation Subseries, Springer Verlag, to appear

List of Participants

1. Afzal Shehzad
   shehzad.afzal@stud.unileoben.ac.at
2. Alabau-Boussouira Fatiha **(lecturer)**
   alabau@math.univ-metz.it
3. Ancona Fabio
   ancona@math.unipd.it
4. Antonelli Paolo
   P.Antonelli@damtp.cam.ac.uk
5. Benedetti Irene
   irene.benedetti@dmi.unipg.it
6. Beauchard Karine
   Karine.Beauchard@cmla.ens-cachan.fr
7. Bonfanti Giovanni
   g.bonfanti3@campus.unimib.it
8. Bright Ido
   ido.bright@weizmann.ac.il
9. Brockett Roger **(lecturer)**
   brockett@scas.harvard.edu
10. Cannarsa Piermarco **(editor)**
    cannarsa@mat.uniroma2.it
11. Caponigro Marco
    marco.caponigro@inria.fr
12. Cazacu Cristian-Mihai
    cazacu@bcamath.org
13. Cellina Arrigo
    arrigo.cellina@unimib.it
14. Chaves Silva Felipe Wallison
    chaves@bcamath.org
15. Coron Jean-Michel (editor)
    coron@ann.jussieu.fr

F. Alabau-Boussouira et al., *Control of Partial Differential Equations*,
Lecture Notes in Mathematics 2048, DOI 10.1007/978-3-642-27893-8,
© Springer-Verlag Berlin Heidelberg 2012

16. Court Sébastien
    sebastien.court@math.univ-toulouse.fr
17. Di Meglio Florent
    florent.di_meglio@mines-paristech.fr
18. Ervedoza Sylvain
    ervedoza@math.univ-toulouse.fr
19. Fahroo Fariba
    fariba.fahroo@afosr.af.mil
20. Floridia Giuseppe
    floridia@mat.roma2.it
21. Gaggero Mauro
    gaggero@diptem.unige.it
22. Galan Ioana Catalina
    galan_ioana@yahoo.com
23. Garavello Mauro
    mauro.garavello@mfn.unipmn.it
24. Guglielmi Roberto
    guglielm@mat.uniroma2.it
25. Laurent Camille
    camille.laurent@math.u-psud.fr
26. Léautaud Matthieu
    leautaud@ann.jussieu.fr
27. Lequeurre Julien
    julien.lequeurre@math.univ-toulouse.fr
28. Le Rousseau Jérôme **(lecturer)**
    jlr@univ-orleans.fr
29. Lukasiak Renata
    renatka@mat.umk.pl
30. Loheac Jerome
    Jerome.Loheac@iecn.u-nancy.fr
31. Loreti Paola
    loreti@dmmm.uniroma1.it
32. Mamadou Diagne
    diagne@lagep.univ-lyon1.fr
33. Marahrens Daniel
    d.o.j.marahrens@damtp.cam.ac.uk
34. Marica Aurora Mihaela
    marica@bcamath.org
35. Marigonda Antonio
    antonio.marigonda@univr.it
36. Marson Andrea
    marson@math.unipd.it
37. Martinez Patrick
    martinez@mip.ups-tlse.fr

38. Mascolo Elvira
    mascolo@math.unifi.it
39. Maurizi Amelio
    amelio.maurizi@gmail.com
40. Marson Andrea
    marson@math.unipd.it
41. Mazzola Marco
    m.mazzola7@campus.unimib.it
42. Mironchenko Andrii
    andmir@math.uni-bremen.de
43. Moncef Mahjoub
    moncef_mahjoub@yahoo.fr
44. Nguyen Khai
    khai@math.unipd.it
45. Olive Guillaume
    golive@cmi.univ-mrs.fr
46. Ortiz-Robinson Norma
    nlortiz@vcu.edu
47. Ozer Ahmet Ozkan
    oozer@iastate.edu
48. Panasenko Elena
    panlena_t@mail.ru
49. Perrollaz Vincent
    perrollaz@ann.jussieu.fr
50. Rocchetti Dario
    dario.rocchetti@ensta.fr
51. Rosier Lionel
    rosier@iecn.u-nancy.fr
52. Rykaczewski Krzysztof
    mozgun@mat.umk.pl
53. Salem Ali
    a3li.salem@gmail.com
54. Shang Peipei
    peipeishang@hotmail.com
55. Sharon Yoav
    ysharon2@illinois.edu
56. Silva Cristiana
    cjoaosilva@ua.pt
57. Sonja Veelken
    veelken@mathc.rwth-aachen.de
58. Thi Nhu Thuy Nguyen
    nthuyc8@yahoo.com
59. Tort Jacques
    jacques.tort@math.univ-toulouse.fr

60. Tosques Mario
    mario.tosques@unipr.it
61. Van Loi Nguyen
    loinv14@yahoo.com
62. Wang Zhiqiang
    wang@ann.jussieu.fr
63. Zair Ouahiba
    wahzair@gmail.com
64. Zhong Peng
    zhong@math.utk.edu
65. Zuazua Enrique **(lecturer)**
    zuazua@bcamath.org

LECTURE NOTES IN MATHEMATICS  Springer

Edited by J.-M. Morel, B. Teissier; P.K. Maini

Editorial Policy (for Multi-Author Publications: Summer Schools / Intensive Courses)

1. Lecture Notes aim to report new developments in all areas of mathematics and their applications - quickly, informally and at a high level. Mathematical texts analysing new developments in modelling and numerical simulation are welcome. Manuscripts should be reasonably selfcontained and rounded off. Thus they may, and often will, present not only results of the author but also related work by other people. They should provide sufficient motivation, examples and applications. There should also be an introduction making the text comprehensible to a wider audience. This clearly distinguishes Lecture Notes from journal articles or technical reports which normally are very concise. Articles intended for a journal but too long to be accepted by most journals, usually do not have this "lecture notes" character.

2. In general SUMMER SCHOOLS and other similar INTENSIVE COURSES are held to present mathematical topics that are close to the frontiers of recent research to an audience at the beginning or intermediate graduate level, who may want to continue with this area of work, for a thesis or later. This makes demands on the didactic aspects of the presentation. Because the subjects of such schools are advanced, there often exists no textbook, and so ideally, the publication resulting from such a school could be a first approximation to such a textbook. Usually several authors are involved in the writing, so it is not always simple to obtain a unified approach to the presentation.

   For prospective publication in LNM, the resulting manuscript should not be just a collection of course notes, each of which has been developed by an individual author with little or no coordination with the others, and with little or no common concept. The subject matter should dictate the structure of the book, and the authorship of each part or chapter should take secondary importance. Of course the choice of authors is crucial to the quality of the material at the school and in the book, and the intention here is not to belittle their impact, but simply to say that the book should be planned to be written by these authors jointly, and not just assembled as a result of what these authors happen to submit.

   This represents considerable preparatory work (as it is imperative to ensure that the authors know these criteria before they invest work on a manuscript), and also considerable editing work afterwards, to get the book into final shape. Still it is the form that holds the most promise of a successful book that will be used by its intended audience, rather than yet another volume of proceedings for the library shelf.

3. Manuscripts should be submitted either online at www.editorialmanager.com/lnm/ to Springer's mathematics editorial, or to one of the series editors. Volume editors are expected to arrange for the refereeing, to the usual scientific standards, of the individual contributions. If the resulting reports can be forwarded to us (series editors or Springer) this is very helpful. If no reports are forwarded or if other questions remain unclear in respect of homogeneity etc, the series editors may wish to consult external referees for an overall evaluation of the volume. A final decision to publish can be made only on the basis of the complete manuscript; however a preliminary decision can be based on a pre-final or incomplete manuscript. The strict minimum amount of material that will be considered should include a detailed outline describing the planned contents of each chapter.

   Volume editors and authors should be aware that incomplete or insufficiently close to final manuscripts almost always result in longer evaluation times. They should also be aware that parallel submission of their manuscript to another publisher while under consideration for LNM will in general lead to immediate rejection.

4. Manuscripts should in general be submitted in English. Final manuscripts should contain at least 100 pages of mathematical text and should always include

   - a general table of contents;
   - an informative introduction, with adequate motivation and perhaps some historical remarks: it should be accessible to a reader not intimately familiar with the topic treated;
   - a global subject index: as a rule this is genuinely helpful for the reader.

   Lecture Notes volumes are, as a rule, printed digitally from the authors' files. We strongly recommend that all contributions in a volume be written in the same LaTeX version, preferably LaTeX2e. To ensure best results, authors are asked to use the LaTeX2e style files available from Springer's web-server at

   ftp://ftp.springer.de/pub/tex/latex/svmonot1/ (for monographs) and

   ftp://ftp.springer.de/pub/tex/latex/svmultt1/ (for summer schools/tutorials).

   Additional technical instructions, if necessary, are available on request from:

   lnm@springer.com.

5. Careful preparation of the manuscripts will help keep production time short besides ensuring satisfactory appearance of the finished book in print and online. After acceptance of the manuscript authors will be asked to prepare the final LaTeX source files and also the corresponding dvi-, pdf- or zipped ps-file. The LaTeX source files are essential for producing the full-text online version of the book. For the existing online volumes of LNM see:

   http://www.springerlink.com/openurl.asp?genre=journal&issn=0075-8434.

   The actual production of a Lecture Notes volume takes approximately 12 weeks.

6. Volume editors receive a total of 50 free copies of their volume to be shared with the authors, but no royalties. They and the authors are entitled to a discount of 33.3 % on the price of Springer books purchased for their personal use, if ordering directly from Springer.

7. Commitment to publish is made by letter of intent rather than by signing a formal contract. Springer-Verlag secures the copyright for each volume. Authors are free to reuse material contained in their LNM volumes in later publications: a brief written (or e-mail) request for formal permission is sufficient.

Addresses:

Professor J.-M. Morel, CMLA,
École Normale Supérieure de Cachan,
61 Avenue du Président Wilson, 94235 Cachan Cedex, France
E-mail: morel@cmla.ens-cachan.fr

Professor B. Teissier, Institut Mathématique de Jussieu,
UMR 7586 du CNRS, Équipe "Géométrie et Dynamique",
175 rue du Chevaleret,
75013 Paris, France
E-mail: teissier@math.jussieu.fr

For the "Mathematical Biosciences Subseries" of LNM:

Professor P. K. Maini, Center for Mathematical Biology,
Mathematical Institute, 24-29 St Giles,
Oxford OX1 3LP, UK
E-mail : maini@maths.ox.ac.uk

Springer, Mathematics Editorial I,
Tiergartenstr. 17,
69121 Heidelberg, Germany,
Tel.: +49 (6221) 4876-8259
Fax: +49 (6221) 4876-8259
E-mail: lnm@springer.com